Randomness
Through
Computation
Some Answers, More Questions

Randomness
Through
Computation
Some Answers, More Questions

editor

Hector Zenil
Wolfram Research Inc., USA

W **World Scientific**

NEW JERSEY · LONDON · SINGAPORE · BEIJING · SHANGHAI · HONG KONG · TAIPEI · CHENNAI

Published by

World Scientific Publishing Co. Pte. Ltd.

5 Toh Tuck Link, Singapore 596224

USA office: 27 Warren Street, Suite 401-402, Hackensack, NJ 07601

UK office: 57 Shelton Street, Covent Garden, London WC2H 9HE

British Library Cataloguing-in-Publication Data
A catalogue record for this book is available from the British Library.

ISBN-13 978-981-4327-74-9
ISBN-10 981-4327-74-3

1006375913
Printed in Singapore by B & Jo Enterprise Pte Ltd

To Ray Solomonoff
(1926–2009)

From left to right: Ray Solomonoff, Mary Silver,
Rollo Silver, Marvin Minsky, *unknown*, and Ed Fredkin
at the Silver's home in Brookline, MA (1960 or early 1961).
(with kind permission of Grace Solomonoff
and information details from Ed Fredkin.)

Douglas Hofstadter and Ray Solomonoff
at the NKS Midwest Conference 2005, Bloomington, IN.
(with kind permission of Adrian German.)

Preface

Everything existing in the
universe is the fruit of chance
Democritus (ca. 460–370)

Chance, too, which seems to rush along with
slack reins, is bridled and governed by law
Boethius (ca. 480–525)

The notion of chance has been used for games and has puzzled thinkers for centuries. Questions about determinism and free will are among the oldest, having been broached in one form or another since ancient times.

(Dado di Bucchero) Etruscan Bucchero die (7th century BC). Castel Sant'Angelo National Museum special exhibition *Archeological and Artistic Treasure of the Castiglion Fiorentino Museum*, Rome 2010.

For many centuries randomness was viewed as a nuisance by science. It wasn't until the last century that mathematics made significant progress in formally describing and studying the concept.

More recently still, connections and applications to other areas of human knowledge have been created. It was in the twentieth century, thanks to probability theory, that randomness was first formulated in terms of the likelihood of a potential event happening according to a certain distribution.

Yet it wasn't until the latter half of the twentieth century, with the development of computer science and information theory, that a consensus was reached that in the emerging field of algorithmic complexity, with its fertile connections to recursion theory, something fundamental about randomness had been grasped in the language of mathematics.

Scientists began to realize that random sampling was a powerful tool for making predictions about the real world based on statistical inference. Researchers of all stripes began using randomness—for simulations, for testing hypotheses, and for carrying out all manner of experiments.

Around the same time, computer scientists began to realize that the deliberate introduction of randomness into computation could be an effective tool for designing faster algorithms, that is, the possibility that a source of random bits might be used to reduce the time required by an algorithm to compute a set of problems.

Given that the intuitive concept of randomness conveys the idea of things having no cause, of being unpredictable, of showing no pattern at all, how dare mathematics try to capture its elusive essence? Beyond its philosophical appeal, chance has come to occupy a central role in our everyday lives, whether we are aware of it or not. Nowadays, many procedures based on the simulation of random processes lie at the core of practical applications of banking, the stock market, television games and even the generation of the encrypted password you might have used to check email today, just to cite a few uses.

Because of its role in games—particularly of the sort played in casinos—it has also become a source of profit. So much so that there are devices being marketed using the pitch that they are superior random number generators.

In communication theory, randomness in a signal is referred to as *noise* and is identified with the concept of *meaninglessness*.

In economics, it is often encountered in the form of random walks regarding the evolution of prices, and plays an important role in the conceptualization of financial markets.

In biology, randomness in the guise of mutation, the driving force behind natural selection, is an essential feature of the theory of evolution.

The concept of free will has been connected to randomness since the beginning, but it has played the role of devil's advocate, which has meant that it has sometimes been construed as the cause of randomness and sometimes as its consequence.

It could be that what we experience when we exercise what we think of as free will is comparable to what we experience when we encounter the digits of a number, such as the constant π without knowing it is π. The digits of π are deterministically generated by simply calculating the ratio of any circle's circumference to its diameter, but to an observer the digits look random and unpredictable.

There seems to be no notion more essential or inherent to randomness than unpredictability, for if one can predict the outcome of an event, its random character is immediately compromised. The kind of randomness that is predictable is usually called pseudo-randomness, as opposed to what may be called "true" indeterministic randomness.

What appears to be a matter of chance in our everyday macroscopic life is a consequence of lack of knowledge, knowledge of the microscopic states of every particle and force involved. Were we to grasp all particles and forces with absolute precision, then we would see that there is no such thing as chance but pseudo-randomness only.

One may be able to disclose pseudo-randomness in seemingly random processes in nature by ascertaining whether an experiment yields the same results when the initial conditions are repeated, just as sowing a certain kind of seed always yields the same crop. What may prevent us from doing so in the real world is our inability to reproduce the exact initial conditions.

Determinism does not, however, imply practical predictability, just as non-determinism does not imply no sense of "freedom of choice".

Yet, seen through the lens of mathematics, randomness has managed to hold on to its most enigmatic features, without which it would lose its essence. Even when mathematically captured, under all hypotheses, randomness remains tied to the most rigorous sense of unpredictability through the formal concept of uncomputability.

Due to these connections that have been forged between the concept of randomness and the idea of computation, some of its deepest properties, properties that vindicate the elusiveness of the concept, have been unveiled.

Algorithmic information theory, the mathematical arm for tackling the concept of randomness through the study of bit sequences and computing, suggests that there must be an infinite amount of information for full randomness to exist. The subject has generated its own paradoxical name *algorithmic randomness*, since what is random is actually non-algorithmic.

Using the tools provided by Algorithmic Information Theory, unpredictability has been found to be connected to concepts such as compressibility, for if one can compress an event by cutting short its outcome, its unpredictable essence evidently breaks down. Unpredictability, incompressibility and other key features of randomness have been found to be deeply related.

These mathematical characterizations have turned out to imply a chain of properties, with each of them capturing an essential feature of randomness. Since every attempt to seize upon an essential part of the concept leads to a convergence with other definitions capturing other features, each of these characterizations may be considered to have objectively captured essential attributes of randomness.

Nevertheless, the concept of randomness remains untamed in a formal sense. For even when approached using Algorithmic Information Theory, it is not possible to generate or verify "true" randomness by calculation. Indeed the study of randomness has formalized and made plain its untamed character, yielding a long hierarchy of infinitely random wildness in the form of degrees of uncomputability, and thereby connecting it even more profoundly to certain areas of mathematical logic and recursion theory.

According to the theory, no finite object may be declared to be truly random as long as it can be shown to be part of a larger non-random object. One is allowed only to say whether a finite string looks or does not look random. Yet, tools for measuring the degree of apparent randomness of a finite string have been successfully developed, and have been proven to be of great use, with applications to several fields both inside and outside computer science.

Hitherto, something has been deemed completely random or not to the degree that it possesses some of the properties of randomness, but in point of fact the most intuitive notion of physical randomness is bias because no physical object is perfect.

A related problem concerns the extraction of randomness from a physical process, and the methods developed for stripping away biased randomness and arriving at a notion of fair randomness. Here statistical tests come to the fore again. They have been proven to be more useful for testing *good* randomness than for testing *true* randomness.

Using statistics, one can determine whether something that is random looking is good enough for particular purposes and applications. In a sense statistical tests only tell us whether a sequence looks random enough, rather than whether or not it is truly random, because one can only devise finite tests for local (useful) randomness, even when the randomness in question appears to be "true" randomness, as is the case with certain phenomena from quantum mechanics.

Quantum mechanics is believed to be fundamentally non-deterministic. The contingency of quantum mechanics may suggest that (uncaused) randomness actually operates at some underlying level of our physical reality, but the very foundations of the theory have yet to be agreed upon, and its impact on macroscopic events determined. Some interpretations of quantum mechanics do not postulate quantum randomness but instead posit hidden underlying variables which would undermine the no-causality of quantum mechanics.

Quantum mechanics somehow undermines the sturdy relationship between randomness and unpredictability. What makes quantum randomness random is its set of true inherent indeterministic properties, by virtue of which, for instance, there is no reason why a particle should be in a particular state at any given moment unless observed.

Nevertheless quantum mechanics reconnects unpredictability to randomness through the concept of irreversibility, because the determination of the state of a superposed particle involves the loss of the superposed state, which makes the process irreversible.

In its superposed state a particle does not carry information; it does so minimally only when disengaged. Hence the superposed state is in some sense a state of randomness because the determination of the state when observed has no cause. Which is to say that even if one were to entangle the particle again, its future states would be disconnected and independent of each other. Under the mainstream (Copenhagen) interpretation, what happens in quantum mechanics has no past and happens for no evident reason.

Some have claimed that the individual measurement result in quantum mechanics remains objectively random because of the finiteness of information, and have proceeded to suggest that the randomness of the individual event is the strongest indication of a reality independent of the observer. Note however that it is not clear whether a truly random source exists because its existence in the real world, has by no means been conclusively established beyond the interpretation of the results of the experiments and the current theoretical understanding.

Everything said so far may seem to suggest that any attempt to capture the nature of (finite) randomness is subjective at the end, since it always seems to depend on the observer, for instance, on how locally random something may look, or on whether the observer takes a measurement.

On the other hand, one may ask why nature manifests any order at all, if, the underlying processes at the quantum level are truly random. As it happens, the study of randomness has helped us toward a better grasp of the concept of complexity, and it may even explain the patterns that we find in a world comprising innumerable sophisticated structures, life being the most sophisticated.

Using the concept of Algorithmic Probability, it may be possible to explain why there is something rather than nothing. So not only are computer programs capable of great diversity and sophistication, but they may produce the order and complexity we see in the world out of pure randomness.

Though in making randomness into a field of study much has been accomplished, not all that can be said on the subject has been said, with many questions still remaining to be formulated and perhaps someday answered.

The present volume participates in this rather paradoxical quest by offering a series of essays presenting the views of some of the founders and leading thinkers in the field, authored by themselves. Taken together, they represent an attempt to present the intelligible side of randomness — what has been deciphered so far through the use of pure reason — with due attention paid to its historic and philosophical dimension, its relevance to the real world, and to what still remains to be figured out. They provide an up-to-date introduction to the subject matter and an in-depth overview of what I have only briefly outlined here. The distinguished contributors use their own personal stories to offer broad critical insights. Their essays discuss in significant detail a large spectrum of research, ranging from probability to physics, from statistics to computability.

The regular contributions are followed by two transcriptions, where pioneers in different areas engage each other's ideas, discussing whether the world is random and in what sense, whether the universe (only) computes and if so how, among other related topics.

Among the contributors to this volume is Ray Solomonoff, one of the founding fathers of the theory of algorithmic information particularly the concept of algorithmic probability and its connection to machine learning and artificial intelligence. Ray unfortunately passed away during the preparation of this volume. He submitted his contribution only six months before leaving us, and sadly this may be one of the last, if not the last article he prepared. I feel privileged to have had the opportunity to work on this book and honor Ray by dedicating it to his memory.

I hope you enjoy reading this volume as much as I enjoyed editing it.

Hector Zenil

Paris, October 2010

http://www.mathrix.org/experimentalAIT/RandomnessThroughComputation.htm

Acknowledgments

I approached the task of editing this book with great enthusiasm, though I fully expected it would be a challenge. And a challenge it was, though in some rather surprising ways as well as in the ways I had anticipated.

Convincing people that they had something to share beyond their technical work was at times unexpectedly difficult. I want to thank the authors who accepted the invitation to contribute to this volume. Without them it would quite simply not have been possible. They have been generous enough to devote the time to write up their contributions and share with us their views and original ideas.

I want to thank Adrian German for the great work he did with the panel discussion transcriptions included at the end of this volume. Thanks also to Cris Calude, Jean-Paul Delahaye, Stephen Wolfram, Greg Chaitin, Jean Mosconi and Eric Allender for their support and advice, to Carlos Gershenson who suggested a first publisher and encouraged me to undertake a previous project that evolved into this one, and to the editors and the helpful staff at World Scientific.

Contents

Contents

Part VI. Panel Discussions (Transcriptions)

PART I

Stochastic Randomness and Probabilistic Deliberations

Chapter 1

Is Randomness Necessary?

Ronald Graham

Department of Computer Science and Engineering,
University of California, San Diego
graham@ucsd.edu

WHAT DREW ME TO THE STUDY OF COMPUTATION AND RANDOMNESS

As long as I can remember, I have been interested in the search for structure in whatever I encounter. In essence, mathematics can be thought of as the science of patterns, so finding them and proving they persist is the bread and butter of mathematicians.

WHAT WE HAVE LEARNED

There are many instances where apparent randomness is actually in the eye of the beholder. An example from my own experience is the following. A fellow researcher once came upon a curious sequence in connection with his work on a certain sorting algorithm. It was defined recursively as follows. The first term, call it x(1), was equal to 3. In general, to get the $(n + 1)$st term $x(n + 1)$, you take the square root of the product 2 times $x(n)$ times $(x(n) + 1)$ and round it down to the next integer value. Thus, the sequence $x(1)$, $x(2)$, $x(3)$, ..., begins 1,2,3,4,6,9,13,19,27,38,54,.... Now, form a new sequence of 0's and 1's by replacing each term in the sequence by the remainder you get when you divide the term by 2. This gives us a new sequence 1, 0, 0, 1, 1, 1, 1, 0, 0, 1, 1, 0, 0, 1, 1, 0, 0, 0, 0, 0, 0, 0, 0, 0, 0, 1, 1, ... Suppose now you take every other term of this sequence, starting with the first term. This would result in the sequence (omitting the commas) S = 10110101000001100111100110...The question was whether anything sensible could be said about S, e.g., would it eventually repeat, are there roughly as many 0's as 1's in it, do all fixed 0/1 blocks occurs about equally often, etc. S certainly appeared "random", although it was produced by a fixed deterministic rule. It turns out that we were able to prove that in fact, if you place a decimal point after the first 1, and interpret the result as the binary number $1.01101010000010011110010...$, then this is exactly the binary representation of the square root of 2! As a consequence, we know that the sequence is not periodic, for example. Of course, no one can yet prove that the number of 0's is asymptotically equal to the number of 1's as you take longer and longer runs of the bits, and from this point of view, this expansion seems to be behaving like a random sequence. A similar phenomenon may be occurring with Wolfram's "Rule 30". This is a deterministic rule he devised for certain cellular automata which operate on a linear tape which appears to generate sequences with no discernible structure. In fact, it is used in *Mathematica* for generating random numbers, and appears to work quite satisfactory. However, it may be that no one (e.g., no human being) has the intelligence to perceive the structure in what it produces, which actually might well be quite striking.

WHAT WE DON'T YET KNOW

A fundamental technique pioneered in the 1950's by Paul Erdös goes under the name of the probabilistic method. With this technique, one can prove the existence of many remarkable mathematical objects by proving that the probability that they exist is positive. In fact, many of the sharpest results on the sizes of such objects are only obtained by using the probabilistic method. However, this method gives absolutely no clue as to how such objects might actually constructed. It would certainly be wonderful if someone could make progress in this direction.

A recent theme appearing in the mathematical literature is the concept of "quasirandomness". This refers to a set of properties of objects, (I'll use graphs as an example), which are all shared by truly random graphs, and which are equivalent, in the sense that any graph family possessing any one of the properties, must of necessity have all the other quasirandom properties as well. It turns out to be relatively easy to give explicit constructions of such quasirandom graphs, which makes it quite useful in many situations in which an explicit construction of a random-like object is desired. One mathematical area where this is especially apparent is in an area of combinatorics called Ramsey theory. They guiding theme in this subject can be described by the phrase "Complete disorder is impossible". Basically, it is the study of results which assert that a certain amount of structure is inevitable no matter how chaotic the underlying space appears. For example, it can be shown that for any choice of a positive number N, there is a least number $W(N)$ so that no matter how the numbers from 1 to $W(N)$ are colored red or blue, in at least one of the colors we must always have an arithmetic progression of equally spaced numbers in a single color. It is of great interest to estimate the size of $W(N)$. The best upper bound known is due to Field's Medalist Tim Gowers and states that $W(N) < 2^{2^{2^{2^{2^{(N+9)}}}}}$. The best lower bound known is of the form $W(N) > N2^N$. I current offer $1000 for a proof (or disproof) that $W(N) < 2^{(N^2)}$.

THE MOST IMPORTANT OPEN PROBLEMS

An important trend in computer science is the use of so-called randomized algorithms, i.e., those that have some access to a source of true randomness. It appears that this increases the range of which problems can be efficiently solved, although it has recently been shown that if there really are computationally intractable problems then randomness cannot be all that helpful. A major open problem is to resolve this uncertain situation. Of course, another important problem would be to show how to construct objects now known to exist only with of use of the probabilistic method.

THE PROSPECTS FOR PROGRESS

I am optimistic that we will be making great progress in tackling many of these questions although it is always hard to predict the timing of such advances. For example, when will the celebrated P versus NP problem be resolved? In our lifetimes? Perhaps. And then again, perhaps not!

Chapter 2

Probability is a Lot of Logic at Once:
If You Don't Know Which One to Pick, Take 'em All

Tommaso Toffoli

Electrical and Computer Engineering Department
Boston University
tt@bu.edu

How I got into it

For our doctoral thesis in physics we* designed a device to help tell the rare muons that were to be produced by the first storage-ring particle accelerator (ADONE) from the innumerable muons continually originated by cosmic-ray showers. Determining the performance of this finder of "needles in a haystack," as it were, required statistics of a very conventional sort; it certainly wasn't this that fired me up to investigate probability and randomness. However, it provided an empirical basis that must have helped keep my feet solidly on the ground whenever I felt the pull of theory. Many years later Ed Fredkin, who was to be my longtime boss and friend at MIT, told me after my first job interview there that my application had been lying around for a while, with no action taken because "The theoreticians thought you were too practical; the practitioners, too theoretical!"

Then a friend asked me if I wanted to take over (he'd promised to do it but couldn't find the time) reviewing Bruno de Finetti's autobiographical sampler, *A Mathematician's Affair with Economics*, for a science magazine.[18] Let's draw a merciful curtain over this. Not only did I not realize what I was getting into; I still hadn't realized it by the time I was done! I didn't know anything about the history of probability and didn't suspect —

---

*This is not the expected "plural of modesty." My colleague Aldo Santonico and I surprised our prospective advisor, Prof. Marcello Conversi, with the idea that the two of us *as a team* would work *jointly on two theses*; thus, I literally mean 'we designed'. Aldo grew into a distinguished particle physicist; I partly morphed into a computer scientist.

blessed are the ignorant! — that de Finetti's ideas were then so original and controversial; to me, they had appeared natural and obvious. In fact, at that time I didn't feel that anything special had happened to me. In any event, de Finetti's passionate arguments left me with no doubt that science can still offer live conceptual questions as well as educational causes worth fighting for.

Fast forward a few years — I was now working on my thesis in computer sciences. This time it was a case of one doctoral student with two advisors, John Holland — he of genetic algorithms — and Arthur Burks — a collaborator of the late John von Neumann in the self-replicating automata project. For my formation this was my most productive period, especially because I had been given *carte blanche* to do whatever I wanted; so I spent practically the whole time furiously teaching myself so many things I needed to know. With this schooling format I was forced to structure facts and concepts myself; try to find out what certain ideas were "really for" (and whether I truly needed them, with my limited time and mental equipment); and find ways to externalize and document what had cost me so much to explain to myself. The economics of this process turned me into a rather demanding customer: I've become a compulsive "knowledge structurer." So, with any concept or discipline I approach, I feel the need to handle it, look at it from different angles, take it apart and reassemble it or reconstruct it in a different way, and I'm not happy until the result

- It's so simple that even I can understand it.
- It's so sincere that I can believe it.
- It's so generally sweeping that it must be the expression of a tautology.

If we doggedly insist on structuring and restructuring knowledge until complex things finally appear trivial — until our students ask themselves, "Is that all there was to it?" — then we'll have fulfilled one of the goals of the educator. Moreover, by hierarchically compressing knowledge in this way, we'll have room in our minds for more of it and *more levels* of it.

At this stage of the game two things happened that played a determining role in the evolution of my interests and my strategies.

(a) Probability: Practice, "Foundations", "Interpretation"

I had much enjoyed studying what goes under the name of "elementary probability" — basically, what is covered by vol. 1 of Feller's classic.[5] That is a mixture of motivational commonsense; applied combinatorics, labeling and enumeration strategies, and a little asymptotics; a practical introduction to organizing concepts useful in a number of common situations: one goes through various basic distributions — say, uniform, normal, and Poisson — much as through roasting, frying, and baking in Rombauer's *Joy of Cooking*;[14] and, finally, life, death, and miracles of the Random Walk in several incarnations (in one or more dimensions, with absorbing or reflecting boundaries, with expected time of first transit or of return, etc.). If you remove the folkloristic attire, this all boils down to what Ivan Niven felicitously calls "how to count without counting[†]".[11]

So it was with much anticipation that I took a course in Advanced Probability, based on Breiman's *Probability* classic.[2] I suffered through it always expecting that, after interminable mathematical preliminaries, with the next lecture we'd finally start doing probability. *We never did!* The whole thing was about *measure theory* — on its own a perfectly legitimate business, but with no new probability contents whatsoever. Measure theory, and much of the axiomatic apparatus that goes into what is often called the "foundations" of probability, is just about developing more refined *accounting techniques* for when the outcome space becomes so large (viz., *uncountably infinite*) that simple-minded techniques lead to paradoxes: "If a line consists of points, and a point has no length, how come a line has length?" As a matter of fact, Breiman's preface (but how many young people read prefaces?) gives the whole scheme away: "[As a prerequisite] roughly the first seven chapters of *Measure Theory* by Halmos is sufficient background. *No prior knowledge of probability is assumed* [italics mine], but browsing through an elementary book such as the one by Feller,[5] with its diverse and vivid examples, gives an excellent feeling for the subject." Note the condescending tone ("Let Feller give his colorful examples;") and the implicit promise ("in the end you'll have to come to us for the real thing!") Regrettably, from that course I didn't get any "*posterior*" knowledge of probability either.

So, if not in Feller or in Breiman, where is *bona fide* probability to be found? Surely we must turn to the doctrine of "interpretation" of probability — doesn't that claim to study the connections between the theory

---

[†]To know how many cubic centimeters in a liter, we do not literally count 1, 2, 3, ... all the way up to one thousand; in our mind we shape the liter as a cube of 10 cm side and quickly multiply $10 \times 10 \times 10$. Let us also recall of the story of Gauss who, asked by his grade-school teacher to add up all the numbers from 1 to 100 (so that he could read the newspaper in peace), surprised — and irritated — said teacher by coming up with the answer in a few moments: $(100 + 1) \times 50$.

and real life? I will try to do justice to this topic in the next section, when the "me" of this narrative will have accumulated a little more scientific maturity.

(b) Reversibility, The "Blessing and Curse of Immortality"

Back to my computer science thesis,[20] what had started as an incidental detail was to shape my entire scientific career. I was studying a class of "solitaire" computer games called *cellular automata*, in which tokens are placed on or removed from the "sites" (the squares) of a checkerboard according to definite rules — there is some similarity here with the Japanese game GO. Since there is "space" (the board itself with its mesh of sites), "matter" (the tokens), a local "dynamics" (the specific rule which tells how to update the contents of a site depending on the token pattern that surrounds it), and "time" (the sequencing of board configurations, move after move, as they describe a "trajectory"), the whole thing can be viewed as a *toy universe*. Basically, I would choose a rule and study what would happen starting from a number of different initial configurations. Then I would explore another rule, and so on, systematically.[‡]

A board of finite-size,[§] say, 32×32, will have an astronomically large yet *finite* number of configurations. Thus, *any* trajectory must eventually enter a cycle. In a computer simulation, to know when the system had entered a cycle one ought to check whether the current state matches any of the previously encountered ones. This was of course rather expensive; so I made the computer memorize, along the simulation, a sparse number of breakpoint states, and compare each new state only with the breakpoints. If I found a match, I would know that I was on a cycle; but to know exactly when the cycle had been entered I had to recompute part of the trajectory, from a breakpoint forward or from the current state *backwards*. The issue thus arose of how to determine the "backward law" corresponding to a certain "forward law." For a given rule, how could one "take a move back,"

[‡] A few years later Stephen Wolfram, with similar motivations, embarked on a remarkably similar project using a more sophisticated conceptual background and heavier computing guns.[29]

[§] Like quadrille paper, a cellular automaton is characterized by a local structure that may extend indefinitely in all directions. If, in order to have a finite universe to experiment with, we delimited this structure with boundaries, then we'd have to introduce artificial, ad hoc rules to justify the special behavior required at the boundaries. But we can turn a finite swatch of quadrille paper into a *boundaryless* structure if we "wrap it around" into a *torus* by joining opposite edges.

i.e., construct the predecessor of a given board configuration? Might it have more than one predecessor? or none at all?

This led me to investigating *invertible* cellular automata — those in which every configuration has one and only one predecessor. And this opened a Pandora box of challenges and at the same time a treasure chest of rewards. Our own physical world appears to follow, at the microscopic level, *strictly invertible* laws (this applies to both classical and quantum mechanics). While conventional cellular automata may offer a good approach to modeling certain macroscopic phenomenologies (e.g., urban traffic, biological growth), invertible cellular automata offer a direct shortcut to modeling aspects of fundamental and emergent physics — thence the title, *Cellular Automata Mechanics*, of my thesis. Part of the ensuing story is told in a quite readable early review.[21] Exciting new discoveries, which in a certain sense take a long line of inquiry to a natural completion, are reported in Ref. 9 and a burst of recent papers;[10,24–26] the latter show that *thermodynamically efficient information processing* in a uniform microphysical medium is in principle possible, but at the cost of *huge* investments in computational infrastructure.

In sum, it was invertible cellular automata that led me to developing an intimate interest in the "new" thermodynamics — that of information, communication, and computation, classical and quantum. More of this in the third section ("Hot and cold").

There is a strict parallel here between the subject of the above theoretical investigations and the concrete problems faced today by our society as it strives to achieve a "green" approach to production — where energy and materials would be used as efficiently as possible. Being efficient is certainly a lofty aspiration, but basic models of the kind I've been studying prove that there are *intrinsic, irreducible* reasons that make the design of efficient processes hard to achieve. Worse than that, even when feasible solutions are found on paper, their material implementation turns out to depend on large and complex infrastructure. Essentially because of thermodynamics' second principle, the "green" target of lowering the *daily operating expenses* of a production process can only be achieved at the cost of *huge capital outlays*. Thus, beyond a certain point (and actually rather soon), the amortization costs of the infrastructure will wipe out any savings in operating expenses. Recycling a first small fraction (of materials, energy, or information) is easy; recycling additional fractions grows exponentially harder.

2.1. "Probability Doesn't Exist"

What I've learned by restructuring knowledge for my own uses is that everybody keeps telling you what is supposed to be "*good for you.*" You may bet that whatever they suggest is most likely good *for them.* Whether and to what extent it is also good for you — only *you* can be the judge! This is not as simple a matter as freely exercising your *will* ("From what is offered I'll just pick whatever I want"); the real issue is "But what is it that I should want?" For this, even and especially if you hope to benefit from other people's wisdom, you have to work hard at developing and maintaining *your own* model of the world — one that *you* can trust. It is in this spirit that I read Leonardo's dictum, "Woe to the disciple that does not go past his master!"

The above attitude is particularly helpful for making sense of and not getting lost into so much that has been written about probability and randomness — not to mention entropy, the second principle of thermodynamics, inference, and decision. While preparing this essay I happened on a sentence from von Mises' seminal paper[27] that made me shudder:

> "The following account is based on the conception of probability theory as a special science of the same sort as geometry or theoretical mechanics."

In other words, "We shall try to cast in abstract mathematical form, like Euclid and Lagrange did, certain empirical constructs and relations that the tangible world out there confronts us with." This was reiterated and clarified a few years later:[28]

> "... just as the subject of geometry is the study of space phenomena, so probability theory deals with mass phenomena and repetitive events."

Note the term 'phenomena'.

Now let us compare that with de Finetti's preface to his treatise on probability:[4]

> "My thesis, paradoxically, and a little provocatively, but nonetheless genuinely, is simply this:
>
> PROBABILITY DOES NOT EXIST.
>
> The abandonment of superstitious beliefs about the existence of Phlogiston, of Cosmic Ether, Absolute Space and Time, ... , of Fairies and Witches, was an essential step along the road to scientific thinking. Probability too, if regarded as something

endowed with some kind of *objective existence* [italics mine],
is no less a misleading misconception, an illusory attempt to
exteriorize or materialize our true probabilistic beliefs."

Is probability then something *objective* (von Mises), that has to do with
facts out there, and that we try to internalize, or something *subjective*
(de Finetti), that has to do with our beliefs, which we then project onto
the outside world?

Enter Jaynes (see Refs. 8 and 19), best known as a pioneer and cham-
pion of the Maximum Entropy Principle (MAXENT for insiders): If, in a
given context, you need to *formulate* a probability distribution on which to
base your initial bets, choose, among all possible distributions that agree
with what *you know* about the problem, the one having *maximum entropy*.
Why? Is this guaranteed to be the "real" (whatever that may mean) probability distribution? Of course not! In fact you will most likely replace it
with a new one as soon as you see the outcome of the first trial — because
by then you'll have one more piece of information. Why, then? Because
any other choice would be equivalent to (a) *throwing away* some of the
information you have or (b) *assuming* information you *don't* have. If you
had to program a robot to autonomously cope with the uncertainties of
Mars' environment, you'd want its choices to be weighed by the MAXENT
principle; anything else would be *indefensible*.

This is real progress towards resolving the objective-vs-subjective
dilemma. According to Jaynes, a probability distribution is created by
me *on the spot* on the basis of the information I currently have, but there
should be nothing *subjective* about the *creation process* itself. If the robot
had different information, I would want it to use *that* information rather
than the one I have (that's why I made an autonomous robot in the first
place!), but using the *same process* I would have used — the process which,
for that matter, *anyone else* should have used! The information may be
subjective but the process of honest inference should be objective and universal — that's the way science works.

As Jaynes stresses with his book's title, *Probability Theory: The Logic
of Science*, the above process is nothing but *ordinary logic* — plain syllogism — only run *in parallel* over a number of cases. You know the story that
"All men are mortal; Socrates is a man; therefore Socrates is mortal." That
is not, of course, what Logic says; she only says that "*If* all men are mortal,
and *if* Socrates is a man, *then* Socrates is mortal" — note the 'if's. In a
similar way, *without having to know anything about Mac Donald's farm*,

probability theory can nevertheless correctly affirm, for instance, that "*If* Mac Donald's menagerie consists of the farmer, his daughter, a cow, and two chickens, and *if* (nonhuman) animals don't pay taxes, then no more than 40% of McDonald's menagerie pay taxes."

From this example you also see how probability theory yields a number between 0 and 1 rather than True or False: we want it to return a value that applies to a collection of (actual or potential) cases *as a whole* rather than a single case, and so we get the *fraction* of the individual cases for which a certain predicate ("pays taxes") is true. So far, our fraction has come out of the ratio of two integers — two *counts*. Now you may ask how one should handle the situation where cases may have different *weights* — as with the faces of a (literally) weighted die. Let me give you a hint. You will agree that if in listing the farmer's menagerie I had mentioned "three brace of pheasants" that should have counted as *six* animals, because 'brace' is just a quaint way of saying 'two' when you are counting pheasants; similarly, if I had mentioned "a litter of piglets," you would not treat 'a litter' as a single animal: in order to return a definite value you would ask *me* (certainly not probability theory; it's not its job) how many piglets there were in a litter.

Now, I go beyond de Finetti and Jaynes (remember Leonardo's saying?) by extrapolating the spirit if not the letter of their approach. Let us have it that *any* probability is, in principle, the ratio of two counts ('good' cases over total cases). Suppose the tax assessor asks me, "how many piglets?" and I reply "I don't remember precisely; maybe three or four." To which he may reply, "Well, let me write down 3.5; do you mind?" But if each individual case still ought to be a logical case, what sense should one make of 0.5 piglets? The law says that a piglet, as an animal, doesn't pay taxes; but half a piglet???

My reply — we are talking now about the *interpretation* of probability — is this. We envisage *two* possible universes, one with three piglets and one with four, and we take them *both* — as two distinct cases. For each of these two cases all other counts remain the same and so will appear two times (two farmers, two daughters, etc.) in our overall head count — which will also include 7 piglets, 3 from one universe and 4 from the other. The number 3.5 the assessor wrote down is really an abbreviation — a convenient *normalization* — for "7 piglets out of the total head count from 2 very similar universes." Here, again, we are running ordinary logic in parallel over two universes.

Indeed, in the matter of interpretation of probability theory I am "more royalist than the king." According to von Mises, a probability distribution captures certain numerical aspects of an objective state of affairs — of a *fact*. For de Finetti and Jaynes, it captures aspects of a subjective state of knowledge, of *information* one has. What I propose is that a probability distribution should be associated, more neutrally and more generally, with the *description* of a system — typically, an *incomplete* description that by its very nature matches *any of a number of different complete descriptions*. Here 'complete' will mean anything that is appropriate to the nature of the system and of the idealization of it that we entertain, such as a 'microscopic state' for a physical system, or 'a specific combination' for a hand of cards.

In this vein, I would be tempted to change Jaynes' book's title to "Probability Theory: The Logic of *Incomplete Description*," where "incomplete" simply means that I don't know — or, more generally, *I don't care to specify* — which precisely of cases a, b, c we actually have, and therefore all I can do is give you the whole "bag" $\{a, b, c\}$ and say, "Sorry, but you'll have to run the same logic argument on scenarios a, b, and c. If your argument is such as to lead to a yes/no outcome in each case, then the overall answer instead of a "Yes" or "No" will be a number, namely, the relative fraction of "Yes's;" in other words, a *probability*.

An (incomplete) description could be a conventional thermodynamical characterization of a physical system, such as "one liter of Helium at standard temperature and pressure," that a frequentist like von Mises would be quite happy with. Note that the latter description is a "wildcard" matched by $\approx 10^{5 \cdot 10^{22}}$ distinct microscopic states [15, p. 101] — that is, a number of cases equal to 1 followed by 50,000 billion billion zeros! When the numbers of "good" cases and of total cases are so super-astronomically large, it is certainly forgivable to handle their ratio as if it were a real number rather than a rational, and thus do all probabilistic accounting in terms of measure theory even though ordinary arithmetic would in principle always suffice. But note that my claim that probability theory is simply "wholesale logic" is tenable only if these real numbers are *in principle* reducible to ratios of counts.

For descriptions that would please subjectivists like de Finetti and Jaynes, let's take two scenarios. In Scenario 1, on the day after the drawing of the British National Lottery (I haven't read the papers yet) a stranger in Trafalgar Sq. offers me ticket #038823604021 at one-tenth the nominal price — should I accept the offer? Scenario 2 has the same day, the same ticket number, the same discount, the same dilemma; but this time the offer comes from a ship mate — we've been stranded on a desert island for a week; no communication with the outside world has been possible, but we have reasonable hopes to be located and rescued. Clearly, the two scenarios describe different states of affairs — even though the *mechanics* of the Lottery is quite insensitive to them[¶] — and the probability distribution that any one should use should be based on the description provided for the two scenarios. The probability that the ticket I'm offered is the winning one is certainly not, à la von Mises, a physical-like quantity attached to the ticket or residing in the lottery urn, and in fact one doesn't need to know any of the lottery's details to refuse the first offer and accept the second.

My "descriptivist" interpretation of probability claims that, even though subjective information may provide ingredients used to *formulate* a certain description, information *per se* is a red herring. Take the case of a corporation officer who wants to make a routine adjustment to his personal securities portfolio, and also happens to have access to critical insider's (i.e., non-public) information about his company. To stay within the law, his trading decisions should be based on a probability distribution that deliberately *ignores* important information he may have, and thus violates MAXENT. Different scenarios ("If I want to stay clean," "If I want to maximize my gains," "If I want to maximize my overall expected benefit, including the the negative benefit of a term in jail") will lead to different probability distributions, just as different premises will lead to different conclusions in a syllogism. What is true in a syllogism is not the *conclusion*, but a certain *relationship with its premises*. What a probability expresses is certain numerical aspects of a situation *as described*, i.e., of a particular scenario. The latter could be a totally fictitious yarn; a snapshot of somebody's knowledge; or a conventional, impersonal characterization of a physical setup; in general, the enumeration (explicit or implicit) of a countable 'universe' of cases and of a number of subsets of it called 'events'. All the same, as Jaynes stresses, the conceptual analysis of what we do when we introduce a probability distribution stops here; passage to the continuum or to more sophisticated measure-theoretical accounting does not need — and, I add, *should not* — entail any conceptual or interpretational novelties.

[¶] "The farmer prays to Zeus for rain, the potter for dry days," or, as in Voltaire's *Candide*, "After the two armies had thoroughly massacred one another, each of the two kings thanked God for a victory with a *Te Deum*."

In this interpretation, a probability distribution simply "projects" over a relatively small number of numerically-valued bins a possibly astronomical number of truth-valued syllogisms, as per the following commutative diagram:

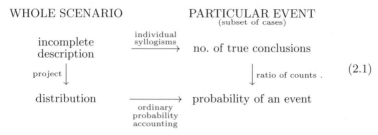

$$(2.1)$$

If, out of convenience or necessity, the projection is somewhat sloppy or approximate — as when we do roundoff in ordinary arithmetic — then the probability obtained by following, from the START square ("incomplete description") to the FINISH ("probability of an event"), the lower of the two paths (passing through "distribution") will be an *approximation* of that which would be arrived at by following the upper path (passing through "no. of true conclusions").‖

The reason why the above diagram commutes is that an event is by definition just a collection of outcomes, and thus any accounting may ignore the *order* of the outcomes within an event. This is what makes a probability theory *possible* — indeed, it is the essence of any such theory. In the next section we'll ask what makes probability theory useful *in our world*.

2.2. "Hot and Cold" is the Only Game in Town

It will not have escaped the astute reader that my "descriptivist" approach is just like the (neo-)Bayesian one — "all probability is conditional probability" (just like "the truth of any syllogism's conclusion is a conditional truth") — but with the added stipulation that probability DOES NOT EXIST (cf. de Finetti above) as an independent quantity: it is just the fraction of oucomes encompassed by an event. Note that substantially the same approach to interpretation had already been entertained by Laplace, as early as 1779, as a matter of commonsense.

‖This is exactly what happens, for instance, when in the Boltzmann equation one approximates the collision term by that resulting solely from two-body collisions between particles assumed to be uncorrelated prior to the collision (*Stosszahl Ansatz* or "molecular chaos" assumption). That, of course, disregards the correlations introduced by all previous collisions, but doing this makes little difference near equilibrium.

With the approach discussed here, how do we know the *prior probabilities*? Of course we *don't*, exactly as we don't *know* a syllogism's premises. Rather, we *propose* them as working hypotheses and thus just play a game of "What if?" or "Let's pretend!" As Jaynes clearly states on a number of occasions,[8] a proposed prior becomes *useful* precisely when experimental results (or later information, or a more complete description) invalidate the forecasts based on that prior. If our gambling partner gets 75 heads out of 100 coin tosses we start suspecting that something is amiss (perhaps the coin is asymmetric or the tossing doctored) *precisely* because that result departs from our 50% or "null" prior. Only a mismatch between expectation and outcome will make us feel the need to revise the canonical "fair coin" prior — that is, consider different (typically more detailed) scenarios and ask more questions ("Does that also happen when he uses *my* coin?" or "What if I try *his* coin?")

In the end, all we ever actually *know* is *our own scenarios*, because *we made them ourselves*. Yet, through the feedback provided by the above "Hot or Cold" game we attempt to converge on scenarios for which we can say not so much "This model of the world works well" (as if we had been testing an uncertain model against a known world), as "The world [which we don't know] *appears* to work pretty much like our model [which we happen to know perfectly]." Like in a *phase-locked loop* demodulator of FM broadcasting, in the end all we listen to is not the external signal-plus-noise voltage that reached us, but the variable voltage which *we ourselves* generate and constantly adjust in order to keep a voltage-controlled oscillator in phase with the received signal.

In sum, the world doesn't directly answer questions from us such as "*What* are you *like?* — Please describe yourself." But we can make a *correlator C* that takes in an aspect W of the world and a proposed model M for it and answers the question "*To what extent* do W and M match (according to the specific criterion embodied by C)?" The output f_C from the correlator is not a description of the world, but merely a *number* ranging, say, from 1 ("perfect match") to 0 ("no response whatsoever"). Those numbers, i.e., sample values of the function $f_C(W, M)$ — are all we ever get to know about the world. In the words of Charles Darwin,** "*How odd it is that anyone should not see that all observation must be for or against some view if it is to be of any service!*"

** In a letter to his friends Henry Fawcett.

Remark that that is precisely what *evolution*, which has successfully been in the inferring business for four billion years, gets away with, as a strategy to perform its wonders by! There, the correlator between a proposed genotype and the world is played by letting loose in the world the corresponding phenotype; the number ("goodness of match") returned as feedback is simply the survival rate.[12,23]

Now our contract with probability theory is clear. This theory shall (a) help us *design the above correlators* and suggest reasonable approximations to their logic and their physics so that they can actually be built;[††] and (b) help us *design scenarios*, seeking those for which the system under investigation might provide an ever better match.

At the end of the first section, *invertibility* was announced as a lead actor in this play. Here is where it makes its entrance.

What worth is a probability — a count or a number — that may come out of our theory, if this number changes all the time? — if last month our securities were worth \$1,000,000 and today we discover that they are worth only half as much? The problem is that our descriptions typically refer to objects from the real world, and the world and the objects therein evolve according to a nontrivial dynamics — so that the descriptions themselves keep changing! Are there any properties of these evolving descriptions that nonetheless remain *constant*?[‡‡] In the present case, it is the *invertibility of the microscopic dynamics* that guarantees that the *fine-grained entropy* associated with an initial description remains *constant* as the system — and accordingly its description — evolves in time.

With a given dynamics τ, at every "turn of the crank" (here for simplicity we assume a discrete dynamics) the current state of the world, s_t, advances to a new state, $s_{t+1} = \tau(s_t)$. In Figure 2.1 the numbered circles represent states and the arrows lead from one state to the next. All the parameters that enter in the premises of a syllogism will advance to a corresponding new value, and the result of the syllogism itself will accordingly

---

[††]Suppose we want to determine if a certain mixture with given concentrations of H_2 and O_2 is explosive. We could model the individual collisions of some 10^{24} molecules of hydrogen and oxygen and explicitly count their number. Alternatively, using schema (2.1), we can estimate the *average number* of collisions of H_2 with the walls of the container by measuring its partial pressure p_H — and similarly for p_O. By multiplying these two pressures together we get a good idea of the number of collisions between the two species. Thus we replace molecular microphysics with simple *macro*physics and the ANDing of 10^{48} logic equations with the multiplication of two numbers; again, we manage to "count without counting."

[‡‡]In physics, we are familiar, for instance, with the importance of *conservation of energy*.

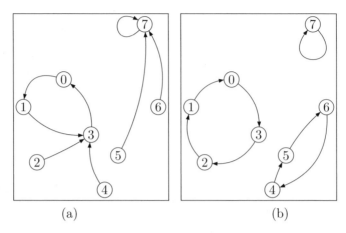

(a)   (b)

Fig. 2.1. (a) A noninvertible dynamics on eight states, (b) An invertible dynamics on the same state set: every state has *exactly one* predecessor. All trajectories are then loops with no side branches flowing into them.

change. The count of outcomes favorable to a given event will inexorably be updated at every "tick."

Take the system of Figure 2.1(a); this specific dynamics is *not* invertible. Suppose I am told that the system is in one of microstates 2, 3, or 4; my "state of incomplete knowledge," or *macrostate*, is described at time $t = 0$ by the 8-tuple of occupation numbers for the eight states 01234567, namely $\langle 00111000 \rangle$. (With only two possible values, 0 and 1, for an occupation number, a tuple is but a way to represent a macrostate by its *indicator set* — which tells us which microstates belong to it and which don't.) After one step of the dynamics state 3 will have gone into 0 while both states 2 and 4 will have gone to state 3; this is reflected by the new macrostate 8-tuple, $\langle 10020000 \rangle$, which records an occupation number of 1 for state 0 and of 2 for state 3. (Don't worry about a value of 2 or more for an occupation number: as explained below, in the present context a tuple with larger occupation numbers can be viewed as just a convenient shorthand for one with only 0s and 1s.) Continuing in this way, I can trace the whole "trajectory" of my state of knowledge, namely,

$$\langle 00111000 \rangle, \langle 10020000 \rangle, \langle 21000000 \rangle, \langle 02010000 \rangle, \langle 10020000 \rangle, \ldots$$

which eventually becomes cyclic with period 3. Weights move along step by step, together with the respective microstates, and are added together at every merger of trajectories. The occupation octuple is none but an ordinary probability distribution, if we normalize each entry by dividing it by the number of initial cases (3, here). In this system, the fine-grained entropy $\sum_i p_i \log \frac{1}{p_i}$ of a macrostate, which started at $\log 3$, *drops* at the first step to an amount $\frac{2}{3} \log 2$ (and thereafter remains constant as the system enters a cycle). For a deterministic system, microscopy entropy is in general *monotonically decreasing*.

A special case is the dynamics of Figure 2.1(b), which is *invertible*. Here trajectories never merge, occupation numbers hop from one state to its successor together with the state itself but never merge — they are merely *permuted* by the dynamics. As a consequence, the fine-grained entropy of any macrostate remains *strictly constant* along its trajectory — since a sum like the above $\sum_i$ is of course insensitive to the order of its terms. Such a system is, as it were, "informationlossless."

Remark that, in the invertible case, not only does the entropy of a distribution remain *constant in time*, but also the *probability* of an individual microstate (represented, as we've seen, by its occupation number) does — provided that we attribute an identity to "a specific occupation number" even as it hops from microstate to microstate according to the dynamics. To keep track of such a "traveling" occupation number we can use, as coordinates for it, the microstate to which it initially belonged and the current time t; then the dynamics τ will tell us on which microstate that occupation number is "parked" at present. (Such a duality between "*what object* is at a given place" and "at *what place* a given object is" is analogous to the duality between Eulerian and Lagrangian viewpoint in fluid dynamics, and Schrödinger vs Heisenberg picture in quantum mechanics, and in all these cases comes to us as a gift of the underlying invertibility.)

With this coordinatization, the occupation tuples used above become genuine indicator sets, with only 0 and 1 values, whether the system is invertible or not. It may still be convenient to make use of larger occupation values, with the understanding that they count the "nanostates" of an even finer-grained underlying dynamics; we keep track of the number — but not individual identity — of these nanostates in the form of a *weight* attached to a microstate. In this way, a microstate-cum-weight becomes, so to speak, a "light" macrostate — what in conventional probability theory would be called an "elementary event," indivisible but provided with a weight.

But why spend time arguing about the interpretation of probability theory, if in practice almost all people, even though attending different "churches" (or no church at all!), end up performing, for the same problem, *essentially the same calculations*? I will just mention, without discussion or proof, a conclusion on which I stumbled following the above trail of ideas, and that has freed much room in my mind—taking away a lot of junk left over by two centuries of debates and replacing it with a small jewel. Namely, with an appropriate, quite pragmatic definition of 'entropy', it turns out that

> "The *second law of thermodynamics* is logically equivalent to the *invertibility of the dynamics!*"

In other words, if a system is microscopically reversible, then its macroscopic behavior will strictly obey the second law of thermodynamics (incidentally, displaying macroscopic *irreversibility* and all that); conversely, if it obeys this law, then it *must* be microscopically reversible.

As simple as that. Wasn't it worth it?

2.3. Maximum Entropy was Easy; Now, How about Least Action?

In the early 1870's Boltzmann had proposed to identify entropy with a *count of microstates*, according to the well-known formula $S = k \log W$. This proposal set in motion a cathartic process, which took a good century, of turning thermodynamics into statistical mechanics, and at the same time convincing most of us that randomness, probability, and entropy are human-made categories rather than physical quantities or parameters independently existing "out there" and just waiting to be measured.

At just about the same time (1877), by proposing the idea (picked up and developed by Planck in 1900) that the energy levels of a physical system could be discrete, Boltzmann himself also set in motion an extraordinarily disruptive process—*quantum theory*—which has picked up new speed in the past two decades with "quantum information, communication, computation, and all that"—and still looks far from settling down (cf. Ref. 7). As earlier with probability theory, (almost) everyone agrees on the formulas and the predictions of quantum mechanics, no matter what "church" (if any) they come from, but they can't seem to agree on a common "theology"—an intuitively satisfactory *model*. Richard Feynman, who delighted at being an *enfant terrible*, famously quipped "I think I can safely say that nobody understands quantum mechanics".[6] According to Paul Quincey[13], "The greatest mystery of quantum mechanics is how its ideas have remained so weird while it explained more and more about the world around us." For comic relief,

"Erwin [Schrödinger] with his ψ can do
Calculations quite a few.
But one thing has not been seen:
Just what does ψ really mean?"

[Erich Hückel]

The best minds on Earth have been mulling over the interpretation of quantum mechanics for a century. Some, like Bell with his inequality, sharpened the horns of the question in a helpful way; others proposed contrived philosophical "explanations" that didn't help much; others abandoned the quest and declared victory; others maintain that there is no problem — quantum mechanics is just the way the world is; others (including me) feel that the problem is a serious one but hasn't been solved yet. Among the latter was Jaynes, whose lifelong detour into probability theory had started as just a minor side step for reconnoitering the quantum mechanics fortress before attempting to breach it. He kept his hopes high to the end:

"A century from now the true cause of microphenomena will be known to every schoolboy and, to paraphrase Seneca, they will be incredulous that such clear truths could have escaped us throughout the 20th (and into the 21st) century....

This conviction *has affected the whole course of my career.* I had intended originally to specialize in Quantum Electrodynamics, but this proved to be impossible [because] whenever I look at any quantum-mechanical calculations, the basic craziness of what we are doing *rises in my gorge*[!] and I have to try to find some different way of looking at the problem that makes physical sense. Gradually, I came to see that the foundation of probability theory and the role of human information have to be brought in, and I have spent many years trying to understand them in the greatest generality."

Tantalizingly, on first sight quantum mechanics appears to be centered on issues of probability. Even more, the principles of quantum mechanics are an expression, as Feynman showed with his "path integral" formulation, of the so called "least action" principle, and the latter is an Enlightenment-time incarnation of a deep-seated desire of humanity to identify Providence (or at least recognize efficiency) in Nature:

"No natural action can be abbreviated....
Every natural action is generated by nature in the shortest possible way she can find. From given causes, effects are generated by nature in the most efficient way."

[Leonardo da Vinci]

More recently science has developed just the opposite reflex — the conviction that Nature is as lazy and sloppy as can be, not because she is perverse but because she just doesn't (or *can't possibly*) care. By now we all know about Darwinian evolution, mindless bureaucracies, and the loose cannon of the free market. In fact, in most cases statistical mechanics explains away "optimal" solutions (usually a constrained maximum or minimum) in terms of *combinatorial tautologies*. A loose chain tied at its ends and given a shove, in a vacuum and in the absence of gravity, just flaps about chaotically and restlessly, visiting or coming close to all of its accessible states. But, after you introduce air in the container, in a short time the energy of the chain will be found shared and traded, quite democratically, with all of the gas particles as well; these are order of 10^{24}, so that the chain's average share will be *very close to zero* — the chain will have virtually come to rest. If we now turn gravity on, the energy budget will come to include potential energy. It turns out that the *overwhelming majority* of the allowed configurations of the whole system (chain plus gas particles) — those having the same energy as at the start — comes very close to showing a chain shaped as a barely quivering *catenary* (that figure, similar to a parabola, that so pleased and intrigued Galileo, Leibniz, Christiaan Huygens, and Johann Bernoulli). A catenary is indeed the figure of *least potential energy*, but not because some frugal manager is minding the "energy shop;" on the contrary, because no one tries to prevent the countless gas molecules from helping themselves as they please. Since the system configurations in which the chain displays a shape close to a catenary happen in this case to be in the majority, the macrostate in which the chain has "catenary shape" is that of *maximum entropy*. That state is preferred not because it is most orderly — though it so appears if one only sees the chain and not the gas molecules — but because indeed it is *most random*. With a tautology of evolutionary flavor, that shape *succeeds* because it is the *most successful*, comprising, as it does, *almost all* microscopic states. Almost any of the system's accessible states that one may pick at random would display a catenary, even though once in a billion billion billion years one may perhaps see the chain twitch for a moment away from the catenary ideal.

That is the typical shape of a MAXENT argument — to just bring to the surface an underlying combinatorial tautology. Now, just as maximum and minimum are often interchangeable, can't we presume that action will work just as well as entropy in an argument of the same flavor, and so hope to explain away the mysteries of quantum mechanics as some specialization, extension, or transliteration of MAXENT?

Before even *suggesting* trying to interpret quantum mechanics as a more complete version of what statistical mechanics should have been in the first place — which, by the way, I believe is a worthy endeavor — one must list a number of serious obstacles. We are comparing apples with oranges, and there is no hope that turning an orange into an apple will be as simple a job as painting it red.

- Naive statistical explanations for "wave mechanics," as it was called then, were tried without success as soon as quantum effects emerged as a permanent feature of physics. Jaynes himself must have been pursuing this rainbow, spending an entire lifetime at developing more mature and sophisticated tools, and yet ever seeing it recede towards the horizon.

- While the variational principles of statistical mechanics have to do with system *states*, and probability has, as we've seen, an obvious interpretation in terms of "the size of a bunch of microstates," the variational principles of analytical mechanics have to do with system *trajectories*. How should "a bundle of trajectories" be counted, and what statistical quantities analogous to probability or entropy should this count give rise to?

- More specifically, entropy has to do with a distribution of *states*, that is, with a partition of the whole set of microstates; it doesn't need to know anything about a dynamics. Action, on the other hand, is assigned to a virtual trajectory (a sequence of microstates obeying general kinematic constraints) by integrating over it a local quantity (the *Lagrangian*) as assessed by a *specific dynamics*; one cannot speak of action if not in the context of a given dynamics.

- Even more specifically, maximum entropy, as a relation of a distribution vis-à-vis neighboring ones, only cares about the structure of the set of distributions as a multidimensional simplex; again, no dynamics has to be mentioned. On the other hand, stationary action, as a relation of a piece of virtual trajectory vis-à-vis neighboring ones, has to do with the manifold of all virtual trajectories and the particular Lagrangian function imposed on this manifold by a specific dynamics.

- The conservation of fine-grained entropy is purely a consequence of the invertibility of the dynamics. The conservation principles of an-

alytical mechanics and their relation to a dynamical system's symmetries (Noether's theorem) seem to be dependent on a much more
richly structured form of invertibility, namely, *symplectic structure*
in classical mechanics, and *unitarity* in quantum mechanics.

- Finally, the verbal analogy between the "least action" of the epony-
mous principle and the "maximum entropy" of Jaynes' principle is
a red herring. The Lagrange–Euler–Hamilton principle is one of
stationary (not *least*) action. Totally different variational principles enter into play.

All these differences between statistical mechanics[§§] and dynamics (classical or quantum), should make one despair of finding a reductionistic *deus
ex machina* that will bring a resolution to the quantum mechanical conun-
drum.

If the task appears so hard, one may ask, "Wait a moment — before
trying this reductionistic program on quantum mechanics, shouldn't I try
it first on ordinary analytical mechanics? Suppose quantum mechanics
had never been discovered, shouldn't someone want to try to explain the
beautiful variational principles of analytical mechanics in terms of emergent
features of an underlying discrete, 'nanoscopic' dynamics?[¶¶] Or instead it
was truly necessary to wait for quantum mechanics in order to have a non-
anthropomorphic explanation of this teleological 'least action' principle?"

In sum, much as we did for the "optimal" shape naturally taken by
soap bubbles and hanging chains, can't we explain the providential shape
of trajectories obeying the variational principles of analytical mechanics as
emerging from the statistics of a combinatorial solitaire played with a large
number of discrete tokens? We have no problem with making use of probabilities — since these real numbers arise quite intuitively and tangibly as
a way to summarize ratios of big counts — but we'd like to avoid introducing ad hoc *complex amplitudes* unless these too can be given an intuitive
combinatorial embodiment and shown to be emergent features of the game.

The "energetics" school of the late 1800's, which included Wilhelm Ost-
wald and Ernst Mach, had worked hard at building a rational, self-contained

---

[§§]Which should really be called "statistical *quasi-statics*," since there dynamics is used,
if at all, in a qualitative way, to tell which way a scale will tilt, not when or how fast.
[¶¶]I gave this experiment a first try in "Action, or the fungibility of computation",[22]
arguing that *action* is a good candidate for an interpretation as a macroscopic measure
of the *computational capacity* of an underlying fine-grained computing structure, much
as *entropy* is a measure of the *information capacity* of a communication channel. I intend
to pick up that trail again with better preparation and equipment.

science of thermodynamics—what Mach called, in fact, *Pure Thermodynamics*—mostly expressed in terms of partial derivatives like those that still decorate thermodynamics textbook. It was easy for them to believe that those equations, which look so commanding and coherent, directly captured the essential reality of physics. They had no need for atoms and other such "old wives' tales" (at that time, atoms had been postulated by many but hadn't been seen or counted yet; only later in life was Mach forced to grudgingly accept their existence). Eventually, that "Sleeping Beauty"-like fairy tale was shattered by the like of Boltzmann, Gibbs, and Einstein. Those lofty equations could be derived, as convenient large-number approximations, by applying purely logical and combinatorial arguments—not new principles of physics—to ordinary particulate mechanics: the phenomena they described were but *accounting epiphenomena.*

In this context, a brief "elevator speech" for our quest could be, "Does Schrödinger's equation capture irreducible physics or just accounting epiphenomena?"

2.4. Quantum Mechanics: Mystery, No Problem, or Sour Grapes?

The way I have depicted the conceptualization and the use of probability, it is clear that most of the actual work goes into generating a variety of working hypothesis—even while accepting that most of them will be found wanting. Nature (but even man-made mathematics, for that matter) can only speak to us by muttering "Cold, cold, cold, hot, cold again, hot, ...," as explained before. It's up to us to imagine and propose *what scenarios* to subject to this primitive trial by fire and water. As they say, winning the lottery may not be easy, but how can you hope to win if you haven't even bought one ticket?

Since, in this game, we can submit as many tickets as we wish, fantasy is more important than technique. Concerning the interpretation of quantum mechanics, my feeling is that we've been too conservative in imagining models, perhaps fearing that we may embarrass ourselves in the eyes of our peers. Here is an edifying story.

In addressing the issue of the "spring of air" (what today we call Boyle's law for ideal gases, $pV = $ const), Boyle (1660) compared air particles to coiled-up balls of wool or *springs* (what else?) which would resist compression and expand into any available space. Newton toyed with this idea and showed that first neighbors *repelling* one another (after all, we want a spring that will push out, not suck in, don't we?) *inversely as the dis-*

tance would give the required macroscopic behavior. What became popular next is models viewing air as a swarm of discrete particles freely jostling about and interacting with one another in some prescribed way, and Newton himself eventually considered one with short range *attractive* forces (wasn't he the world expert in gravitation?), using inverse *fifth-power* law for its computational advantages. Bernouilli considered elastic collisions between hard spheres ("billiard-ball" model), while Maxwell considered inverse fifth-power *repulsion* (he got the same result as Newton with inverse fifth-power *attraction*; what would that suggest to you?). Much more recently (1931), taking into account the quantum-mechanical electrical structure of orbitals, Lennard–Jones established, for simple nonpolar molecules, an inverse seventh-power attractive law.

The moral of this story is that, if one is looking for a microscopic model for the laws of ideal gases, there is no dearth of *plausible* candidates. On the contrary, the same generic macroscopic behavior embodied by these laws will emerge from *almost any* microscopic model — using attractive or repulsive forces, power-law or exponential or whatever — that displays certain basic symmetries. In fact, even a model *with no interactions whatsoever* between particles — only elastic collisions with the container's walls — will do perfectly well. The "spring of air" is the most generic macroscopic expression of certain conservation laws (basically, of *microscopic invertibility*) — not of the technical details of a particular microscopic dynamics. Conversely, the fact that many power-law repulsive models gave the right macroscopic results was no guarantee that the interparticle forces are repulsive, power-law, or, least of all, fifth-power. If there was an embarrassment with the "spring of air," it was one of *riches*. Far from there being no plausible candidates, there were too many to choose from!

Given the generality of application and the *genericity* of design of the least-action principle — and similarly for the structure of quantum mechanics — I'm amazed that the same predicament — having many plausible interpretation and at worst not knowing how to pick the right one — hasn't occurred yet. Certain areas of research — materials science, economics, operations research, origin and extinction of species — have learned to look almost by reflex for an underlying fine-grained combinatorics every time they encounter phenomenology obeying variational principles. On the other hand, virtually all books that I have seen that deal with the calculus of variations from a mathematical viewpoint — and, specifically, the books on analytical mechanics — do not even *entertain the possibility* of a combinatorial origin. What's more serious, they do not even tell the reader that most variational principles happen to be of a *mathematical form* which, whatever its actual origin in each specific case, would most naturally *emerge* from an underlying fine-grained combinatorics.

What are the prospects for progress? To break the current impasse on the interpretation of quantum mechanics it might be useful to explore a detour first, namely, try to devise fine-grained combinatorial models of ordinary analytical mechanics — Noether's theorem and all that. (The latter shows that conserved quantities arise from certain symmetries of a system's dynamics. Edward Fredkin and Norman Margolus have been throwing in the ring the suggestion that, in a combinatorial model, it may be the conserved quantities — say, a quantity specifying a number of indestructible tokens — that generate symmetries, rather than the other way around.) If we get interesting combinatorial models that way, we may get inspiration for novel models of quantum mechanics. If, on the other hand, we can prove (for instance, by arguments similar to those based on Bell's inequalities) that no such model can exist unless it already incorporates quantum-mechanical principles (as does the path-integral formulation), then we'll know that already classical mechanics displays features as mysterious and irreducible as those of quantum mechanics, and so we might as well tackle directly the latter.

When will enough be enough? Operations theory swears by the *branch-and-bound* approach (and it appears that humans, and even animals, intuitively follow a similar strategy). Say, you want to know whether it is the case that A. First (a) you try to prove A; if after spending a lot of resources you haven't attained that goal, (b) you switch to trying to prove $\overline{A}$ (the negation of A); after having spent even more resources on that, (c) you go back to trying to prove A; and so forth. In the long run, (q) you may decide to invest part of your budget into trying to prove that A is undecidable. In other words, you hedge your bets and invent new bet options. When your wallet is almost empty, you may even start asking yourself, "Am I really *that* interested in this problem?"

In the matter of "prospects for progress," I will conclude with a topical snapshot. In 1994 Peter Shor invented an algorithm that, on a quantum computer, would factor integers in *polynomial* time — while the current best factoring algorithms for a classical machine take *exponential* time.[16] This event stimulated an outburst of research and a whole "academic industry" around quantum computing — the dangling Holy Grail being the prospect of exponential speedup over the entire computing front. This goal has so far proven elusive. Ten years later, Shor commented on this with a review paper, "Progress in quantum algorithms",[17] whose abstract sincerely reports, "We discuss the progress *(or lack of it)* that has been made in discovering algorithms for computation on a quantum computer. Some possible reasons are given for the paucity of quantum algorithms so far discovered."

Now that the best minds have been tackling this problem for fifteen years, shouldn't they switch to stage (b) of the branch-and-bound strategy, and ask the question, "Precisely what feature of quantum mechanics is supposed to grant it the godly power to speed every computation up exponentially? and are there any tradeoff? Here, the issue of interpretation — of what irreducible mechanisms might lie at the base of quantum mechanics — comes to the forefront. After all, evolution does perform design miracles even with no brains, but the conceptual simplicity of its design loop is bought by an exponential amount of waste, pain, and senseless cruelty. We are caught in a Faustian dilemma. The more we wish for quantumness to support miraculously fast computation (and who can rule that out yet?), the more we wish it to be miraculous and thus uninterpretable and irreducible. Conversely, if we root for quantum mechanics being a mere epiphenomenon — like statistical mechanics — we shouldn't at the same time bet that any apparently miraculous performance of it in the computing domain will not have to be paid for by astronomical outlays of resources at the fine-grained level, just as in the (otherwise quite legitimate) "DNA computing" business.[1]

So, we shall clench our teeth and keep proceeding in parallel on the (a), (b), and (why not?) (q) fronts. As Candide replied to Pangloss' confession of faith in an explainably providential world, "Excellently observed, my friend; but let us go on cultivating our garden."

References

1. L. Adleman, "Molecular computation of solutions to combinatorial problems," *Science* **266** (1994), 1021–1024.
2. L. Breiman, *Probability*, Addison–Wesley 1968.
3. B. de Finetti, *Un matematico e l'economia*, Angeli, F. Milano 1969. This is an edited, autobiographically-minded selection of de Finetti's papers from 1935 to 1968.
4. B. de Finetti, *Theory of Probability — A critical introductory treatment*, vol. 1 and 2, Wiley & Sons 1974, 1975, translated by Machí A. and Smith, A. from *Teoria della Probabilità*, vol. 1 and 2, Einaudi 1970.

5. W. Feller, *An Introduction to Probability Theory and Its Applications*, Vol. 1, 2nd ed., Wiley 1957.
6. R. Feynman *The Character of Physical Law* (1964), Modern Library 1994.
7. G. Jaeger, *Entanglement, Information, and the Interpretation of Quantum Mechanics*, Springer 2009.
8. E. T. Jaynes, *Probability Theory: The Logic of Science* (Larry Bretthorst, ed.), Cambridge 2003.
9. J. Kari, "Theory of cellular automata: A survey," *Theor. Comp. Sci.* **334** (2005), 3–33.
10. Lev B. Levitin and T. Toffoli, "Thermodynamic cost of reversible computing," *Phys. Rev. Lett.* **99** (2007), 110502.
11. I. Niven *The Mathematics of Choice: How to count without counting*, Math. Assoc. Am. 1965.
12. H. Plotkin *Darwin Machines and the Nature of Knowledge*, Harvard 1993.
13. P. Quincey "Why quantum mechanics is not so weird after all," *Skeptical Inquirer* **30**:4 (Jul.–Aug. 2006).
14. I. Rombauer and M. Rombauer Becker *The Joy of Cooking: The all-purpose cookbook*, Plume 1973.
15. D. Ruelle, *Chance and Chaos*, Princeton 1993.
16. P. Shor, "Polynomial-time algorithms for prime factorization and discrete logarithms on a quantum computer," *SIAM J Sci. Statist. Comput.* **26** (1997), 1484–1509.
17. P. Shor, "Progress in quantum algorithms," *Quantum Information Processing* **3** (2004), 5–13.
18. T. Toffoli, book review of Ref. 3, for the magazine *Sapere, Italy*, circa 1972.
19. T. Toffoli, "Honesty in inference" (a review of Ref. 8, *Am. Scientist* **92** (2004), 182–185. A longer version, "Maxwell's daemon, the Turing machine, and Jaynes' robot," is available as `arxiv.org/math.PR/0410411`.
20. T. Toffoli, "Cellular Automata Mechanics," PhD Thesis, The Univ. of Michigan, Comp. Comm. Sci. Dept. 1977; microfilm: `proquest.umi.com/pqdlink?did=758535201&Fmt=7&clientId=3740&RQT=309&VName=PQD`
21. T. Toffoli and M. Norman, "Invertible Cellular Automata: A review," *Physica D* **45** (1990), 229–253.
22. T. Toffoli, "Action, or the fungibility of computation," *Feynman and Computation: Exploring the limits of computers* (Anthony Hey ed.), Perseus 1998, 348–392.
23. T. Toffoli, "Nothing makes sense in computing except in the light of evolution," *Int. J Unconventional Computing* **1** (2004), 3–29.
24. T. Toffoli, P. Mentrasti and S. Capobianco, "A new inversion scheme, or how to turn second-order cellular automata into lattice gases," *Theor. Comp. Sci.* **325** (2004), 329–344.
25. T. Toffoli, S. Capobianco and P. Mentrasti, "When — and how — can a cellular automaton be rewritten as a lattice gas?" *Theor. Comp. Sci.* **403** (2008), 71–88.
26. T. Toffoli, "Lattice-gas vs cellular automata: the whole story at last," *J Cellular Automata* **4** (2009), 267–292.

27. R. von Mises, "Grundlagen der Wahrscheinlichkeitsrechnung," *Math. Z* **5** (1919), 52–99.

28. R. von Mises, *Wahrscheinlichkeit, Statistik und Wahrheit*, Springer 1928.

29. S. Wolfram, "Statistical mechanics of cellular automata," *Rev. Mod. Phys.* **55** (1983), 601–644.

Chapter 3

Statistical Testing of Randomness: New and Old Procedures

Andrew L. Rukhin

Statistical Engineering Division
National Institute of Standards and Technology
andrew.rukin@nist.gov

WHAT DREW ME TO THE STUDY OF COMPUTATION AND RANDOMNESS

The study of randomness testing discussed in this chapter was motivated by attempts to assess the quality of different random number generators which have widespread use in encryption, scientific and applied computing, information transmission, engineering, and finance. The evaluation of the random nature of outputs produced by various generators has became vital for the communications and banking industries where digital signatures and key management are crucial for information processing and computer security.

A number of classical empirical tests of randomness are reviewed in Knuth.[13] However, most of these tests may pass patently nonrandom sequences. The popular battery of tests for randomness, Diehard,[16] demands fairly long strings (2^{24} bits). A commercial product, called CRYPT-X,[7] includes some of tests for randomness. L'Ecuyer and Simard[15] provide a suite of tests for the uniform (continuous) distribution.

The Computer Security Division of the National Institute of Standards and Technology (NIST) initiated a study to assess the quality of different random number generators. The goal was to develop a novel battery of stringent procedures. The resulting suite[22] was successfully applied to pseudo-random binary sequences produced by current generators. This collection of tests was not designed to identify the best possible generator, but rather to provide a user with a characteristic that allows one to make an informed decision about the source. The key selection criteria for inclusion in the suite were that the test states its result as a numerical value (P-value) indicating "the degree of randomness", that the mathematics behind the test be applicable in the finite sequence domain, and that there be no duplication among the tests in the suite. All of the tests are applicable for a wide range of binary strings size and thus exhibit considerable flexibility. While an attempt was made to employ only procedures which are optimal from the point of view of statistical theory, this concern was secondary to practical considerations.

In the next sections we review some of the tests designed for this purpose. Most of them are based on known results of probability theory and information theory, a few of these procedures are new. Before doing this, however, we discuss one of the first applications of the test suite.

3.1. Testing Block Ciphers: Statistical Theory and Common Sense

One application of the tests of randomness is block ciphers. These ciphers are widely used in cryptographic applications. Ten years ago NIST carried out a competition for the development of the "Advanced Encryption Standard (AES)". Its goal was to find a new block cipher which could be used as a standard. One of the requirements was that its output sequence should look like a random string even when the input is not random.

Indeed, one of the basic tests for the fifteen AES candidates was "Randomness Testing of the AES Candidate Algorithms," whose aim was to evaluate these candidates by their performance as random number generators.[28] The winner of the competition, the Rijndael algorithm, as well as other finalists, Mars, RC6, Serpent and Twofish, were used in the experiment involving randomness testing of their bit output by using statistical procedures in the NIST test suite.

To measure their performance, a numerical characteristic of the degree of randomness was required. For a given test, such a characteristic is provided by the P-value which quantifies the likelihood of a particular, observed data sequence under the randomness assumption. We discuss the P-values, their interpretation, and the difficulties of assigning them in the next section. Meantime, it should be mentioned that although different from the probability of the randomness hypothesis being true, P-values bear the same interpretation: small P-values indicate a disagreement between the hypothesis and the observed data. Although larger P-values do not imply validity of the hypothesis at hand, when a string is being tested by a number of tests, they are necessary to continue the study of multi-

faceted aspects of non-randomness. A very important property of P-values
is that they are uniformly distributed over the unit interval when the tested
null hypothesis is correct. Thus, about 10 P-values should be expected in
the interval $(0, 0.01)$ if the true hypothesis is being tested $1,000$ times.

In a simplified description, in the AES evaluation experiment each algorithm generated a long string (about 2^{20} bits) stemming from the data type
and the keysize Altogether 300 different data sequences were generated by
each algorithm under combinations of data type and keysizes. Each of these
sequences was subject to a randomness test from the suite resulting in a
pass/fail decision. This decision was based on comparison of the P-value
and a significance level which was chosen as 0.01. The P-values obtained
were tested for uniformity. The sum of overall number of pass/fail decisions
over all sequences was used as the statistic to judge randomness according
to the particular test: if this sum is below a certain bound, the data set
is deemed to have passed this statistical test. Thus, each algorithm under a particular test of randomness generated three hundred decisions with
regard to agreement of the output and the randomness hypothesis. Both
the sum of pass/fail decisions and the characteristic of uniformity (based
on χ^2-test discussed in the next section) were used in the final assessment.
If none of the P-values fell below 0.0001, the sample was believed to have
satisfied the criterion for being random from the point of view of the given
statistical test.

This procedure has been criticized[19] from several perspectives. According to principles of statistical inference it is preferable to have one summary statistic on the basis of a long sequence rather than a number of such
statistics obtained from shorter subsequences. But testing of encryption
algorithms is not a purely statistical exercise. The validation of uniform P-
values does not enter into the calculation of the power of a particular test,
yet it can be seriously recommended from a practical point of view. The
whole problem of testing randomness is not as unambiguous as the parametric hypothesis testing problems of classical mathematical statistics.

The same common sense approach led to concatenation of cipherblocks
derived from random keys and different plaintexts. Murphy[19] compares this
process to interleaving or decimating an underlying message, so that either
some signal is removed or some noise is added. Exploration and validation
of various transmission regimes may be a more apt comparison. Besides,
from the cryptographic point of view, the entropy of the concatenated text
is larger than that of the original sequence. The concept of *randomization* in
statistical design theory presents a similar idea. Randomized designs do not

lead to better performance of statistical procedures if the postulated model is correct. However, they provide a commonly recommended safeguard against violations of the model assumptions.

Equally misguided seem to be arguments in favor of tests which are invariant to data transformations. In the statistical inference context this principle is violated by virtually all proper Bayes procedures. In AES testing context symmetrization over all data permutations is unpractical if not impossible. The use of random plaintext/random 128-bit key data type employed in the preliminary testing of AES candidates resulted in a mere permutation of the plaintext blocks. This data category was abandoned at later stages.

The concept of admissibility of tests (cf. Ref. 21) was not very helpful when testing randomness. Indeed, practical considerations led to inclusion in the suite not only the frequency (monobit or balance) test, but also the frequency test within a block. From the theoretical point of view the latter is superfluous, from the pragmatic perspective it is a useful check-up.

3.2. P-values, One-sided Alternatives versus Two-sided Alternatives and χ^2-tests

One of the principal difficulties of studying tests of randomness in statistical hypothesis formulation is that the null hypothesis, according to which the observed bits represent independent Bernoulli random variables with the probability of success $1/2$, is typically false. Indeed, this is certainly the case in the previous example of block ciphers, and more generally for all pseudorandom number generators which are based on recursive formulas. In view of this fact, one may expect only a measure of randomness to be attested to by a given string.

To explain, we recall the basic definitions of the classical hypothesis testing which traditionally involve the so-called parameter space Θ. This set indexes all possible probability distributions of the observed data. A subset Θ_0 of Θ corresponding to special parametric values of interest is the null hypothesis $H_0 : \theta \in \Theta_0$. In many problems one can specify the alternative hypothesis $H_1 : \theta \in \Theta_1$, often taking by the default $\Theta_1 = \Theta - \Theta_0$. However this specification is not straightforward and this is the case of randomness testing. For example, if Θ is formed by real numbers and the null hypothesis specifies a particular probability distribution for the data, the alternative could be all probability distributions different from that one, or perhaps all probability distributions which are stochastically larger or smaller. (Think of the life-time distribution of a device, or of the distribution of defective items in a lot.) What if elements of Θ are vectors or even more complicated objects?

In any case, assume that the particular alternative hypothesis leads to rejection of the null hypothesis for large values of a *test statistic* T, say when $T > T_0$. How can one find the cut-off constant T_0? The traditional (but somewhat dated) approach is to specify a significance level α (a smallish probability, like 0.01 or 0.05) so that under H_0 the probability of its false rejection is α (or does not exceed α.) Then the probability of the event $T > T_0$ evaluated under the alternative represents the chance of the correct rejection and is called the *power* of the test.

A body of classical statistical literature deals with the problem of finding tests of a given significance level which have the largest power. To accomplish that typically a distribution from the alternative is to be fixed. An additional difficulty of the randomness hypothesis is that its amorphous alternative is enormous and cannot be fully described by a sensible finite-dimensional parameter set Θ_1.

A more modern approach (implemented in almost all statistical software packages) suggests that the main characteristic of a test rejecting the null hypothesis for large values of a statistic T, is the empirical significance level, i.e., the probability of the random variable T exceeding its observed value $T(obs)$ evaluated under the null hypothesis, $P(T > T_n(obs)|H_0)$. One of the immediate benefits of this concept, called the P-value, is that the classical test of level α obtains if the null hypothesis is rejected when the P-value is smaller that α. The power function of a classical test can be expressed in terms of the P-value, but it involves calculation for $\theta \in \Theta_1$.

Since we believe that the P-value (not the significance level and not the power) is the most important practical characteristic of these procedures, each test in the suite results in such a value, and the collection of P-values from all the tests are consolidated into a vector reported to the consumer. If the distributions of two test statistics under the randomness hypothesis coincide, we consider them to be equivalent as the P-values are the same for both of these statistics.

Let n be the length of the string under testing. Each of tests in the suite is based on its statistic $T = T_n$ which, under the randomness assumption, has a desirably continuous distribution function $G(t) = G_n(t) = P(T \leq t)$ whose tail probabilities can be numerically evaluated. If a one-sided alternative corresponds to distributions of T which are stochastically larger than the distribution of T under the null hypothesis, then the P-value is

$1 - G(T(obs))$. Its large values are indicative of the fact that the null hypothesis is false, i.e., they support the alternative hypothesis.

For example, in the classical goodness-of-fit test, T is the chi-squared statistic, and H_0 postulates the probabilities of a multinomial distribution as calculated from the randomness condition. As discussed later, the P-value can be obtained from the incomplete gamma-function, and its small values lead one to believe in the falsity of the null hypothesis. This type of statistic is common in the suite. On the other hand, statistics distributed as a mixture of chi-squared distributions with different degrees of freedom were deemed to be too inconvenient to work with.

For some tests the alternative to our randomness hypothesis may not necessarily be restricted to distributions of T which are stochastically larger (or smaller) than the distribution of T evaluated under this hypothesis. Then the two-sided alternatives can be more appropriate with the validity of the null hypothesis being in doubt for small values of $\min[G(T(obs)), 1 - G(T(obs))]$. An interpretation of P-values in this case is as a "degree of agreement" between the statistic and its "typical" value measured by the median $\hat{T}$ of its distribution (see Refs. 5 and 22). Section 3.4 gives an example.

When G is a discrete distribution and the alternative is one-sided, the P-value with the continuity correction is defined as $\frac{1}{2}P(T = T(obs)) + P(T > T(obs))$. Under the randomness hypothesis, these P-values have an approximate uniform distribution on the interval $(0, 1)$ (exactly uniform in the continuous case.) This can be tested, for example, by the classical Kolmogorov-Smirnov test.

To achieve P-values with a uniform distribution on the interval $(0, 1)$, when a discrete-valued statistic is used, the original string of length $n = NM$ is partitioned into N substrings each of length M. For each of these substrings the frequencies, $\nu_0, \nu_1, \ldots, \nu_K$, of values of the corresponding statistic within each of $K + 1$ chosen classes, $\nu_0 + \nu_1 + \ldots + \nu_K = N$, are evaluated. The theoretical probabilities $\pi_0, \pi_1, \ldots, \pi_K$ of these classes are determined from the distribution of the test statistic.

The frequencies are aggregated by the χ^2-statistic,

$$\chi^2 = \sum_0^K \frac{(\nu_i - N\pi_i)^2}{N\pi_i},$$

which under the randomness hypothesis has an approximate χ^2-distribution with K degrees of freedom. The reported P-value cab be written as the incomplete gamma function.

This example (in which the exact distribution of T is a complicated sum of multinomial probabilities) demonstrates another difficulty of our testing problem. The exact distribution of T is usually difficult to find or it is too involved from the practical point of view. However the approximate (limiting as $n \to \infty$) distribution may be (more) tractable and available. By replacing G_n by this distribution one obtains approximate P-values. For insufficiently large n, these may lack accuracy when compared to the exact probabilities.

WHAT HAVE WE LEARNED: STATISTICAL TESTS WHICH WORK WELL AND SOME WHICH DO NOT

This section illustrates the concepts discussed in the first section by using some basic tests in the suite.

3.3. Tests based on the Properties of a Random Walk

Denote by $\epsilon_k, k = 1, 2, \ldots, n$ the underlying series of bits taking values 0 and 1 which is to be tested for randomness. In many situations it is more convenient to deal with the sequence $X_k = 2\epsilon_k - 1, k = 1, 2, \ldots, n$, with X_k taking values $+1$ or -1.

Quite a few statistical tests can be derived from the well-known limit theorems for the random walk, $S_n = X_1 + \cdots + X_n$. Under the randomness hypothesis, $(S_n + n)/2$ has the binomial distribution with parameters n and $p = 1/2$, which is not convenient to use when say, $n \geq 200$. However the classical Central Limit Theorem, according to which

$$\lim_{n \to \infty} P\left(\frac{S_n}{\sqrt{n}} \leq z\right) = \Phi(z) = \frac{1}{\sqrt{2\pi}} \int_{-\infty}^{z} e^{-u^2/2} \, du,$$

provides a useful approximation, which forms the foundation for the most basic monobit test of the null hypothesis that in a sequence of independent random variables X's or ϵ's the probability of ones is $1/2$.

More tests of randomness can be derived on the distribution of the maximum of the absolute values of the partial sums, $\max_{1 \leq k \leq n} |S_k|$, and from the distribution of the number of visits within an excursion of the random walk S_k to a certain state. However, not all tests based on the probabilistic properties of random walk are equally suitable for randomness testing. For example, the limiting distribution of the proportion of time U_n that the sums S_k are non-negative, leads to a fairly weak test.

3.4. Discrete Fourier Transform (Spectral) Test

The spectral test which appeared in the suite turned out to be trouble-
some. As it happened, it was not properly investigated, which resulted in
a wrong statistic and a faulty constant. This fact was duly noticed by the
cryptographic community.[8,10,11] Now the original version is replaced by the
following modification.

Let $X_k = \pm 1, k = 1, \ldots, M$, be a sequence of random bits. Denote

$$f_j = \sum_{k=1}^{M} X_k \exp\left\{\frac{2\pi(k-1)j\mathbf{i}}{M}\right\},$$

$j = 0, \ldots, M/2 - 1$. Then $Ef_j = 0, Ef_j\bar{f}_{j'} = \delta_{jj'}M$. Here $\bar{z} = a - b\mathbf{i}$ is the
complex conjugate of a complex number $z = a + b\mathbf{i}$. For a fixed m (i.e., m
which does not depend on M) the joint distribution of the complex vectors
$M^{-1/2}(f_1, \ldots, f_m)$ for large M is approximately the multivariate complex
normal distribution. It follows that under the randomness hypothesis, $W =$
$2\sum_{k=1}^{m} mod_k^2/M, mod_k^2 = f_k\bar{f}_k$, has an approximate χ^2-distribution with
$2m$ degrees of freedom.

These facts lead to the following procedure. Partition a string of length
n, such that $n = MN$ into N substrings, each of length M. For each
substring evaluate $mod_k^2, k = 1, \ldots, m$ (as in the original version of this
test but for a much smaller m than $M/2 - 1$). For each $j = 1, \ldots, N$,
calculate the statistic $W = W_j = 2\sum_{k=1}^{m} mod_k^2/M$. Reject the randomness
hypothesis if the χ^2 goodness-of-fit test does not accept the χ^2-distribution
with $2m$ degrees of freedom.

More exactly, choose a number $K + 1$ of disjoint intervals (classes),
evaluate the theoretical probabilities $\pi_i, i = 0, 1, \ldots, K$ of the intervals
according to this distribution, and form the familiar χ^2-statistic, $\chi^2 =$
$\sum_{i=0}^{K}(\nu_i - N\pi_i)^2/(N\pi_i)$ with ν_i denoting empirical frequencies of W-values
in i-th interval, $\nu_0 + \cdots + \nu_K = N$. The P-value can be given through the
incomplete gamma-function,

An alternative way to calculate the P-value is to follow a suggestion in
Section 3.2. Namely, determine the median, $\hat{W}$, of the χ^2-distribution with
$2m$ degrees of freedom, $Q(m, 0.5\hat{W}) = 0.5$. Then $0.5\hat{W} \approx m - 1/3$. The
P-value corresponding to j-th substring is

$$\text{P-value} = \begin{cases} Q(m, 0.5W_j) + 1 - Q(m, \hat{W} - 0.5W_j) & W_j \geq \hat{W}, \\ 1 - Q(m, 0.5W_j) + Q(m, \hat{W} - 0.5W_j) & W_j < \hat{W}. \end{cases}$$

3.5. Non-overlapping and Overlapping Template Matchings

Most conventional pseudo random number generators, such as the linear congruential generators and lagged-Fibonnaci generators used in IMSL, C++, and other packages, tend to show patterning due to their deterministic recursive algorithms. Because of this patterning, it is natural to investigate statistical tests based on the occurrences of words (patterns or templates).

We start here with the tests which utilize the observed numbers of words or the frequency of a given word in a sequence of length M. Let $\imath = (i_1, \ldots, i_m)$ be a given word (template or pattern, i.e., a fixed sequence of zeros and ones) of length m.

An important role belongs to the set $\{j, 1 \leq j \leq m, i_{j+k} = i_k, k = 1, \ldots, m - j\}$, which is the set of periods of $\imath$. For example, when $\imath$ corresponds to a run of m ones, $\{1, \ldots, m - 1\}$ is the set of all periods. For *aperiodic* words $\imath$, this set is empty. Such words cannot be written as $\ell\ell \ldots \ell\ell'$ for a pattern ℓ shorter than $\imath$ with ℓ' denoting a prefix of ℓ. In this situation occurrences of $\imath$ in the string are necessarily non-overlapping.

Denote by $W = W(m, M)$ the number of occurrences of the given aperiodic pattern. If $(M - m + 1)/2^m = \lambda$, then $EW = (M - m + 1)2^{-m} = \lambda$. When both M and m tend to infinity, W has a Poisson limiting distribution with the parameter λ, i.e., $P(W = k) \to e^{-\lambda}\lambda^k/k!$ $k = 0, 1, \ldots$ (see Ref. 2). When the length m is fixed, the limiting distribution of standardized statistic W is normal. Each of these facts can be used for randomness testing.

This statistic is defined also for periodic patterns, but the accuracy of Poisson approximation is good only when $\imath$ does not have small periods. A test of randomness can be based on the number of possibly overlapping occurrences of templates in the string. If $U = U(m, M)$ is this number for a periodic word of length m then the asymptotic distribution of U is the compound Poisson distribution(the so-called Polya-Aeppli law). The probabilities of this distribution can be expressed in terms of confluent hypergeometric function. So to implement the test of randomness based on overlapping patterns, partition the string into N substrings and evaluate the empirical frequencies of occurrences of aperiodic or periodic patterns within each substring of length M comparing them to the theoretical probabilities via the χ^2-statistic.

Notice that M must be sufficiently large for validity of this test. For example, when $M = 1032$ and $\imath$ is a run of $m = 9$ ones, so that $\lambda =$

1.9980468750, the comparison of Polya-Aeppli probabilities and the exact probabilities (due to K. Hamano) is given in the Table 3.1.

Table 3.1. The exact probabilities and the Polya-Aeppli law probabilities when $M = 1032$ and $m = 9$.

	exact probabilities	Polya-Aeppli probabilities
$P(U = 0)$	0.367879	0.364091
$P(U = 1)$	0.183940	0.185659
$P(U = 2)$	0.137955	0.139381
$P(U = 3)$	0.099634	0.100571
$P(U = 4)$	0.069935	0.070431
$P(U = 5)$	0.140657	0.139865

WHAT WE DO NOT KNOW YET: TESTS BASED ON PATTERNS, PERIODIC OR NOT

For a given set of words (patterns), it is of interest to determine the probability of the prescribed number of (overlapping) occurrences of these patterns in the text. This problem appears in different areas of information theory such as source coding and code synchronization. It is also important in molecular biology, in DNA analysis, and for gene recognition.

It is convenient now to consider a random text formed by realizations of letters chosen from a finite (not necessarily binary) alphabet. This setting can be used for a binary sequence if its substrings of a given length p represent the new letters, so that there are $q = 2^p$ such letters. Then the length n of q-nary sequence is related to the length n' of the original binary string by the formula, $n = n'/p$. This extension opens the possibility of choosing q in an optimal way.

To find words with prescribed frequencies one can use asymptotically normal estimates of word probabilities or the exact distributions obtained from generating functions (see, for example, Ref. 21). These results suggest that the probability for a given word $\imath$ to appear exactly r times in the string of length n can be approximated by the Poisson probability of the value r, when the Poisson parameter is $nP(\imath)$. Thus, the distribution of the number of words with given r can be expected approximately equal to that of the sum of Bernoulli random variables whose success probability is this Poisson probability. However, the more detailed structure of this distribution, in particular, the covariance of several such random variables, needed in the study of large sample efficiency, is less intuitive.

The approximate Poisson distribution for the number of missing words ($r = 0$) is alluded to in Marsaglia and Zaman.[17] It forms the basis of the so-called OPSO test of randomness in the Diehard Battery.[16] This test takes non-overlapping substrings formed by zeros and ones of length $p = 10$ to represent the letters of the new alphabet, so that there are $q = 2^{10}$ new letters, which in general is not the optimal choice. In the OPSO test one counts the number of two-letter patterns (the original substrings of length $2p = 20$) which never occur. We consider the case of arbitrary m in the next section.

3.6. Tests of Randomness Based on the Number of Missing Words

Let $\epsilon_1, \ldots, \epsilon_n$ denote a sequence of independent discrete random variables each taking values in the finite set $\mathbb{Q}$, say, $\mathbb{Q} = \{1, \ldots, q\}$, $P(\epsilon_i = k) = p_k, k = 1, \ldots, q$. Thus, the probability of the word $\imath = (i_1 \ldots i_m)$ is $P(\imath) = p_{i_1} \cdots p_{i_m}$. The situation when $p_k \equiv q^{-1}$ corresponds to the randomness hypothesis.

To determine efficient tests for randomness, we look at the alternative distributions of the alphabet letters which are close to the uniform in the sense that $p_k = q^{-1} + q^{-3/2}\eta_k$, $k = 1, \ldots, q$, $\sum_{k=1}^{q} \eta_k = 0$. We assume that as $n \to \infty$, $q \to \infty$ so that $n/q^m \to \lambda$ and for a positive κ, $\sum_k \eta_k^2/q \to \kappa$. Then $nP(\imath) \to \lambda$.

The first object of interest is the probability $\pi_\imath(n)$ that a fixed pattern $\imath$ is missing in the string of length n. To find this probability one can use the correlation polynomial of two patterns which was introduced by Guibas and Odlyzko.[6] It plays an important role in the study of the distribution of their frequencies.

For a complex z, let $F_\imath(z) = \sum_n \pi_\imath(n) z^{-n}$ be the probability generating function. Then in can be expressed as a ratio of two polynomials closely related to the correlation polynomials, and for any word $\imath$ the probability $\pi_\imath(n)$ can be found by comparing the coefficients in the series expansions of $F_\imath(z)$.

For example when $m = 2$, $\jmath = (i, k), i \neq k$, then $F_\jmath(z) = z^2/(p_i p_k + (z - 1)z)$. With $s = 1/2 + \sqrt{1/4 - p_i p_k}$, $t = 1/2 - \sqrt{1/4 - p_i p_k}$, $\pi_\jmath = (s^{n+1} - t^{n+1})/(s - t)$. These formulas lead to very accurate answers for the expected value and the variance. For example when $n = 2^{21}$, $q = 2^{10}$ (so that $\lambda = 2$),

$$\pi_{(i,i)} = 0.13559935200020, \quad \pi_{(i,k)} = 0.13533502510527.$$

The asymptotic approximation for the probabilities $\pi_j^t(n)$ for a word ȷ of length m and of period t is $\pi_j^t(n) \approx e^{-\lambda}(1 + \lambda q^{-t})$. For aperiodic words, $\pi_j^\infty(n) \sim e^{-\lambda}(1 - (2m - 1)\lambda/(2q^m) + (m - 1)/q^m)$. To get the formula for the expected value of the number of missing m-words, X^0 let $N_t = N_t(m), t = 1, \ldots, m - 1, \infty$, denote the total number of words of the period t ($t = \infty$ corresponds to aperiodic words). Then $\sum N_t = N^m$, and as $q \to \infty$ for $t = 1, \ldots, m - 1$, $N_t \sim q^t, N_\infty \sim q^m$. One gets

$$EX^0 = \sum_t N_t \pi_j^t(n) \approx e^{-\lambda} q^m + e^{-\lambda}\left[m - 1 - \frac{\lambda}{2}\right].$$

In the example when $m = 2, n = 2^{21}$, $q = 2^{10}$, the exact value of the mean is $EX^0 = 141909.3299555$, and the approximate formula gives 141909.3299551. The formula for the variance can be obtained from the probabilities $\pi_{1\jmath}^{00}(n) = P\,(\text{words } \imath \text{ and } \jmath \text{ are missing})$ which can be found from the probability generating function technique. See Refs. 22 and 23 for details.

After the number X^0 of missing two letter words in the string of length n has been determined, one evaluates the ratio, $(X^0 - EX^0)/\sqrt{Var(X^0)}$, which leads to the P-value obtained from the standard normal distribution.

3.7. Testing Randomness via Words with a Given Frequency

More powerful tests can be derived by using the observed numbers of words which appear in a random text a prescribed number of times (i.e., which are missing, appear exactly once, exactly twice, etc.) In practice these statistics are easier to evaluate than the empirical distribution of occurrences of all m-words.

It turns out that such tests can be obtained by techniques of the previous section which lead to the formula for the expected value of the number of m-words, which occur exactly r times in a random string of length n, $X^r = X_n^r$, A surprising fact is that the asymptotic behavior of the expected value and of the covariance matrix is the same for overlapping and non-overlapping occurrences, i.e., when the word occurrences are counted in the non-overlapping m-blocks. Therefore, the form of the following optimal test coincides with that in Kolchin, Sevastyanov and Chistyakov[14] who give the formulas for the first two moments of the joint distribution of the number of words appearing a prescribed number of times when the frequencies of these words are independent.

To derive the optimal test of the null hypothesis $H_0 : \eta_i \equiv 0$, we look at the class of linear test statistics of the form $S = \sum_{r=0}^{R} w_r(X^r - EX^r)$ for a fixed positive integer R. The (Pitman) efficiency of this statistic can be obtained from the fact that S is asymptotically normal both under the null hypothesis and the alternative $H_1 : \kappa > 0$. This efficiency is determined by the normalized distance between the means under the null hypothesis and under the alternative, divided by the standard deviation (which is common to the null hypothesis and the alternative). This test is asymptotically optimal not only within the class of linear statistics, but in the class of all functions of $X^0, \ldots, X^R$.

The following table for $R = 0, \ldots, 9$, gives the value of $\lambda = \lambda^\star$, which maximizes the efficiency and the corresponding optimal weights $\tilde{w} = w / \sum w_r$ normalized so that their sum is equal to one.

Table 3.2. Optimal weights and the optimal $\lambda^\star$.

R	$\lambda^\star$	$\tilde{w}$
0	3.59	1
1	4.77	$[0.62, 0.38]$
2	5.89	$[0.47, 0.33, 0.20]$
3	6.98	$[0.39, 0.29, 0.20, 0.14]$
4	8.06	$[0.33, 0.25, 0.19, 0.14, 0.09]$
5	9.13	$[0.29, 0.23, 0.18, 0.14, 0.09, 0.07]$
6	10.17	$[0.25, 0.21, 0.18, 0.14, 0.09, 0.07, 0.06]$
7	11.21	$[0.23, 0.19, 0.16, 0.14, 0.09, 0.07, 0.06, 0.05]$
8	12.24	$[0.21, 0.18, 0.16, 0.14, 0.09, 0.07, 0.06, 0.05, 0.04]$
9	13.26	$[0.19, 0.17, 0.16, 0.14, 0.09, 0.07, 0.06, 0.05, 0.04, 0.03]$

To implement this test on the basis of a string of binary bits for a fixed R, as discussed in the beginning of this section, choose a positive integer p, such that $n \approx 2^{mp}\lambda^\star$, and take all substrings of length p formed by zeros and ones to represent the letters of the new alphabet of the size $q = 2^p$. The numbers X^r of m-letter patterns (the original non-overlapping consecutive substrings of length $2m$), which occurred r times with the weights from the table lead to the asymptotically optimal test. In particular, the most efficient test based on the number of missing words $m = 2, R = 0$ arises when $\lambda^\star = 3.594..$, which means that the best relationship between q and n, is $n \approx 3.6q^2$ Extensions of these results to Markov dependent alternatives are in Ref. 25.

WHAT ARE THE MOST IMPORTANT OPEN PROBLEMS: DATA COMPRESSION
AND RANDOMNESS TESTING

It is desirable to develop tests based on patterns suggested by the data
themselves. A powerful heuristic idea is that random sequences are those
that cannot be compressed or those that are most complex. However its
practical implementation is limited by scarcity of relevant compression code
based statistics whose (approximate) distributions can be evaluated.

3.8. Linear Complexity for Testing Randomness

Here we look at linear complexity which is related to one of the main
components of many keystream generators, namely, Linear Feedback Shift
Registers (LFSR). Such a register consists of L delay elements each having
one input and one output. If the initial state of LFSR is $(\epsilon_{L-1}, \ldots, \epsilon_1, \epsilon_0)$,
then the output sequence, $(\epsilon_L, \epsilon_{L+1}, \ldots)$, satisfies the following recurrent
formula for $j \geq L$

$$\epsilon_j = (c_1 \epsilon_{j-1} + c_2 \epsilon_{j-2} + \cdots + c_L \epsilon_{j-L}) \bmod 2.$$

Here $c_1, \ldots, c_L$ are coefficients of the so-called connection polynomial corresponding to a given LFSR. The *linear complexity* $L = L_n$ of a given
sequence $\epsilon_1, \ldots, \epsilon_n$, is defined as the length of the shortest LFSR that generates it as first n terms. The possibility of using the linear complexity
characteristic for testing randomness is based on the Berlekamp-Massey
algorithm, which provides an efficient way to evaluate the connection polynomial for finite strings.

When the binary n-sequence is truly random, the formulas for the mean,
$\mu_n = EL_n$, and the variance are known. However, the asymptotic distribution as such does not even exist; one has to treat the cases, n even, and
n odd, separately with two different limiting distributions. Both of these
distributions can be conjoined in a discrete distribution obtained via a mixture of two geometric random variables (one of them taking only negative
values).

The monograph of Rueppel[20] gives the distribution of the random variable L_n, the linear complexity of a random binary string, which can be used
to show that

$$T_n = (-1)^n [L_n - \xi_n] + \frac{2}{9}, \quad \xi_n = \frac{n}{2} + \frac{4 + r_n}{18},$$

can be used for testing randomness. The sequence T_n converges in distribution to the random variable T whose distribution is skewed to the

right. While $P(T = 0) = 1/2$, $P(T = k) = 2^{-2k}$, for $k = 1, 2, \ldots$, and $P(T = k) = 2^{-2|k|-1}, k = -1, -2, \ldots$, which provide easy formulas for the P-values.

In view of the discrete nature of this distribution one can use the strategy described in Section 3.2 for a partitioned string. A more powerful test which efficiently uses the available data was suggested by Hamano, Sato and Yamamoto.[9] It is based on the statistic $\sum |j/2 - L_j|$, which can be interpreted as the sum of areas of triangles formed by vertexes $(j, L_j), (j + 1, L_{j+1}), j = 0, 1, 2, \cdots$, around the line $L_j = j/2$.

3.9. Tests based on Data Compression

The original suite attempted to develop a randomness test based on the Lempel-Ziv algorithm[30] of data compression via parsing of the text. Let W_n represent the number of words in the parsing of a binary random sequence of length n according to this algorithm. Aldous and Shields[1] have shown that $EW_n/(n/\log_2 n) \to 1$, so that expected compression can be asymptotically approximated by $n/\log_2 n$. Moreover,

$$P\left(\frac{W_n - EW_n}{\sqrt{Var(W_n)}} \leq w\right) \to \Phi(w).$$

The behavior of $Var(W_n)$ was elucidated by Kirschenhofer, Prodinger, and Szpankowski[12] who proved that $Var(W_n) \sim (n[C + \delta(\log_2 n)])/\log_2^3 n$, where $C = 0.26600$ (to five significant places) and $\delta(\cdot)$ is a slowly varying continuous function with mean zero, $|\delta(\cdot)| < 10^{-6}$.

One of the tests in the original version of the suite compressed the sequence using the Lempel-Ziv algorithm. If the reduction is statistically significant when compared to the expected result, one can declare the sequence to be nonrandom. It was expected that the ratio $(W - n/\log_2 n)/\sqrt{0.266n/\log_2^3 n}$, where W is the number of words obtained, would provide the P-value corresponding to the two-sided alternative. Unfortunately, this test failed because the normal approximation was too poor, i.e., the asymptotic formulas are not accurate enough for values of n of the magnitude encountered in testing random number generators.

More practical turned out to be the so-called "universal" test introduced by Maurer.[18] The test requires a long (of the order $10 \cdot 2^L + 1000 \cdot 2^L$ with $6 \leq L \leq 16$) sequence of bits which it divides into two stretches of L-bit blocks: D, $D \geq 10 \cdot 2^L$, initialization blocks and K, $K \approx 1000 \cdot 2^L$ test blocks. The test looks back through the entire sequence while inspecting the test segment of L-bit blocks, checking for the nearest previous exact match and recording the distance (in number of blocks) to that previous match. The algorithm computes the logarithm of all such distances for all the L-bit templates in the test segment (giving effectively the number of digits in the binary expansion of each distance), and averages over all the expansion lengths by the number of test blocks K to get the test statistic F_n. A P-value is obtained from the normal error function based on the standardized version of the statistic, with the test statistic's mean EF_n equal to that of $\log_2 G$, where G is a geometric random variable with the parameter $1 - 2^{-L}$.

The difficult part is determination of the variance $Var(F_n)$. There are several versions of empirical approximate formulas, $Var(F_n) = c(L, K)Var(\log_2 G)/K$, where $c(L, K)$ represents the factor which takes into account dependent nature of the occurrences of templates, The latest of the approximations belonging to Coron and Naccache[3] has the form $c(L, K) = 0.7 - 0.8/L + (1.6 + 12.8/L)K^{-/L}$. However, these authors report that "the inaccuracy due to [this approximation] can make the test to be 2.67 times more permissive than what is theoretically admitted."

The prospects for better approximations, in particular for the exact variances $Var(W_n)$ or $Var(F_n)$ do not look very good. In view of this fact, it may be advisable to test the randomness hypothesis by verifying normality of the observed values F_n assuming that the variance is unknown. This can be done via a classical statistical technique, namely the t-test. The original sequence must be partitioned in a number N (say $N \leq 20$) of substrings on each of which the value of the universal test statistic is evaluated (for the same value of parameters K, L and D). The sample variance is calculated, and the P-value is obtained from the t-distribution with $N - 1$ degrees of freedom.

The most interesting randomness test would be based on Kolmogorov's definition of complexity which is the length of the shortest (binary) computer program that describes the string. One of the universal Turing machines is supposed to represent the computer which could use this description to exhibit this string after a finite amount of computation. As was argued, the random sequences are the most complex ones, so if Kolmogorov's complexity were computable, a randomness test would reject the null hypothesis for its small values. Unfortunately, this complexity characteristic is not computable,[4] and there is no hope for a test which is directly based on it.

Pursuing the idea of using data compression codes as randomness testing statistics, let $\mathbb{Q}$ be a finite alphabet of size q. A data compression method consists of a collection of mappings ϕ_n of $\mathbb{Q}^n$, $n = 1, 2, \ldots$ into a set of all finite binary sequences, such that for $\imath, \jmath \in \mathbb{Q}^n$, $\imath \neq \jmath$, one has $\phi_n(\imath) \neq \phi_n(\jmath)$. This means that a message of any length n can be both compressed and decoded.

For a given compression code, the randomness hypothesis is accepted on the basis of the string $\epsilon_1, \ldots, \epsilon_n$ if the length T_n of $\phi_n(\epsilon_1, \ldots, \epsilon_n)$ is large enough. Ryabko and Monarev[27] show that the choice of the cut-off constant $T_0 = n \log q + \log \alpha - 1$ leads to a test of significance level α. A code is universal if for any ergodic stationary source Γ, the ratio T_n/n converges with probability one to the entropy of Γ. This entropy equals to $\log q$ if the randomness hypothesis is true, and it is smaller than $\log q$ under any alternative that can be modeled by an ergodic stationary process. For universal codes the power of the corresponding test tends to one as n increases. However, finding non-trivial compression codes with known distributions of the codewords length (so that P-value can be evaluated) is quite difficult.

WHAT ARE THE PROSPECTS FOR PROGRESS: CONCLUDING REMARKS

To sum up, there are major challenges in the area of empirical randomness testing. It may be a bit surprising that so many available procedures are based on the one hundred years old χ^2-test. Since this area is so important, one can expect more stringent methods based on new ideas. In particular, a study of overlapping spatial patterns is of great interest, as it may lead to such procedures.

References

1. Aldous, D. and Shields, P. A diffusion limit for a class of randomly-growing binary trees. *Probability Theory and Related Fields* 79, 1988, 509–542.
2. Barbour A. D., Holst, L. and Janson, S. *Poisson Approximation*, Clarendon Press, Oxford, 1992.
3. Coron, J-S. and Naccache, D. An accurate evaluation of Maurer's universal test. Proceedings of SAC'98, Lecture Notes in Computer Science, Springer, Berlin, 1998.
4. Cover, T. and Thomas, J. Elements of Information Theory, J. Wiley, New York, NY, 1991,
5. Gibbons, J. and Pratt, J. P-values: interpretations and methodology. *American Statistician* 29, 1975, 20–25.

6. Guibas, L. J. and Odlyzko, A. M. Strings overlaps, pattern matching and nontransitive games. *J. Comb. Theory* A 30, 1970, 183–208.

7. Gustafson, H., Dawson, E., Nielsen, L., and Caelli, W. A computer package for measuring the strength of encryption algorithms. *Computers and Security* 13, 1994, 687-697.

8. Hamano, K. The distribution of the spectrum for the discrete Fourier transform test included in SP800-22. *IEICE Trans. Fundamentals* E88, 2005, 67-73.

9. Hamano, K., Sato, F., and Yamamoto, H. A new randomness test based on linear complexity profile. *IEICE Trans. Fundamentals* E92, 2009, 166-172.

10. Killmann, W., Schüth, J., Thumser, W. and Uludag, I. A note concerning the DFT test in NIST special publication 800-22.T-Systems Integration, Technical Report, 2004.

11. Kim, S.-J., Umeno, K. and Hasegawa, A. Corrections of the NIST statistical suite for randomness, IEICE Technical Report, ISEC2003-87, 2003.

12. Kirschenhofer, P., Prodinger, H., and Szpankowski, W. Digital Search Trees Again Revisited: The Internal Path Length Perspective. *SIAM Journal on Computing* 23, 1994, 598-616.

13. Knuth, D. E. *The Art of Computer Programming*, Vol. 2, 3rd ed. Addison-Wesley Inc., Reading, MA. 1998.

14. Kolchin, V. F., Sevast'yanov, B. A., and Chistyakov, V. P. *Random Allocations*. Whinston Sons, Washington, DC, 1978.

15. L'Ecuyer, P. and Simard, R. TestU01: A C library for empirical testing of random number generators, *ACM Transactions of Mathematical Software.* 33, 4, Article 22, 2007.

16. Marsaglia, G. *Diehard: A battery of tests for randomness.* http://stat.fsu.edu/geo/diehard.html 1996.

17. Marsaglia, G. and Zaman, A. Monkey tests for random number generators. *Computers & Mathematics with Applications* 9, 1993, 1–10.

18. Maurer, U. M. A universal statistical test for random bit generators. *Journal of Cryptology* 5, 1992, 89-105.

19. Murphy, S. The power of NIST's statistical testing of AES candidates. 2000. http://www.cs.rhbnc.ac.uk/sean/StatsRev.ps

20. Rueppel, R. *Analysis and Design of Stream Ciphers.* Springer, Berlin, 1986.

21. Rukhin, A. L. Admissibility: survey of a concept in progress. *International Statistical Review* 63, 1995, 95–115.

22. Rukhin, A. L. Testing randomness: a suite of statistical procedures. *Theory of Probability and its Applications* 45, 2000, 137–162.

23. Rukhin, A. L. Pattern correlation matrices and their properties. *Linear Algebra and its Applications* 327, 2001, 105–114.

24. Rukhin, A. L. Distribution of the number of words with a prescribed frequency and tests of randomness. *Advances in Applied Probability* 34, 2002, 775–797.

25. Rukhin, A. L. Pattern correlation matrices for Markov sequences and tests of randomness. *Theory of Probability and its Applications* 51, 2006, 712–731.

26. Rukhin, A. L., Soto, J., Nechvatal, J., Smid, M., Levenson, M., Banks, D., Vangel, M., Leigh, S., Vo, S., Dray, J. A statistical test suite for the validation of cryptographic random number generators. Special NIST Publication, NIST, Gaithersburg, 2000.

27. Ryabko, B. Ya. and Monarev, V. A. Using information theory approach to randomness testing. *Journal of Statistical Planning and Inference* 133, 2005, 95–110.

28. Soto, J. and Bassham, L. Randomness testing of the advanced encryption standard finalist candidates. *Proceedings of AES Conference*, 2001, http://csrc.nist.gov/publications/nistir/ir6483.pdf

29. Szpankowski, W. *Average Case Analysis of Algorithms on Sequences*. Wiley-Interscience, New York, 2001.

30. Ziv, J. and Lempel, A. A universal algorithm for sequential data compression. *IEEE Transactions on Information Theory*, 23, 1997, 337-343.

Chapter 4

Scatter and Regularity Imply Benford's Law... and More[*]

Nicolas Gauvrit and Jean-Paul Delahaye

Laboratoire André Revuz, University Paris VII, France
adems@free.fr
Université des Sciences et Techniques de Lille
Laboratoire d'Informatique Fondamentale de Lille
delahaye@lifl.fr

WHAT DREW US TO THE STUDY OF THE APPARENT (AWKWARD) RANDOMNESS OF BENFORD'S LAW

A random variable (r.v.) X is said to follow Benford's law if $\log(X)$ is uniform mod 1. Many experimental data sets prove to follow an approximate version of it, and so do many mathematical series and continuous random variables. This phenomenon received some interest, and several explanations have been put forward. Most of them focus on specific data, depending on strong assumptions, often linked with the log function.

Some authors hinted–implicitly–that the two most important characteristics of a random variable when it comes to Benford are regularity and scatter.

In a first part, we prove two theorems, making up a formal version of this intuition: scattered and regular r.v.'s do approximately follow Benford's law. The proofs only need simple mathematical tools, making the analysis easy. Previous explanations thus become corollaries of a more general and simpler one.

These results suggest that Benford's law does not depend on properties linked with the log function. We thus propose and test a new general

---

[*]This is an updated version of the results published in Gauvrit, N. and Delahaye, J.-P., *La loi de Benford générale*, in Math. & Sci. hum. / Mathematics and Social Sciences, 47, n. 186, 2009(2), 5-15 available online at http://www.ehess.fr/revue-msh/pdf/N182R1280.pdf.

version of Benford's law. The success of these tests may be viewed as an *a posteriori* validation of the analysis formulated in the first part.

WHAT WE HAVE LEARNED

First noticed by Newcomb[19] and again later by Benford[1] the so-called Benford's law states that a sequence of "random" numbers should be such that their logarithms are uniform mod 1. As a consequence, the first non-zero digit of a sequence of "random" numbers is d with probability $\log\left(1 + \frac{1}{d}\right)$, an unexpectedly non-uniform probability law. log here stands for the base 10 logarithm, but an easy generalization follows: a random variable (r.v.) conforms to *base b* Benford's law if its base b logarithm $\log_b(X)$ is uniform mod 1. Lolbert[17] recently proved that no r.v. follows base b Benford's law for all b.

Many experimental data roughly conform to Benford's law (most of which no more than roughly). However, the vast majority of real data sets that have been tested do not fit this law with precision. For instance, Scott and Fasli[24] reported that only 12.6 of 230 real data sets passed the test for Benford's law. In his seminal paper, Benford[1] tested 20 data sets including lakes areas, length of rivers, populations, etc., half of which did not conform to Benford's law.

The same is true of mathematical sequences or continuous r.v.'s. For example, binomial arrays $\binom{n}{k}$, with $n \geq 1$, $k \in \{0, ..., n\}$, tend toward Benford's law,[5] whereas simple sequences such as $(10^n)_{n \in \mathbb{N}}$ obviously don't.

In spite of all this, Benford's law is actually used in the so-called "digital analysis" to detect anomalies in pricing[26] or frauds, for instance in accounting reports[6] or campaign finance.[4] Faked data indeed usually depart from Benford's law more than real ones.[11] However, advise caution,[10] arguing that real data do not always fit the law.

Many explanations have been put forward to elucidate the appearance of Benford's law on natural or mathematical data. Some authors focus on particular random variables,[7] sequence,[14] real data,[3] or orbits of dynamical systems.[2] As a rule, other explanations assume special properties of the data. Hill[13] and Pinkham[22] show that scale invariance implies Benford's law. Base invariance is another sufficient condition.[12] Mixtures of uniform distributions[9] also conform to Benford's law, and so do the limits of some random processes.[25] Multiplicative processes have been mentioned as well.[21] Each of these explanations accounts for some appearances of data fitting Benford's law, but lacks generality.

While looking for a truly general explanation, some authors noticed that data sets are more likely to fit Benford's law if they were scattered enough. More precisely, a sequence should "cover several orders of magnitude", as Raimi[23] expressed it. Of course, scatter alone is no sufficient condition. The sequence 0.9, 9, 90, 900... indeed covers several orders of magnitude, but is far from conforming to Benford's law. The continuous random variables that are known to fit Benford's law usually present some "regularity": exponential densities, normal densities, or lognormal densities are of this kind. Invariance assumptions (base-invariance or scale-invariance) lead to "regular" densities and so do central limit-like theorem assumptions of mixture.

Some technical explanations may be viewed as a mathematical expression of the idea that a random variable X is more likely to conform to Benford's law if it is regular and scattered enough. Mardia and Jupp[18] linked Benford's law to Poincaré's theorem in circular statistics, and Smith[27] expressed it in terms of Fourier transform and signal processing. However, a non expert reader would hardly notice the smooth-and-scattered implications of these developments.

Though scatter has been explicitly mentioned and regularity allusively evoked, the idea that scatter and regularity (in a sense that will be made clear further) may actually be a *sufficient* explanation for Benford's phenomenon related to continuous r.v.'s have never been formalized in a simple way, to our knowledge, except in a recent article by Fewster.[8] In this paper, Fewster hypothesizes that *"any distribution [...] that is reasonably smooth and covers several orders of magnitude is almost guaranteed to obey Benford's law."* She then defines a smoothing procedure for a r.v. X based on $\left[\pi^2\left(x\right)\right]''$, π being the probability density function (henceforth *p.d.f.*) of $\log\left(X\right)$, and illustrates with a few eloquent examples that under smoothness and scatter constraints, a r.v. cannot depart much from Benford's law. However, no theorem is given that would formalise this idea.

WHAT WE DON'T (YET) KNOW

Benford's law has been always considered either mysterious or too complicated. In the following, however, we prove a theorem from which it follows that scatter and regularity can be modelled in such a way that they, alone, imply *rough* compliance to Benford's law (again: real data usually do not perfectly fit Benford's law, irrespective of the sample size).

It is not surprising that many data sets or random variables samples are scattered and regular hence our explanation of Benford's phenomena corroborates a widespread intuition. The proof of this theorem is straight-forward and requires only basic mathematical tools. Furthermore, as we shall see, several of the existing explanations can be understood as corollaries of ours. Our explanation encompasses more specific ones, and is far simpler to understand and to prove.

Scatter and regularity do not presuppose any log-related properties (such as the property of log-normality, scale-invariance, or multiplicative properties). For this reason, if we are right, Benford's law should also admit other versions. We set that a r.v. X is u-Benford for a function u if $u(X)$ is uniform mod 1. The classical Benford's law is thus a special case of u-Benford's law, with $u = \log$. We test real data sets and mathematical sequences for "u-Benfordness" with various u, and test a second theorem echoing the first one. Most data conform to u-Benford's law for different u, which is an argument in favour of our explanation.

4.1. Scatter and Regularity: A Key to Benford

The basic idea at the root of theorem 1 (below) is twofold.

First, we hypothesize that a continuous r.v. X with density f is almost uniform mod 1 as soon as it is scattered and regular. More precisely, any f that is non-decreasing on $]-\infty, a]$, and then non-increasing on $[a, +\infty[$ (for regularity) and such that its maximum $m = \sup(f)$ is "small" (for scatter) should correspond to a r.v. X approaching uniformity mod 1. Figure 4.1 illustrates this idea.

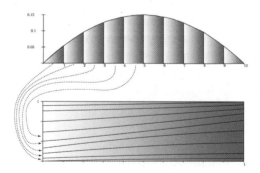

Fig. 4.1. Illustration of the idea that a regular p.d.f. is almost bound to give rise to uniformity mod 1. The stripes–restrictions of the density on $[n, n+1]$–of the p.d.f. of a r.v. X are stacked to form the p.d.f. of X mod 1. The slopes partly compensate, so that the resulting p.d.f. is almost uniform. If the initial p.d.f. is linear on every $[n, n+1]$, the compensation is perfect.

Second, note that if X is scattered and regular enough, so should be $\log(X)$. These two ideas are formalized and proved in theorem 1.

Henceforth, for any real number x, $\lfloor x \rfloor$ will denote the greatest integer not exceeding x, and $\{x\} = x - \lfloor x \rfloor$. Any positive x can be written as a product $x = 10^{\lfloor \log x \rfloor}.10^{\{\log x\}}$, and the Benford's law may be rephrased as the uniformity of the random variable $\{\log(X)\}$.

Theorem 4.1 *Let X be a continuous positive random variable with p.d.f. f such that $Id.f : x \longmapsto xf(x)$ conforms to the two following conditions : $\exists a > 0$ such that (1) $\max(Id.f) = m = a.f(a)$ and (2) $Id.f$ is nondecreasing on $]0, a]$, and nonincreasing on $[a, +\infty[$. Then, for any $z \in]0, 1]$,*

$$|P(\{\log X\} < z) - z| < 2\ln(10)m. \tag{4.1}$$

In particular, (X_n) being a sequence of continuous r.v.'s with p.d.f. f_n satisfying these conditions and such that $m_n = \max(Id.f_n) \longrightarrow 0$, $\{\log(X_n)\}$ converges toward uniformity on $[0, 1[$ in law.

Proof. We first prove that for any continuous r.v. Y with density g such that g is nondecreasing on $]-\infty, b]$, and then nonincreasing on $[b, +\infty[$, the following holds:

$$\forall z \in]0, 1], \ |P(\{Y\} < z) - z| < 2M, \tag{4.2}$$

where $M = g(b) = \sup(g)$.

We may suppose without loss of generality that $b \in [0, 1[$. Let $z \in]0, 1[$ (the case $z = 0$ is obvious). Put $I_{n,z} = [n, n+z[$. For any integer $n \leq -1$,

$$\frac{1}{z}\int_{I_{n,z}} g(t)dt \leq \int_n^{n+1} g(t)dt. \tag{4.3}$$

Thus

$$\frac{1}{z}\sum_{n \leq -1}\int_{I_{n,z}} g(t)dt \leq \int_{-\infty}^0 g(t)dt. \tag{4.4}$$

For any integer $n \geq 2$,

$$\frac{1}{z}\int_{I_{n,z}} g(t)dt \leq \int_{n-1+z}^{n+z} g(t)dt, \tag{4.5}$$

so

$$\frac{1}{z}\sum_{n \geq 2}\int_{I,z} g(t)dt \leq \int_{1+z}^{+\infty} g(t)dt. \tag{4.6}$$

Moreover, $\int_{I_{0,z}} g \leq zM$ and $\int_{I_z} g \leq zM$. Hence,

$$\frac{1}{z}\sum_{n\in\mathbb{Z}}\int_{I_{n,z}} g \leq \int_{-\infty}^{\infty} g + 2M. \tag{4.7}$$

We prove in the same fashion that

$$\frac{1}{z}\sum_{n\in\mathbb{Z}}\int_{I_{n,z}} g \geq \int_{-\infty}^{\infty} g - 2M. \tag{4.8}$$

Since $\sum_{\mathbb{Z}}\int_{I_{n,z}} g = P\left(\{Y\} < z\right)$, $z < 1$ and $\int_{-\infty}^{\infty} g = 1$, the result is proved. Now, applying this to $Y = \log\left(X\right)$ proves theorem 1.   □

Remark 4.1 *The convergence theorem is still valid if we accept f to have a finite number of monotony changes, provided this number does not exceed a previously fixed k. The proof is straightforward.*

Remark 4.2 *The assumptions made on $Id.f$ may be seen as a measure of scatter and regularity for X, adjusted for our purpose.*

EXAMPLES

Type I Pareto

A continuous r.v. X is type I Pareto with parameters α and x_0 (α, $x_0 \in \mathbb{R}_+^*$) iff it admits a density function

$$f_{x_0,\alpha}\left(x\right) = \frac{\alpha x_0^\alpha}{x^{\alpha+1}}\mathbb{I}_{[x_0,+\infty[}. \tag{4.9}$$

Besides its classical use in income and wealth modelling, type I Pareto variables arise in hydrology and astronomy.[20]

The function $Id.f = g : x \longmapsto \frac{\alpha x_0^\alpha}{x^\alpha}\mathbb{I}_{[x_0,\infty[}$ is decreasing. Its maximum is

$$\sup\left(Id.f\right) = Id.f\left(x_0\right) = \alpha. \tag{4.10}$$

Therefore, X is nearly Benford-like, in the extent that

$$|P(\{\log X\} < z) - z| < 2\ln(10)\alpha. \tag{4.11}$$

Type II Pareto

A r.v. X is type II Pareto with parameter $b > 0$ iff it admits a density function defined by

$$f_b(x) = \frac{b}{(1+x)^{b+1}} \mathbb{I}_{[0,+\infty[} . \qquad (4.12)$$

It arises in a so-called *mixture model*, with mixing components being gamma distributed r.v.'s sequences.

The function $Id.f_b = g_b : x \longmapsto \frac{bx}{(1+x)^{b+1}} \mathbb{I}_{[0,+\infty[}$ is $C^\infty(\mathbb{R}_+)$, with derivative

$$g_b'(x) = \frac{b(1-bx)}{(x+1)^{b+2}}, \qquad (4.13)$$

which is positive whenever $x < \frac{1}{b}$, then negative. From this result we derive

$$\sup g_b = g_b\left(\frac{1}{b}\right) = \frac{1}{\left(1+\frac{1}{b}\right)^{1+b}} = \left(\frac{b}{1+b}\right)^{b+1}, \qquad (4.14)$$

since

$$\ln\left[\left(\frac{b}{1+b}\right)^{b+1}\right] = (b+1)\left[\ln b - \ln(b+1)\right], \qquad (4.15)$$

which tends toward $-\infty$ when b tends toward 0,

$$\sup g_b \xrightarrow[b \to 0]{} 0. \qquad (4.16)$$

Theorem 1 applies. It follows that X conform toward Benford's law when $b \longrightarrow 0$.

Lognormal distributions

A random variable X is lognormal iff $\log(X) \sim N(\mu, \sigma^2)$. Lognormal distributions have been related to Benford.[16] It is easy to prove that whenever $\sigma \longrightarrow \infty$, X tends toward Benford's law. Although the proof may use different tools, a straightforward way to do it is theorem 1.

One classical explanation of Benford's law is that many data sets are actually built through multiplicative processes.[21] Thus, data may be seen as a product of many small effects. This may be modelled by a r.v. X that may be written as

$$X = \prod_i Y_i, \qquad (4.17)$$

Y_i being a sequence of random variables. Using the logtransformation, this leads to $\log(X) = \sum \log(Y_i)$.

The *multiplicative central-limit theorem* therefore proves that, under usual assumptions, X is bound to be almost lognormal, with $\log(X) \sim N(\mu, \sigma^2)$, and $\sigma \longrightarrow \infty$, thus roughly conforming to Benford, as an application of theorem 1.

THE PROSPECTS FOR PROGRESS

If we are right to think that Benford's law is to be understood as a consequence of mere scatter and regularity, instead of special characteristics linked with multiplicative, scale-invariance, or whatever log-related properties, we should be able to state, prove, and check on real data sets, a generalized version of the Benford's law were some function u replaces the log.

Indeed, our basic idea is that X being scattered and regular enough implies $\log(X)$ to be scattered and regular as well, so that $\log(X)$ should be almost uniform mod 1. The same should be true of any $u(X)$, u being a function preserving scatter and regularity. Actually, some u should even be better shots than log, since log reduces scatter on $[1, +\infty[$.

First, let us set out a generalized version of theorem 1, the proof of which is closely similar to that of theorem 1.

Theorem 4.2 *Let X be a r.v. taking values in a real interval I, with p.d.f. f. Let u be a C^1 increasing function $I \longrightarrow \mathbb{R}$, such that $\frac{f}{u'} : x \longmapsto \frac{f(x)}{u'(x)}$ conforms to the following: $\exists a > 0$ such that (1) $\max\left(\frac{f}{u'}\right) = m = \frac{f}{u'}(a)$ and (2) $\frac{f}{u'}$ is nondecreasing on $]0, a]$, and nonincreasing on $[a, +\infty[\cap I$. Then, for all $z \in [0, 1[$,*

$$|P(\{u(X)\} < z) - z| < 2m. \tag{4.18}$$

In particular, if (X_n) is a sequence of such r.v.'s with p.d.f. f_n and $\max(f_n/u') = m_n$, and $\lim_{+\infty}(m_n) = 0$, then $\{u(X_n)\}$ converges in law toward $U([0, 1[)$ when $n \longrightarrow \infty$.

A r.v. X such that $\{u(X)\} \sim U([0, 1[)$ will be said u-Benford henceforth.

Sequences

Although our two theorems only apply to continuous r.v.'s, the underlying intuition that log-Benford's law is only a special case (having, however, a special interest thanks to its implication in terms of leading-digits interpretation) of a more general law does also apply to sequences. In this section, we experimentally test u-Benfordness for a few sequences (v_n) and four functions u.

We will use six mathematical sequences. Three of them, namely $(\pi n)_{n \in \mathbb{N}}$, prime numbers (p_n), and$(\sqrt{n})_{n \in \mathbb{N}}$ are known not to follow Benford. The three others, $(n^n)_{n \in \mathbb{N}}$, $(n!)_{n \in \mathbb{N}}$ and $(e^n)_{n \in \mathbb{N}}$ conform to Benford. As for u, we will focus on four cases:

$$x \longmapsto \log\left[\log\left(x\right)\right], \qquad (4.19)$$

$$x \longmapsto \log\left(x\right), \qquad (4.20)$$

$$x \longmapsto \sqrt{x}, \qquad (4.21)$$

$$x \longmapsto \pi x^2. \qquad (4.22)$$

The first one increases very slowly, so we may expect that it will not work perfectly. The second leads to the classical Benford's law. The π coefficient of the last u allows us to use integer numbers, for which $\left\{x^2\right\}$ is nil.

The result of the experiment is given in Table 4.1.

Table 4.1.   Results of the Kolmogorov-Smirnov tests applied on $\{u\left(v_n\right)\}$, with four different functions u (columns) and six sequences (lines). Each sequence is tested through its first N terms (from $n = 1$ to $n = N$), with an exception for $\log \circ \log\left(n^n\right)$ and $\log \circ \log\left(n!\right)$, for which $n = 1$ is not considered. Each cell displays the Kolmogorov-Smirnov z and the corresponding p value.

v_n	$\log \circ \log\left(v_n\right)$	$\log\left(v_n\right)$	$\sqrt{v_n}$	πv_n^2
$\sqrt{n}$ $(N = 10\ 000)$	68.90 (.000)	45.90 (.000)	4.94 (.000)	0.02 (.000)
πn $(N = 10\ 000)$	44.08 (.000)	26.05 (.000)	0.19 (1.000)	0.80 (.544)
p_n $(N = 10\ 000)$	53.92 (.000)	22.01 (.000)	0.44 (.990)	0.69 (.719)
e^n $(N = 1\ 000)$	6.91 (.000)	0.76 (1.000)	0.63 (.815)	0.79 (.560)
$n!$ $(N = 1\ 000)^{(*)}$	7.39 (.000)	0.58 (.887)	0.61 (.844)	0.90 (.387)
n^n $(N = 1\ 000)^{(*)}$	7.45 (.000)	0.80 (.543)	16.32 (.000)	0.74 (.646)

The sequences have been arranged according to the speed with which it converges to $+\infty$ (and so are the functions u). None of the six sequences is $\log \circ \log$-Benford (but a faster divergent sequence such as $\left(10^{e^n}\right)$ would do). Only the last three are log-Benford. These are the sequences going to ∞ faster than any polynomial. Only one sequence (n^n) does not satisfy $\sqrt{x}$-Benfordness. However, this can be understood as a pathological case, since $\sqrt{n^n}$ is integer whenever n is even, or is a perfect square. Doing the same Kolmogorov-Smirnov test with odd numbers not being perfect squares gives $z = 0.45$ and $p = 0.987$, showing no discrepancy with $\sqrt{x}$-Benfordness for (n^n). All six sequences are $\pi^2 x$-Benford.

Putting aside the case of $\sqrt{n^n}$, what Table 4.1 reveals is that the convergence speed of $u(v_n)$ completely determines the u-Benfordness of (v_n). More precisely, it seems that (v_n) is u-Benford whenever $u(v_n)$increases as fast as $\sqrt{n}$, and is not u-Benford whenever $u(v_n)$ increases as slowly as $\ln(n)$. Of course, this rule-of-thumb is not to be taken as a theorem. Obviously enough, one can actually decide to increase or decrease convergence speed of $u(v_n)$ without changing $\{u(v_n)\}$, adding or substracting ad hoc integer numbers.

Nevertheless, this observation suggests that we give a closer look at sequence $f(n)$, where f is an increasing and concave real function converging toward ∞, and look for a condition for $(\{f(n)\})_n$ to converge to uniformity. An intuitive idea is that $(\{f(n)\})_n$ will depart from uniformity if it does not increase fast enough: we may define brackets of integers–namely $[f^{-1}(n), f^{-1}(n+1) - 1[\cap\mathbb{N}$, within which $\lfloor(n)\rfloor$ is constant, and of course $\{f(n)\}$ increasing. If these brackets are "too large", the relative weight of the last considered bracket is so important that it overcomes the first terms of the sequence $f(0), ..., f(n)$ mod 1. In that case, there is no limit to the probability distribution of $(\{f(n)\})$. The weight of the brackets should therefore be small relative to $f^{-1}(n)$, which may be written as

$$\frac{f^{-1}(n) - f^{-1}(n+1)}{f^{-1}(n)} \xrightarrow[\infty]{} 0. \tag{4.23}$$

Provided that f is regular, this leads to

$$\frac{\left(f^{-1}\right)'(x)}{f^{-1}(x)} \xrightarrow[\infty]{} 0, \tag{4.24}$$

or

$$\left[\ln\left(f^{-1}(x)\right)\right]' \xrightarrow[\infty]{} 0. \tag{4.25}$$

Functions $f : x \longmapsto x^\alpha$, $\alpha > 0$ satisfy this condition. Any n^α should then show a uniform limit probability law, except for pathological cases ($\alpha \in \mathbb{Q}$). Taking $\alpha = \frac{1}{\pi}$ gives (with $N = 1000$), a Kolmogorov-Smirnov $z = 1.331$, and a p-value 0.058, which means there is no significant discrepancy from uniformity. On the other hand, the log function which does not conform to this condition is such that $\{\log(n)\}$ is not uniform, confirming once again our rule-of-thumb conjecture.

Real data

We test three data sets for u-Benfordness using a Kolmogorov-Smirnov test for uniformity. First data set is the opening value of the Dow Jones, the first day of each month from October 1928 to November 2007. The second and third are country areas expressed in millions of square-km and the populations of the different countries, as estimated in 2008, expressed in millions of inhabitants. The two last sequences are provided by the CIA[†]. Table 4.2 displays the results.

Table 4.2. Results of the Kolmogorov-Smirnov tests applied on $\{u(v_n)\}$. Dow Jones ($N = 950$), countries areas ($N = 256$) and populations ($N = 242$) were used.

	$\log \circ \log(v_n)$	$\log(v_n)$	$\sqrt{v_n}$	πv_n^2
Dow Jones	5.90 (.000)	5.20 (.000)	0.75 (.635)	0.44 (.992)
Countries areas	1.94 (.001)	0.51 (.959)	0.89 (.404)	1.88 (.002)
Populations	3.39 (.000)	0.79 (.568)	0.83 (.494)	0.42 (.994)

This table confirms our analysis: classical Benfordness is actually less often borne out than $\sqrt{x}$-Benfordness on these data. The last column shows that our previous conjectured rule has exceptions: divergence speed is not an absolute criterion by itself. For country areas, the fast growing $u : x \longmapsto \pi x^2$ gives a discrepancy from uniformity, whereas the slow-growing log does not. However, allowing for exceptions, it is still a good rule-of-thumb.

Continuous random variables

Our theorems apply on continuous r.v.'s. We now focus on three examples of such r.v.'s, with the same u as above (except for $\log \circ \log$, which is not defined everywhere on $\mathbb{R}_+^*$): the uniform density on $]0, k]$ ($k > 0$), exponential density, and absolute value of a normal distribution.

Uniform random variables

It is a known fact that a uniform distribution X_k on $]0, k]$ ($k > 0$) does not approach classical Benfordness, even as a limit. On every bracket $[10^{j-1}, 10^j - 1[$, the leading digit is uniform. Therefore, taking $k = 10^j - 1$ leads to a uniform (and not logarithmic) distribution for leading digits, whatever j might be.

[†]http://www.cia.gov/library/publications/the-world-factbook/docs/
rankorderguide.html

The density g_k of $\sqrt{X_k}$ is

$$g_k(x) = \frac{2x}{k}, \ x \in \left]0, \sqrt{k}\right] \tag{4.26}$$

and $g_k(x) = 0$ otherwise. It is an increasing function on $]-\infty, \sqrt{k}]$, decreasing on $[\sqrt{k}, +\infty[$ with maximum $\frac{2}{\sqrt{k}} \longrightarrow 0$ when $k \longrightarrow \infty$. Theorem 2 applies, showing that X_k tends toward $\sqrt{x}$-Benfordness in law. Now, theorem 3 below proves that X_k tends toward u-Benfordness, when $u(x) = \pi x^2$.

Theorem 4.3 *If X follows a uniform density on $]0, k]$, $\{\pi X^2\}$ converges in law toward uniformity on $[0, 1[$ when $k \longrightarrow \infty$.*

Proof.   Let $X \sim U(]0, k])$. The p.d.f. g of $Y = \pi X^2$ is

$$g(x) = \frac{1}{2a\sqrt{x}}, \ x \in]0, a^2], \tag{4.27}$$

where $a = k\sqrt{\pi}$.

The c.d.f. G of Y is then

$$G(x) = \frac{\sqrt{x}}{a}, \ x \in]0, a^2]. \tag{4.28}$$

Let now $\delta \in]0, 1[$. Call P_δ the probability that $\{Y\} < \delta$.

$$\sum_{j=0}^{\lfloor a^2 - \delta \rfloor} G(j + \delta) - G(j) \le P_\delta \le \sum_{j=0}^{\lfloor a^2 - \delta \rfloor + 1} G(j + \delta) - G(j), \tag{4.29}$$

$$\frac{1}{a} \sum_{j=0}^{\lfloor a^2 - \delta \rfloor} \sqrt{j + \delta} - \sqrt{j} \le P_\delta \le \frac{1}{a} \sum_{j=0}^{\lfloor a^2 - \delta \rfloor + 1} \sqrt{j + \delta} - \sqrt{j}. \tag{4.30}$$

The square-root function being concave,

$$\sqrt{j + \delta} - \sqrt{j} \ge \frac{\delta}{2\sqrt{j + \delta}} \tag{4.31}$$

and, for any $j > 0$,

$$\sqrt{j + \delta} - \sqrt{j} \le \frac{\delta}{2\sqrt{j}}. \tag{4.32}$$

Hence,

$$\frac{\delta}{2a} \sum_{j=0}^{\lfloor a^2-\delta \rfloor} \frac{1}{\sqrt{j+\delta}} \leq P_\delta \leq \frac{1}{a}\left[\sqrt{\delta} + \sum_{1}^{\lfloor a^2-\delta \rfloor+1} \frac{\delta}{2\sqrt{j}} \right], \qquad (4.33)$$

$$\frac{\delta}{2a} \sum_{j=0}^{\lfloor a^2-\delta \rfloor} \frac{1}{\sqrt{j+\delta}} \leq P_\delta \leq \frac{\sqrt{\delta}}{a} + \frac{\delta}{2a} \sum_{1}^{\lfloor a^2-\delta \rfloor+1} \frac{1}{\sqrt{j}}, \qquad (4.34)$$

$x \longmapsto \frac{1}{\sqrt{x}}$ being decreasing,

$$\sum_{j=0}^{\lfloor a^2-\delta \rfloor} \frac{1}{\sqrt{j+\delta}} \geq \int_{\delta}^{\lfloor a^2-\delta \rfloor+1+\delta} \frac{1}{\sqrt{t}}dt \geq 2\left[\sqrt{\lfloor a^2-\delta \rfloor+1+\delta} - \sqrt{\delta} \right] \quad (4.35)$$

and

$$\sum_{1}^{\lfloor a^2-\delta \rfloor+1} \frac{1}{\sqrt{j}} \leq \int_0^{\lfloor a^2-\delta \rfloor+1} \frac{1}{\sqrt{t}}dt \leq 2\left[\sqrt{\lfloor a^2-\delta \rfloor+1} \right]. \qquad (4.36)$$

So,

$$\frac{\delta}{a}\left[\sqrt{\lfloor a^2-\delta \rfloor+1+\delta} - \sqrt{\delta} \right] \leq P_\delta \leq \frac{\delta}{a}\left[\sqrt{\lfloor a^2-\delta \rfloor+1} \right]. \qquad (4.37)$$

As a consequence, for any fixed δ, $\lim_{a \longrightarrow \infty} (P_\delta) = \delta$, and $\{\pi X^2\}$ converges in law to uniformity on $[0, 1[$. $\square$

Exponential random variables

Let X_λ be an exponential r.v. with p.d.f. $f_\lambda(x) = \lambda \exp(-\lambda x)$ $(x \geq 0, \lambda > 0)$. Engel and Leuenberger [2003] demonstrated that X_λ tends toward the Benford's law when $\lambda \longrightarrow 0$.

The p.d.f. of $\sqrt{X_\lambda}$ is $x \longmapsto 2\lambda x \exp(-\lambda x^2)$, which increases on $]0, \frac{1}{2\lambda}]$ and then decreases. Its maximum is $\exp(-\frac{1}{4\lambda})$. Theorem 2 thus applies, showing that X_λ is $\sqrt{x}$ -Benford as a limit when $\lambda \longrightarrow 0$.

Finally, theorem 4 below demonstrates that X_λ tends toward u-Benfordness for $u(x) = \pi x^2$ as well.

Theorem 4.4 *If $X \sim EXP(\lambda)$ (with p.d.f. $f : x \longmapsto \lambda \exp(-\lambda x)$), then $Y = \pi X^2$ converges toward uniformity mod 1 when $\lambda \longrightarrow 0$.*

Proof. Let X be such a r.v. $Y = \pi X^2$ has density g with

$$g(x) = \frac{\mu}{2\sqrt{x}} \exp(-\mu\sqrt{x}), \quad x \geq 0 \qquad (4.38)$$

where $\mu = \frac{\lambda}{\sqrt{\pi}}$. The Y c.d.f. G is thus, for all $x \geq 0$

$$G(x) = 1 - e^{-\mu\sqrt{x}}. \tag{4.39}$$

Let P_δ denote the probability that $\{Y\} < \delta$, for $\delta \in]0, 1[$.

$$P_\delta = \sum_{j=0}^{\infty} \left[e^{-\mu\sqrt{j}} - e^{-\mu\sqrt{j+\delta}} \right]. \tag{4.40}$$

$x \longmapsto \exp(-\mu\sqrt{x})$ being convex,

$$\delta \frac{\mu}{2\sqrt{j+\delta}} e^{-\mu\sqrt{j+\delta}} \leq e^{-\mu\sqrt{j}} - e^{-\mu\sqrt{j+\delta}} \tag{4.41}$$

for any $j \geq 0$, and

$$e^{-\mu\sqrt{j}} - e^{-\mu\sqrt{j+\delta}} \leq \delta \frac{\mu}{2\sqrt{j}} e^{-\mu\sqrt{j}} \tag{4.42}$$

for any $j > 0$. Thus

$$\delta \sum_{j=0}^{\infty} \frac{\mu}{2\sqrt{j+\delta}} e^{-\mu\sqrt{j+\delta}} \leq P_\delta \leq 1 - e^{-\mu\sqrt{\delta}} + \delta \sum_{j=1}^{\infty} \frac{\mu}{2\sqrt{j}} e^{-\mu\sqrt{j}}. \tag{4.43}$$

$x \longmapsto \frac{1}{\sqrt{x}} \exp(-\mu\sqrt{x})$ being decreasing,

$$\delta \sum_{j=0}^{\infty} \frac{\mu}{2\sqrt{j+\delta}} e^{-\mu\sqrt{j+\delta}} \geq \delta \int_{\sqrt{\delta}}^{\infty} \frac{\mu}{2\sqrt{t}} e^{-\mu\sqrt{t}} dt \tag{4.44}$$

$$\geq \delta \left[-e^{-\mu\sqrt{t}} \right]_{\sqrt{\delta}}^{\infty} \tag{4.45}$$

$$= \delta e^{-\mu\sqrt{\delta}}, \tag{4.46}$$

and

$$1 - e^{-\mu\sqrt{\delta}} + \delta \sum_{j=1}^{\infty} \frac{\mu}{2\sqrt{j}} e^{-\mu\sqrt{j}} \leq 1 - e^{-\mu\sqrt{\delta}} + \delta \int_{0}^{\infty} \frac{\mu}{2\sqrt{t}} e^{-\mu\sqrt{t}} dt \tag{4.47}$$

$$\leq 1 - e^{-\mu\sqrt{\delta}} + \delta. \tag{4.48}$$

The two expressions tend toward δ when $\mu \longrightarrow 0$, so that $P_\delta \longrightarrow \delta$. The proof is complete. $\qquad\square$

Absolute value of a normal distribution

To test the absolute value of a normal distribution X with mean 0 and variance 10^8, we picked a sample of 2000 values and used the same procedure as for real data. It appears, as shown in Table 4.3, that X significantly departs from u-Benfordness with $u(x) = \log(x)$, $u(x) = \pi x^2$, but not with $u(x) = \sqrt{x}$.

Table 4.3.   The table displays if uniform distributions, exponential distributions, and absolute value of a normal distribution, are u-Benford for different functions u, or not. The last line shows the results (and p -values) of the Kolmogorov-Smirnov tests applied to a 2000-sample. It could be read as "NO; YES; NO".

	$\log(X)$	$\sqrt{X}$	πX^2
$U\left([0,k[\right) \;\; k \longrightarrow \infty$	NO	YES	YES
$EXP(\lambda) \;\; \lambda \longrightarrow 0$	YES	YES	YES
$\left\lvert \mathcal{N}\left(0,10^8\right)\right\rvert$	14.49 (.000)	0.647 (.797)	28.726 (.000)

As we already noticed, the best shot when one is looking for Benford seems to be the square-root rather than log.

4.2. Conclusions

Random variables exactly conforming the Benford's classical law are rare, although many do roughly approach the law. Indeed, many explanations have been proposed for this approximate law to hold so often. These explanations involve complex characteristics, sometimes directly related to logarithms, sometimes through multiplicative properties.

Our idea–formalized in theorem 1–is more simple and general. The fact that real data often are regular and scattered is intuitive. What we proved is an idea which has been recently expressed by Fewster:[8] scatter and regularity are actually *sufficient* conditions to Benfordness.

This fact thus provides a new explanation of Benford's law. Other explanations, of course, are acceptable as well. But it may be argued that some of the most popular explanations are in fact corollaries of our theorem. As we have seen when studying Pareto type II density, mixtures of distributions may lead to regular and scattered density, to which theorem 1 applies. Thus, we may argue that a mixture of densities is nearly Benford *because* it is necessarily scattered and regular. In the same fashion, multiplications of effects lead to Benford-like densities, but also (as the multiplicative central-limit theorem states) to regular and scattered densities.

Apart from the fact that our explanation is simpler and (arguably) more general, a good argument in its favor is that Benfordness may be generalized–unlike log-related explanations. Scale invariance or multiplicative properties are log-related. But as we have seen, Benfordness is not dependent on log, and can easily be generalized. Actually, it seems that square root is a better candidate than log. The historical importance of log-Benfordness is of course due to the implications in terms of leading digits which bears no equivalence with square-root.

References

1. Benford, F. *The law of anomalous numbers.* Proceedings of the American Philosophical Society, 78, 551-572, 1938
2. Berger, A., Bunimovich L., Hill T., *One-dimensional dynamical systems and Benford's law*, Transactions of the American Mathematical Society 357(1), p. 197-219, 2004.
3. Burke J., Kincanon E., *Benford's law and physical constants: The distribution of initial digits*, American Journal of Physics 59, p. 952., 1991.
4. Cho, W. K. T., and Gaines, B. J. *Breaking the (Benford) law: Statistical fraud detection in campaign finances.* The American Statistician, 61, 218-223, 2007.
5. Diaconis, P. *The distribution of leading digits and uniform distribution mod 1.* The Annals of Probability, 5(1), 72-81, 1977.
6. Drake, P. D., Nigrini, M. J. *Computer assisted analytical procedures using Benford's law.* Journal of Accounting Education, 18, 127-146, 2000
7. Engel, H.-A., Leuenberger, C. *Benford's law for exponential random variables.* Statistics and Probability Letters, 63, 361-365, 2003.
8. Fewster, R. M. *A simple explanation of Benford's law.* The American Statistician, 63(1), 26-32, 2009.
9. Janvresse, E., de la Rue, T. *From uniform distributions to Benford's law.* J. Appl. Probab. Volume 41, Number 4, 1203-1210, 2004.
10. Hales, D. N., Sridharan, V., Radhakrishnan, A., Chakravorty, S. S., Siha, S. M. *Testing the accuracy of employee-reported data: An inexpensive alternative approach to traditional methods.* European Journal of Operational Research, 2007.
11. Hill, T. *Random-number guessing and the first-digit phenomenon*, Psychological Reports 62, p. 967-971, 1988
12. Hill T. *Base-invariance implies Benford's law*, Proceedings of the American Mathematical Society 123, p. 887-895, 1995a.
13. Hill, T. *A statistical derivation of the significant-digit law*, Statistical Science 10, 354-363, 1995b.
14. Jolissaint, P. *Loi de Benford, relations de récurrence et suites equidistribuées*, Elemente der Mathematik 60(1), 2005, p.10-18. http://ww.jura.ch/ijsla/Benford.pdf
15. Kontorovich, A. and Miller, S. J. *Benford's law, values of L-functions and the 3x + 1 problem*, Acta Arith. 120, 269297, 2005.
16. Kossovsky, A. E., 2008. Towards a better understanding of the leading digits phenomena. ArXiv:math/0612627.
17. Lolbert, T. *On the non-existence of a general benford's law.* Mathematical Social Sciences 55, 2008, 103–106.
18. Mardia, K. V. and Jupp, P. E. *Directional statistics.* Chichester: Wiley, Example 4.1. 2000
19. Newcomb, S. *Note on the frequency of use of the different digits in natural numbers.* American Journal of Mathematics, 4, 39-40, 1881.

20. Paolella, M. S. *Fundamental probabilty: A computational approach.* Chichester: John Wiley and Son, 2006
21. Pietronero, L., Tosatti, E., Tosatti, V., Vespignani, A. *Explaining the uneven distribution of numbers in nature: The laws of Benford and Zipf.* Physica A, 293, 297-304, 2001.
22. Pinkham, R. S. *On the distribution of first significant digits,* Ann. Math. Statistics 32, 1223-1230, 1961.
23. Raimi, R. A. *The first digit problem.* The American Mathematical Monthly, 83, 521-538, 1976.
24. Scott, P. D., Fasli, M. *Benford's law: An empirical investigation and a novel explanation.* CSM technical report 349, Department of Computer Science, University of Essex, 2001. http://citeseer.ist.psu.edu/709593.html
25. Shürger, K. *Extensions of Black-Scholes processes and Benford's law.* Stochastic Processes and their Applications, Stochastic Processes and Their Applications, vol. 118, no. 7, pp. 1219–1243, 2008.
26. Sehity, T., Hoelzl, E., Kirchler, E. *Price developments after a nominal shick: Benford's law and psychological pricing after the euro introduction.* International Journal of Research in Marketing, 22, 471–480, 2005
27. Smith, S. W. *Explaining Benford's law.* Chapter 34 in The scientist and engineer's guide to digital signal processing, 2007. Available at http://www.dspguide.com/

PART II

Randomness and Computation in Connection to the Physical World

Chapter 5

Some Bridging Results and Challenges in Classical, Quantum and Computational Randomness

Giuseppe Longo, Catuscia Palamidessi and Thierry Paul

CNRS and Département d'Informatique UMR 8548, École Normale Supérieure

http://www.di.ens.fr/users/longo/

CNRS and Département de Mathématiques et Applications. UMR 8553, École Normale Supérieure

paul@dma.ens.fr

and

INRIA-Saclay and LIX, École Polytechnique

catuscia@lix.polytechnique.fr

WHAT DREW US TO THE STUDY OF COMPUTATION AND RANDOMNESS

We encountered randomness in our different fields of interest, as unpredictable phenomena are omnipresent in natural and artificial processes. In classical physical systems (and by this we mean also relativistic ones) randomness may be defined as 'deterministic unpredictability'. That is, since Poincaré's results (on the Three Body Problem) and his invention of the geometry of dynamical systems, deterministic systems include various forms of chaotic ones, from weak (mixing) systems to ones highly sensitive to border conditions, where random behaviours are part of the deterministic evolutions. Randomness got a new status with the birth of quantum mechanics: access to information on a given systems passes through a non-deterministic process (measurement). In computer sciences, randomness is at the core of algorithmic information theory, all the while nondeterministic algorithms and networks present crucial random aspects. Finally, an extensive use of randomness is made also in biology.

Thus we wondered: all these different sciences refer to a concept of randomness, but is it really the same concept? And if they are different concepts, what is the relation between them?

Let us analyse in more detail the kind of randomness that emerges in the various disciplines.

5.1. Physical Randomness (Classical)

A kind of randomness can be viewed as a property of trajectories within classical dynamical systems, namely as unpredictability in finite time, over approximated physical measure with a given (non reducible) precision.[2,3,7,20,34,35] Moreover, ergodicity (à la Birkhoff) provides a relevant and purely mathematical way to define randomness asymptotically, in the limit of infinite trajectories, which still applies for deterministic systems inspired by physics, but independently of chaotic properties (unpredictability) of physical processes.[18,36] In short, within these deterministic frames, randomness can be derived from a comparison between time averages (along a trajectory starting from a point) and space averages (over the entire space) of the chosen observables.

5.2. Physical Randomness (Quantum)

Randomness in quantum mechanics has a special status, as it is of intrinsic origin. In contrast to other fields, including classical and relativistic physics, quantum randomness (QR) does not appear as a mean of "hiding" a kind of lack of knowledge, like, e.g., in statistical physics. The experimental evidence of the violation of the Bell's inequalities definitely proves that QR really belongs to the paradigm of the theory.

Yet, on the other side, QR fits perfectly well on the axiomatic setting of quantum mechanics, it is part of the 4 axioms of the so-called "Copenhagen" formalism, and the perfect coherence with the other postulates is a great success of the sciences of the last century.

Observe, in particular, that a quantum algorithm cannot avoid randomness, as it provides, as "output", i.e., the result of a final measurement process, a bunch of possibilities, one of them (fortunately the most probable) being the expected result. Therefore (easy) checking of the validity of a result must be part of the handled problem. The spectacular power of quantum algorithms, on the other side, overcomes this double constraint (randomness of the result and necessity of a posteriori validation of the "guess") and fully justifies this new branch of computer sciences. In this perspective, quantum randomness cannot be viewed as a form of (hidden or incomplete) determination.[4,21]

5.3. Algorithmic Randomness

Also recursion theory gave us a proper form of (asymptotic) randomness, for infinite sequences, in terms of Martin-Löf randomness.[39] This has been extensively developed by Chaitin, Schnorr, Calude and many others,[15] also in relation to physics.[51] As a matter of fact, on the grounds of this theory one can give a precise meaning to vague notions such as "easily describable" and "regularities". In short, algorithmic randomness tests enables one to detect those elements whose regularities can be effectively tested, labeling them as "non-random". The elements that remain are called "random". Yet, the existing comparisons within (classical) physical randomness and algorithmic randomness are developed in formal space with insufficient physical generality. Some recent work went beyond this limit.[51]

5.4. Computer Science Randomness

Probabilistic and nondeterministic models of computation have been extensively investigated in the computer science literature, as well as their combination. Still there is no agreement about the precise nature of nondeterminism, and its relation with probability. In different areas the term nondeterminism refers to different concepts, and even in the same area people have different intuition and interpretation of nondeterministic models or phenomena (see for example, Ref. 5, for a good survey). This confusion has brought consequences sometimes dramatic, especially when trying to transfer some theory or methodology from one field to another. One community which is particularly sensible to the problem is that of Computer Security. Although it was discovered only recently, the issue has rapidly become known and recognized as crucial, to the point that the organizers of the 2006 edition of the main forum in Computer Security, the IEEE FCS, set up a panel to discuss about nondeterminism.

In sequential computation the term nondeterminism refers to models in which the transition relation goes from one state to a set of states, like nondeterministic Turing machines. This is a useful device in specification, especially for search problems, in that it allows to explore alternative solutions without having to detail the backtracking policy. A characteristic of this kind of models is that their intended meaning is in terms of may-semantics, in the sense that the computation is considered successful if at least one of the alternative branches is successful. We argue that this has nothing to do with randomness: the re-execution of the system gives always the same result(s), and a deterministic implementation is always possible via backtracking (in a breath-first fashion to avoid that infinite branches would cause a pitfall).

In concurrency theory the situation is very different:[1] nondeterminism is inherent to the model, in the sense that it arises naturally from the concurrent execution and interaction of parallel processes. More precisely, it generates from the different ways in which processes may alternate their execution steps, cooperate with each other, compete for resources, etc. In general, these processes are assumed to run in different nodes of a distributed network, and the intended meaning is in terms of the must-semantics, in the sense that every branch must be successful, because the execution follows only one of the many alternatives, and backtracking is not an option (it would be too expensive, if possible at all, because all the processes would need to be backtracked, in order to ensure consistency). The re-execution of the system gives a different result each time, determined by run-time circumstances. In general we want to abstract from these, and we use the notion of scheduler to represent how the choices are resolved.

5.5. Biology

A common characteristic in the various forms of physical randomness is the predetermination of the spaces of possibilities: random results or trajectories are given among already known possible ones (the six sides of a dice, the spin-up/spin-down of a quanton...). In fact, in quantum physics, even in cases where new particles may be created, sufficiently "large" spaces are provided upstream (the Fock spaces of which Hilbert spaces): Fock's spaces capture all the possible states, infinitely many in general. The classical methods transfer successfully in molecular analysis in Biology, where only physical processes are observed, even though there are meant to happen within cells.

In System Biology, however, phase or reference spaces (that is, the spaces of possible evolutions) are far from being predetermined. Typically, the proper biological observables of Darwinian Evolution, namely phenotypes and species,[26,33] are not pre-given or there is no way to give them in advance within a space of all possible evolutions, in a sound theory. And, of course, there is no way to pre-give the possible molecular interactions (internal and systemic) as well as the feedbacks, from the forthcoming ecosystems onto molecular cascades. An analysis of Species Evolution, in terms of a diffusion equation (thus of underlying random paths) is given in

Ref. 8. Our attention to the problem of randomness in System Biology was stimulated by these analogies and differences w.r. to Physics.

WHAT WE HAVE LEARNED

Randomness is nowadays part of our culture. It appears everywhere. But, as mentioned before, it reflects very different kinds of conceptual (and mathematical) settings. Let us more closely review different forms of knowledge dealing with randomness in modern sciences.

5.6. Physical Randomness (Classical)

In classical dynamics, it is possible to give a notion of individual random element. More precisely, in a dynamical systems, with a transformation T preserving a measure μ, there exists a natural class of (asymptotic) properties allowing a definition of "random state" with respect to the dynamic T and the equilibrium distribution defined by μ. This class is the one associated to the Birkhoff ergodic theorem which states that for each (integrable) function $f : X \to \mathbb{R}$ (representing a quantified observation), the time average of f along the orbit $O(x) = \{x, T(x), T^2(x), ...\}$ converges to the spacial mean $\int f d\mu$ with probability one. Such points are called typical for T.[53] Though given asymptotically, the definition has a robust physical meaning and it may be given for weakly chaotic dynamical systems : the mixing ones (that is, dynamical systems where observable decorrelates with time, see Refs. 23,24).

5.7. Physical Randomness (Quantum)

The appearance of non-determinism in quantum mechanics was a shock. In the '20s Physics was still very rooted in the classical deterministic view of the world. Therefore it took a lot of time to accept this fact, although it appears totally natural now to us that a description on the world could be fully statistical. This randomness flavour of quantum mechanics is usually handled in the so-called field of "conceptual aspects" of the theory.[31] Facing these conceptual aspects with new experimental physics is not so frequent, and the relationship with other fields, as the ones presented here, are even less often considered. At the meantime recent works (see Ref. 46–49) started invoking these links. Let us finally mention a kind of philosophical morality: the high new power of quantum algorithms relies on randomness, that is, releasing determinism (in a way of keeping probability predictions) increases drastically efficiency.

5.8. Algorithmic Randomness

As already mentioned, the notion of individual random infinite sequence has been effectively modeled with the tools of computability theory. The idea is that statistical tests may be "effectively given", that is one can effectively look for regularities along infinite sequences. Then an infinite sequence is random if it passes all effective tests. This algorithmic modelling of randomness has been formalized in different ways and there exist several definitions (the one by Martin-Löf's, in Ref. 39, being the most celebrated, see Ref. 15 for a survey book).

In these approaches, being random yields a high degree of "non-computability", as a random sequence has no infinite recursively enumerable subset. The role of algorithmic randomness in dynamical systems, especially in ergodic ones, has already be the subject of previous research. However, without a more general theory of randomness, all this work is restricted to "symbolic spaces" (spaces of strings), with little physical meaning. The non-obvious results recently obtained proved an equivalence, in very general and "physically meaningful" dynamical spaces, of the (classical) physical randomness, à la Birkhoff, and algorithmic randomness, for infinite trajectories (see the PhD theses by M. Hoyrup and C. Rojas (June 2008) and Refs. 23, 24 and 38).

As for finite time computations, Chaitin, Levin, Calude and many others, following Kolmogorof, deeply analysed also (finite) sequence incompressibility (sequences whose length coincides with their shortest generating program) and showed that for infinite sequences, under suitable conditions, the incompressibility of initial segments yields Martin-Löf asymptotic randomness. But, in our opinion, unless the physical generating process is spelled out, a finite incompressible sequence is not random, it is just algorithmically incompressible. In other words, it is pseudorandom in the strongest way and it is impossible to see in it any regularity whatsoever. Of course, a physical random sequence is incompressible, in principle, but the converse is false or ill defined. That is, if one stays within theory of computation and no physical process is mentioned, there is no other way to give/conceive it but by a program, a formal/linguistic matter, a priori with no physical meaning. A closer analysis of this issue is part of our project, in view of our experience on the relation physical (dynamical) vs. algorithmic randomness.

5.9. Randomness in Computer Science

As argued in previous section, sequential systems are inherently deterministic (despite the use of the adjective 'nondeterministic' for the automata whose transition relation is one-to-many). Concurrency, on the contrary, seems to give rise to a true notion of randomness, due to the unpredictable and unbacktrable nature of interaction between independent and asynchronous agents.

The fact that computation in concurrent systems seems to have features which are basically different from those of sequential computation has inspired some intriguing lines of research, with the common goal of defining a 'good' notion of expressiveness for concurrent formalisms. One of the most successful approaches is based on the notions of 'encoding', and it has produced some interesting results concerning the mechanisms of guarded choice, which are intimately related to the nature of nondeterminism in concurrency. We mention in particular the works by Nestmann and Pierce,[41] and by Palamidessi.[44,45] The first have shown that a certain form of choice (input-guarded choice) can be encoded in parallelism, while the latter has shown that mixed-guarded choice is essentially more expressive.

Another line of investigation has been pursued by Wegner and his collaborators. Starting from the principle that interaction is more expressive than algorithms, Wagner has written an intriguing position paper where he justifies this claim, essentially on the basis of the intuition that interaction can express random computation.[54] In the technical development of this idea in subsequent papers[25,55] the authors considered a sort of interactive Turing machines (persistent Turing machines) and showed that they cannot be reduced to standard Turing machines. However the approach is based on the conventional interpretation of nondeterminism in automata theory.

An interesting result has bee found recently by Busi and her colleagues:[13] they investigate a process calculus (CCS with replication) and show that in this calculus the existence of a divergent path is decidable. Still, in a subsequent paper[14] they show that this formalism can encode Turing machines. The explanation of this apparent paradox is that the nondeterminism of the language is essential to achieve Turing-completeness: any attempt to eliminate it in an effective way is doomed to fail, because otherwise decidability of (existence of) divergence would imply decidability of termination.

In all the above investigations nondeterminism plays a central role: since in concurrency there is no backtracking, it becomes important to control nondeterminism as much as possible, and the expressive power of a concurrent language usually lies on the capability to exert such control.

In a more practical fashion, nondeterminism has been used in concurrency theory as a convenient abstraction from run-time information. Essentially, a concurrent program is nondeterministic because when we write it we do not know yet what will determine at run-time the choice of a particular path, so all the possibilities must be considered.

A common misconception of nondeterminism is to consider it a probabilistic mechanism with uniform distribution. The confusion probably originates from epistemic considerations: the total lack of knowledge about which of the available alternatives will be chosen, in a nondeterministic process, evokes the concept of maximum entropy. But maximum entropy represents the maximum degrees of uncertainty within the probabilistic setting. Nondeterminism is outside the realm of probability and it represents an even higher degree of uncertainty. In any case, confusing nondeterminism with uniform probability has induced wrong approaches. We argue that this due to several aspects: not only a nondeterministic property cannot express the quantitative aspects of a probabilistic one, but also this transformation requires angelic nondeterminism (nondeterminism working in favour of the property), which is a strong assumption usually not guaranteed in a concurrent setting (where nondeterminism is typically demonic).

5.10. Biology

We have no doubt that the issue of randomness in biology is extremely difficult, yet it is amazing to observe that leading biologists, still now and along the lines of Crick and Monod,[40] contrapose determination and randomness according to Laplace's split: deterministic means predictable and random is its opposite (non-deterministic), to be analyzed by statistics and probabilities. Along these laplacian lines, deterministic, as it implies predictability (Laplace's conjecture), yields "programmable", which leads to the the idea that the "DNA is a program" (see Ref. 22 for an history.) Yet, since Poincaré (1890), we know that classical randomness is deterministic unpredictability and that unpredictability pops out almost everywhere in non-linear systems (see Ref. 37 for a critique of the claim that "DNA is a program" from the point of view of Physics and Programming Theory.)

In short, determination as "necessity" in life phenomena, understood in a laplacian way, is far from the frameworks of modern determination in physics, classical or quantum, even if it is supplemented by a few speckles of randomness (le "hasard"). Crick's "central dogma" ("Genetic Information" goes one-way, from DNA to RNA to proteins — and to the phenotype) and the "one gene–one enzyme" hypothesis in Molecular Biology are good examples of this. They guided research for decades and, the first, is still nowadays believed by many, modulo the addition of a few "epigenetic factors" and "norms of reaction" (for an alternative view, see Ref. 27; more discussions and references are in Ref. 37). By their linear causality, these assumptions do not seem to have still integrated the views on the interplay of interactions in XXth century physics, that is the richness of the approaches to physical determination and randomness. In these modern physical frames, causes become interactions and these interactions themselves dynamically constitute the fabric of the processes and of their manifestations; reshaping this fabric modifies the interactions, intervening upon the interactions appears to reshape the fabric.[6]

Our recent work on entropy in System Biology,[8] has been extensively borrowing from the work on far-from-equilibrium thermodynamics, in particular dissipative systems.[42,43] Note that entropy provides a measure for "increasing randomness" also for dynamical systems and this establishes a link, developed by many, with other areas of physics (thermodynamics).

WHAT ARE THE MOST IMPORTANT OPEN PROBLEMS IN THE FIELD?

"Random" is not the opposite of "deterministic", in spite of the opposition of these concepts that is commonly made in Computer Science and Biology. As a matter of fact, the analysis of randomness is part of the proposal for a structure of determination of physical processes, in particular in classical dynamics, where randomness is deterministic unpredictability. But it is so also when it is related the very precise and specific notion of quantum indetermination and quantum measure of "deterministic evolutions of the state function" (determined by Schrödinger equation).

As seen in the previous sections, randomness and nondeterminism appear at least in four thematic fields (mathematics, physics, biology and computer sciences). Yet, they cover different, sometimes antinomic meanings.

5.11. Algorithmic Randomness

Since Chaitin's 1975 construction of the first random infinite word, an entire robust field has been opened concerning the combinatorial properties of randomness in discrete frames. This is leading to an interesting analysis of infinite sequences and cycles in de Bruijn graphs, for example, and a lot more in finite and infinite combinatorics.

A way to stress the difference and the strength of the approach we worked at, is that, given a measurable space of measure μ, classical theorems like "property P holds for μ-almost every point" can be converted into "property P holds for every μ-random point", and this in physical dynamics. Now, the notion of "being (μ-)random", relatively to a measure μ, is fully understood, by the many result in algorithmic randomness and, in particular, by our team's work on asymptotic randomness, relating classical (dynamical) and algorithmic randomness. In short, a major point of these results, is given by the use of discrete, though asymptotic, tools for algorithmic randomness, in applications to continuous dynamics. By this, our approach, by its relation to physically meaningful dynamical systems complements the relevant existing work on algorithmic randomness. Further applications (or correlations between) the recent ideas in algorithmic randomness, like the ones developed by the authors above, and our understanding of classical (and quantum) dynamics is one of the paths to be explored. In particular, a typical example is of cross interest is given by the analysis of normality in the sense of Borel*: a real number is absolutely normal with probability one, but constructing such points is extremely complicate.[12] However, it is widely accepted that computable simulations show the right ergodic behaviour.[50] Is it the case also while modelling relevant physical dynamics? and which ones? We will hint below on how a relation to our approach can be developed.

5.12. Computer Science

In Section 5.2, we mentioned the ambiguities we see in the transfer of notions from algorithmic randomness (an asymptotically well-defined notion) to finite time computations. Some may see our questioning as referring to a terminological nuance, yet too much confusion in computing deserves clarification. Consider, say, a so called "non-deterministic" Turing Machine.

---

*A "normal number" is a real number whose digits in every base show a uniform distribution, with all digits being equally likely, all pairs of digits equally likely, all triplets of digits equally likely, etc.[10]

This is just a formal, deterministic device, associating a set of numbers to a number. Its evolution is determined by an ill-typed input-output function. Indeed, it is a useful device as it allows to speed up computations by a form of basic parallelism. Yet, as long a physical process, choosing, at each step, one or a few of the elements in the output set, is not proposed, one cannot discuss about "determination" nor "randomness": is it classical? Quantum? This is an analogue of the problem posed for finite "random" (incompressible!) sequences mentioned above. A major clarification is needed here, as the relation between the abuses of "non-deterministic", "random" in Computer Science deserve a close analysis in terms of their relation to the underlying physical processes.

We need to tackle this problem within the relevant area which mostly interests, concurrency and networks. In particular, the following three main objectives seem relevant:

Clarify the nature of nondeterminism in concurrency. As discussed in previous sections, there is a lot of confusion about this concept, especially in the way the notions and tools developed in the area of concurrency get applied in other areas. We feel that this is because the concept of nondeterminism is still missing a proper characterization, formalization, and theoretical foundations. The way it is formalized in process calculi, indeed, makes it easy to confuse it with the 'nondeterminism' of automata theory, which, as argued previously, has a radically different nature.

Investigate and formalize the difference of concurrent computations with respect to sequential ones. With respect to Wegner's claim that 'interaction is more powerful than algorithms' on the basis of the fact that interactive (concurrent) systems can create real random sequences, our position is: we actually do not believe that it is a matter of interaction having a superior expressive power (if anything, it should be the reverse — randomization witnesses a lack of control capabilities), but certainly the capability of producing randomness sharply separates interactive (concurrent) computations from algorithmic (sequential) ones. Once this characterization in terms of randomness is given, we can also explore whether we can characterize different classes of random sequences, which would induce a separation in the concurrent formalisms (and mechanisms) that can produce them.

Explore the use of physical randomness (classical and quantum) for computer science applications that use randomized mechanism to protect secret information. Typically randomization is used to obfuscate the link between secrets and observables, and the use of really unpredictable mechanisms (in contrast to the pseudo random ones) is critical. Examples of such applications are protocols for privacy, anonymity, and confidentiality,[52] like Crowds and DCnets. How non-linear systems, reflecting classical randomness, or intrinsic quantum randomness can help in the developments of these areas?

5.13. Biology

A long term ambition would be to obtain, by mathematics if possible, a clarification of the not so much explored problem of randomness in System Biology (again, in Molecular Biology, randomness plays a novel but fundamental role, but the tools and concepts are borrowed from Physics, via Chemistry). As we said, in a systemic approach to Biology, one of the challenges, as for our project, is that species (and phenotypes) are co-constituted with their environment.

To make an analogy with the reasons for chaos in planetary systems, some sort of "resonance effect" takes place in this co-constitutive process. The difference is that in the physical deterministic case, the resonance happens at one (and conceptually simple) level: the gravitational interactions between a few planets, fully determined by Newton-Laplace equations. In Evolution (but also in ontogenesis), the resonance takes place between different levels of organization, each deserving an analysis on terms of an appropriate structure of determination (typically, fractal structures or geodetics in morphogenesis, see Ref. 28, vs. networks dynamics in cellular/neural tissues, see Ref. 30.

That is, a systemic approach requires an analysis of interactions between species, individuals, physical landscapes, but also organs and tissues and, very importantly, by two ways interactions between these levels and molecular activities, starting with DNA expression. Moreover, molecular events belong to microphysics, thus possibly subject to quantum analysis, thus, quantum probabilities. By this, one would need a theory encompassing both classical randomness, which may better fit the description of macroscopic interactions, and quantum randomness, as they may be retroacting one on top of the other. We are far form having such a theory, even in Physics.

In conclusion, Physics has been able to propose two different notions of randomness in finite time: classical deterministic unpredictability and quantum randomness. As we mentioned, we proved that they merge in infinite time. Biology badly needs its own notion, while in search, of course, for unification with physical (molecular?) structures of determination. We observe that the notions of "extended criticality" and "anti-entropy" (the opposite of "increasing disorganization or randomness", yet differing from the classical negative entropy) recently proposed in Refs. 8 and 9 seem pertinent. They deeply involve randomness in Biology, by the role of fluctuations in extended criticality and of random paths in the formation and growths of anti-entropy.

5.14. Quantum Mechanics

Since the beginning of the '80s, and especially the fundamental experiments by Aspect et al, showing the violation of Bell's inequalities, quantum mechanics has got the status of "true" theory of the microscopic world. Roughly at the same time appeared the beginning of what is called now quantum information and quantum computing.

Feasibility of the construction of quantum computers is certainly one of the most challenging perspective in physics. More generally experimental exhibition of single phenomena in quantum mechanics is certainly one of the most striking success of sciences of the last 25 years. In the mean time computer science developed, independently, in a direction where non-determinism took an increasing place. An objective would be to exhibit links between these two independent developments, together with possible oppositions. As a bi-product, incidence of the rich computer network theory on quantum mechanics is also expected to be strengthened.

WHAT ARE THE PROSPECTS FOR PROGRESS?

5.15. Quantum Randomnesses

One of the subjects that should be seriously carried out is the difference, in quantum mechanics, between two different type of randomnesses: the one obtained through decoherence and the one of the measurement process (see Ref. 29). Indeed it is often said that decoherence presents a satisfactory "explanation" of quantum measurement. Let us remind that the phenomenon of decoherence is the action on a "small" quantum system of the big quantum surrounding system (e.g., the apparatus). In the limit of infinite "reservoir" the state of the small system gets a diagonal feature that one can identify with a statistical mixture. This last one is sometimes and wrongly presented as reflecting the quantum randomness, as randomness in statistical mechanics. This point of view, driven by a pure statistical perception of quantum phenomena, must be corrected, as more and more single events in quantum mechanics are nowadays shown experimentally. It seems to us that a clarification of the link ! between this dichotomy between (not intrinsic) randomness and the true (intrinsic) one on the quantum side, and the same kind of distinction in the other fields presented all a long this paper, would be a major achievement.

5.16. Pseudo-randomness and Algorithmic Randomness

Computer simulations (the trajectory given in some initial condition and drawn on the screen) has become a very important component of the analysis of physical dynamics. The problem is that an algorithmically random point is strongly non-computable, and consequently it is impossible to observe the trajectory of such a point in a computer simulation. Worst, the set of computable points has probability 0 ! From the simulation point of view, the fact that a given property holds with probability one says nothing about its observability with a computer. A typical example is (absolute) normality in the sense of Borel, but, as we mentioned, constructing such points is extremely complicate.

In spite of this intrinsic (mathematical) difficulties, it is widely accepted that computable simulations display the right ergodic behaviour. The evidence is mostly heuristic. More precisely, most arguments are based on the various "shadowing" theorems. By these results, it is possible to prove that in a suitable system any "pseudo"-trajectory, that is a trajectory obtained by a simulation with round-off, is approximated by a real (continuous) trajectory of the system (but, in general, not the converse). The main limit of this approach is however that shadowing results hold only in particular systems ("hyperbolic" dynamics), while many physically interesting systems do not need to belong to this class. Can we extend this frame for shadowing to an analysis of randomness for the systems of biological and physical relevance, both for finite and infinite processes? What does this mean in a quantum perspective?

Moreover and as observed in the previous sections, algorithmically random points are those points which have the "generic" behavior, as prescribed by the underlying measure. In a sense, they are physically "randomized", as they do not possess any artificial feature related to the particular supporting space. An open issue would be to turn this into a mathematical frame, where the only role of the measure would be to distinguish the set of random points and give them a topological structure. Once this is done, probabilistic theorems can be formulated and proved using the structure of the random set and topological methods (without referring to the measure).

5.17. Computer Science

The objectives proposed in previous section should be pursued by applying the following approaches and methodologies:

Clarify the nature of nondeterminism in concurrency. The results and perspectives developed in mathematics and physics and quoted above should be more closely used to get new insights that can be used to understand in depth the notion of nondeterminism in concurrency, in its intrinsically random nature. On this basis:

Formalize the difference between concurrent and sequential computation in terms of randomness. What about the applicability of Martin-Löf definition of randomness to give this characterization?

Explore the use of physical randomness (classical and quantum) for computer science applications. Typically, it should be explored the possible use of different kinds of randomness in quantum mechanics (see the paragraph on quantum randomnesses in this section), as well as the extensively studied relations between computational and dynamical randomness in enriching/clarifying randomness in finite computing: how can this affect computing? (on a conceptual basis? by the relevance of hardware?)

5.18. Extended Criticality in Biology

In recent work[9] we proposed to analyze the state of living matter as "extended critical state". The idea is that an organism is in a permanent critical transition, constantly reconstructing its own organization. The well established domain called "physics of criticality"[32] necessarily deals with point-wise critical transitions: this is part of the very definition of phase transition and it is used in an essential way by the main mathematical tool in the approach, the "renormalization methods"[19] In ongoing work, we consider, instead, a set (whose closure is) of null-measure, an extended interval of criticality w.r.t. all pertinent parameters (time, temperature, pressure...). It is as if a snow flake (a "coherence structure", formed at a critical transition) could stand variations within a relatively large interval of its control parameters by continually reconstructing itself, in a permanent "going through" the critical transition (extended criticality applies to far from equilibrium, dissipative and not necessarily steady states). One then has an extended, permanently reconstructed global organization in a dynamic interaction with local structures, as the global/local interaction is proper to critical transitions. The role of randomness in these context is crucial, yet to be explored: the structural stability of the system may be seen as stability w.r.t. perturbations within the margins of criticality. Moreover, the very construction of the coherent state is obtained from fluctuations from an initial relatively stable state. The point now is to turn these conceptually robust ideas into sound mathematics. As the normalization methods cannot work, by principle (non-pointwise nature of the critical "transition"), how to describe and handle rigorously extended criticality and its main properties (the establishment and maintainance of the global coherent structure, its structural stability and complexity, in particular in relation to random events)?

Extended criticality, a notion modeling organisms as "extended coherence structures" (our proposal for structurally stable systems), needs also to be merged with the work on entropy and anti-entropy in Species Evolution, mentioned above, see Ref. 8. The possible technical link may reside in the role of randomness in both analyses. This shows up in fluctuations and resistance to perturbations (average behaviours?), as for extended criticality, in diffusion equations (thus, in underlying random paths) as for entropy and anti-entropy. In this frame, one needs to further analyse the role of time and of diffusion equations (which average random paths) both in phylogenesis (we already carried on a mathematical analysis of this in Ref. 8) and in ontogenesis, an open issue.

5.19. Comparison

It should be clear that the many forms of randomness above differ or, in some cases, are not or ill formalized. Yet, as already mentioned, some recent results of our's prove that they merge, asymptotically. In particular, we recall, one of us proved an asymptotic merging of quantum and classical randomness at the so-called "semi-classical limit"[48] and the team of another of us proved a form of equivalence between Birkhoff physical randomness, a limit notion, and algorithmic randomness (see Ref. 38 for a survey). This poses several open questions, for example in the correlations in *finite time* of classical, quantum and algorithmic randomness, an issue extensively studied by many, since the asymptotic analysis may propose a new perspective. Moreover, as we said, the computational approach (algorithmic randomness) is far from being related to the modern forms of randomness in networks and concurrency. Not to mention the very difficult and rather confusing situation we can witness in Biology, when coming from more mathematized disciplines.

References

1. Aceto L., Longo G., Victor B. (eds.) The difference between Sequential and Concurrent Computations. Special issue, Mathematical Structures in Computer Science, Cambridge U. Press, n. 4-5, 2003.
2. Adler R. L. Topological entropy and equivalence of dynamical systems, American Mathematical Society, 1979.
3. Alligood K., Sauer T., Yorke J. Chaos: an introduction to Dynamical Systems, Springer, New York, 2000.
4. Anandan J. Causality, "Symmetries and Quantum Mechanics". Foundations of Physics Letters, vol.15, no. 5, 415 438, October, 2002.
5. Michal Armoni and Mordechai Ben-Ari. The Concept of Nondeterminism. Journal of Science & Education, Springer, 2008.
6. Bailly F., Longo G., Mathématiques et sciences de la nature. La singularité physique du vivant. Hermann, Paris, 2006 (English introduction, downloadable; ongoing translation).
7. Bailly F., Longo G. "Randomness and Determination in the interplay between the Continuum and the Discrete", Mathematical Structures in Computer Science, vol. 17, n. 2, 2007.
8. Bailly F., Longo G. "Biological Organization and Anti-Entropy", to appear in Journal of Biological Systems, 2008.
9. Bailly F., Longo G. "Extended Critical Situations", in J. of Biological Systems, Vol. 16, No. 2, pp. 309-336, June 2008.
10. Becher V., Figueira S. Picchi R, "Turing's unpublished algorithm for normal numbers", Theoretical Computer Science, Volume 377, 126–138, 2007,
11. Becher V., Dickmann M., "Infinite sequences on de Bruijn graphs", manuscript 2007.
12. Becher V. and Figueira S.. An example of a computable absolutely normal number. Theor. Comput. Sci., 270(1-2):947-958, 2002.
13. Busi N., Gabbrielli M., Zavattaro G. Replication vs. Recursive Definitions in Channel Based Calculi. ICALP 2003: 133-144.
14. Busi N., Gabbrielli M., Zavattaro G. Comparing Recursion, Replication, and Iteration in Process Calculi. ICALP 2004: 307-319.
15. Calude C. Information and Randomness: An Algorithmic Perspective. Springer-Verlag New York, 1994.
16. Calude C. Stay M. "From Heisemberg to Gödel via Chaitin", International J. Theor. Phys. 44 (7), 2005.
17. Chaitin, G. A theory of program size formally identical to information theory. Journal ACM, 22:329–340, 1975.
18. Cornfeld I. Fomin S. and Sinai Ya. G., Ergodic Theory. New York: Springer-Verlag, 1982.
19. Delamotte B. A hint of renormalization, American Journal of Physics 72, pp. 170-184, 2004.

20. Devaney R. L. An introduction to Chaotic Dynamical Systems, Addison-Wesley, 1989.
21. Feynman R. Lectures in Physics. Addison-Wesley, 1966.
22. Fox Keller E. The Century of the Gene, Gallimard, 2000.
23. Galatolo S. Hoyrup M. and Rojas C., "Effective symbolic dynamics, random points, statistical behavior, complexity and entropy", to appear in Information and Computation, 2009.
24. Galatolo S. Hoyrup M. and Rojas C., "A Constructive Borel-Cantelli lemma. Constructing orbits with required statistical properties", Theoretical Computer Science, 410(21-23):2207-2222, 2009.
25. Goldin D., Smolka S., Wegner P. Turing Machines, Transition Systems, and Interaction. Electr. Notes Theor. Comput. Sci. 52(1), 2001.
26. Gould S. J. Wonderful Life, WW. Norton, 1989.
27. Kupiec, J.-J. et al. (eds), Le hasard au coeur de la cellule, Sylleps, Paris, 2009.
28. Jean R. V. Phyllotaxis: a systemic study in plant morphogenesis, Cambridge University Press, 1994,
29. Haroche S., Raimond J.M. Exploring the quantum: atoms, cavities and photons, Oxford U.P. graduate texts 2006.
30. Hertz, J., Krogh, A., Palmer, R. (1991) Introduction to the Theory of Neural Computation, New York: Addison-Wesley.
31. Jammer M. The philosophy of quantum mechanics: the interpretations of quantum mechanics in historical perspective, Willey New York, 1976.
32. Lagues M., Lesne A. Invariance d'echelle, Belin, Paris, 2003.
33. Lecointre G., Le Guyader H. Classification phylogénétique du vivant, Paris, Belin 2001.
34. Lighthill J. The recent recognized failure of predictability in Newtonian dynamics, Proc. R. Soc. Lond. A 407, 35-50, 1986.
35. Laskar J. "Large scale chaos in the Solar System", Astron. Astrophysics, 287, L9 L12, 1994.
36. Longo G., Paul T. "The Mathematics of Computing between Logic and Physics". Invited paper, Computability in Context: Computation and Logic in the Real World, (Cooper, Sorbi eds) Imperial College Press/World Scientific, 2008.
37. Longo G., Tendero P.-E. "The differential method and the causal incompleteness of Programming Theory in Molecular Biology". In Foundations of Science, n. 12, pp. 337-366, 2007.
38. Longo G. Randomness and Determination, from Physics and Computing towards Biology. Invited Lecture at the 5th International Conference on: Current Trends in Theory and Practice of Computer Science, Spindleruv mlyn (Czech Republic), January 24-30, 2009, to appear in Lecture Notes in Computer Science, Springer, 2009.
39. Martin-Löef, P. "The definition of random sequences". Information and Control 9: 602-619, 1966.
40. Monod J. Le Hasard et la Nécessité, PUF, 1973.

41. Nestmann U., Pierce B. Decoding Choice Encodings. Inf. Comput. 163(1): 1–59, 2000.
42. Nicolis G. Prigogine I., Self-Organization in Nonequilibrium Systems, J. Willey, 1977.
43. Nicolis G. "Dissipative systems", Rev. Prog. Phys., IL, p. 873, 1986.
44. Palamidessi C. Comparing The Expressive Power Of The Synchronous And Asynchronous Pi-Calculi. Mathematical Structures in Computer Science 13(5): 685-719, 2003
45. Parrow J. Expressiveness of Process Algebras. Electr. Notes Theor. Comput. Sci. 209: 173-186, 2008.
46. Paul T. "La mécanique quantique vue comme processus dynamique", in "Logique, dynamique et cognition" (dir. J.-B. Joinet), collection "Logique, langage, sciences, philosophie", Publications de la Sorbonne, Paris, 2007.
47. Paul T. "Échelles de temps pour l'évolution quantique à petite constante de Planck", Séminaire X-EDP, École Polytechnique, Palaiseau, 2008.
48. Paul T. "Semiclassical analysis and sensitivity to initial conditions", Information and Computation, 207, p. 660-669 (2009).
49. Paul T. À propos du formalisme mathématique de la Mécanique Quantique, "Logique & Interaction: Géométrie de la cognition" Actes du colloque et école thématique du CNRS "Logique, Sciences, Philosophie" à Cerisy, Hermann, 2009.
50. Pour-El M.B., Richards J.I. Computability in analysis and physics. Perspectives in mathematical logic, Springer, Berlin, 1989.
51. Rojas C. "Computability and Information in models of Randomness and Chaos", Math. Struct. in Computer Science, vol. 18, pp 291-307, 2008.
52. Schneider S., Sidiropoulos A. CSP and Anonymity. ESORICS 1996: 198-218
53. V'yugin, Vladimir V. "Ergodic Theorems for Individual Random Sequences", Theoretical Computer Science, vol. 207, p.343-361, 1998.
54. Wegner P. Interactive Foundations of Computing. Theor. Comput. Sci. 192(2): 315-351, 1998.
55. Wegner P. Goldin D. Coinductive Models of Finite Computing Agents. Electr. Notes Theor. Comput. Sci. 19, 1999.

Chapter 6

Metaphysics, Metamathematics and Metabiology

Gregory Chaitin

*IBM Thomas J. Watson Research Center**

gjchaitin@gmail.com

Introduction

In this essay we present an information-theoretic perspective on epistemology using software models. We shall use the notion of algorithmic information to discuss what is a physical law, to determine the limits of the axiomatic method, and to analyze Darwin's theory of evolution.

6.1. Weyl, Leibniz, Complexity and the Principle of Sufficient Reason

The best way to understand the deep concept of conceptual complexity and algorithmic information, which is our basic tool, is to see how it evolved, to know its long history. Let's start with Hermann Weyl and the great philosopher/mathematician G. W. Leibniz. That everything that is true is true for a reason is rationalist Leibniz's famous *principle of sufficient reason*. The bits of Ω seem to refute this fundamental principle and also the idea that everything can be proved starting from self-evident facts.

6.2. What is a Scientific Theory?

The starting point of algorithmic information theory, which is the subject of this essay, is this toy model of the scientific method:

theory/program/010 $\rightarrow$ **Computer** $\rightarrow$ experimental data/output/110100101.

---

*Honorary professor of the University of Buenos Aires and member of the Académie Internationale de Philosophie des Sciences

A scientific theory is a computer program for producing exactly the experimental data, and both theory and data are a finite sequence of bits, a bit string. Then we can define the complexity of a theory to be its size in bits, and we can compare the size in bits of a theory with the size in bits of the experimental data that it accounts for.

That the simplest theory is best, means that we should pick the smallest program that explains a given set of data. Furthermore, if the theory is the same size as the data, then it is useless, because there is always a theory that is the same size as the data that it explains. In other words, a theory must be a compression of the data, and the greater the compression, the better the theory. Explanations are compressions, comprehension is compression!

Furthermore, if a bit string has absolutely no structure, if it is completely random, then there will be no theory for it that is smaller than it is. Most bit strings of a given size are incompressible and therefore incomprehensible, simply because there are not enough smaller theories to go around.

This software model of science is not new. It can be traced back via Hermann Weyl (1932) to G. W. Leibniz (1686)! Let's start with Weyl. In his little book on philosophy *The Open World: Three Lectures on the Metaphysical Implications of Science*, Weyl points out that if arbitrarily complex laws are allowed, then the concept of law becomes vacuous, because there is always a law! In his view, this implies that the concept of a physical law and of complexity are inseparable; for there can be no concept of law without a corresponding complexity concept. Unfortunately he also points out that in spite of its importance, the concept of complexity is a slippery one and hard to define mathematically in a convincing and rigorous fashion.

Furthermore, Weyl attributes these ideas to Leibniz, to the 1686 *Discours de métaphysique*. What does Leibniz have to say about complexity in his *Discours*? The material on complexity is in Sections V and VI of the *Discours*.

In Section V, Leibniz explains why science is possible, why the world is comprehensible, lawful. It is, he says, because God has created the best possible, the most perfect world, in that the greatest possible diversity of phenomena are governed by the smallest possible set of ideas. God simultaneously maximizes the richness and diversity of the world and minimizes the complexity of the ideas, of the mathematical laws, that determine this world. That is why science is possible!

A modern restatement of this idea is that science is possible because the world seems very complex but is actually governed by a small set of laws having low conceptual complexity.

And in Section VI of the *Discours*, Leibniz touches on randomness. He points out that any finite set of points on a piece of graph paper always seems to follow a law, because there is always a mathematical equation passing through those very points. But there is a law only if the equation is simple, not if it is very complicated. This is the idea that impressed Weyl, and it becomes the definition of randomness in algorithmic information theory.[†]

6.3. Finding Elegant Programs

So the best theory for something is the smallest program that calculates it. How can we be sure that we have the best theory? Let's forget about theories and just call a program *elegant* if it is the smallest program that produces the output that it does. More precisely, a program is elegant if no smaller program written in the same language produces the same output.

So can we be sure that a program is elegant, that it is the best theory for its output? Amazingly enough, we can't: It turns out that any formal axiomatic theory A can prove that at most finitely many programs are elegant, in spite of the fact that there are infinitely many elegant programs. More precisely, it takes an N-bit theory A, one having N bits of axioms, having complexity N, to be able to prove that an individual N-bit program is elegant. And we don't need to know much about the formal axiomatic theory A in order to be able to prove that it has this limitation.

6.4. What is a Formal Axiomatic Theory?

All we need to know about the axiomatic theory A, is the crucial requirement emphasized by David Hilbert that there should be a proof-checking algorithm, a mechanical procedure for deciding if a proof is correct or not. It follows that we can systematically run through all possible proofs, all possible strings of characters in the alphabet of the theory A, in size order,

---

[†] *Historical Note:* Algorithmic information theory was first proposed in the 1960s by R. Solomonoff, A. N. Kolmogorov, and G. J. Chaitin. Solomonoff and Chaitin considered this toy model of the scientific method, and Kolmogorov and Chaitin proposed defining randomness as algorithmic incompressibility.

checking which ones are valid proofs, and thus discover all the theorems, all the provable assertions in the theory A.[‡]

That's all we need to know about a formal axiomatic theory A, that there is an algorithm for generating all the theorems of the theory. This is the software model of the axiomatic method studied in algorithmic information theory. If the software for producing all the theorems is N bits in size, then the complexity of our theory A is defined to be N bits, and we can limit A's power in terms of its complexity $H(A) = N$. Here's how:

6.5. Why Can't You Prove that a Program is Elegant?

Suppose that we have an N-bit theory A, that is, that $H(A) = N$, and that it is always possible to prove that individual elegant programs are in fact elegant, and that it is never possible to prove that inelegant programs are elegant. Consider the following paradoxical program P:

> P runs through all possible proofs in the formal axiomatic theory A, searching for the first proof in A that an individual program Q is elegant for which it is also the case that the size of Q in bits is larger than the size of P in bits. And what does P do when it finds Q? It runs Q and then P produces as its output the output of Q.

In other words, the output of P is the same as the output of the first provably elegant program Q that is larger than P. But this contradicts the definition of elegance! P is too small to be able to calculate the output of an elegant program Q that is larger than P. We seem to have arrived at a contradiction!

But do not worry; there is no contradiction. What we have actually proved is that P can never find Q. In other words, there is no proof in the formal axiomatic theory A that an individual program Q is elegant, not if Q is larger than P. And how large is P? Well, just a fixed number of bits c larger than N, the complexity $H(A)$ of the formal axiomatic theory A. P consists of a small, fixed main program c bits in size, followed by a large subroutine $H(A)$ bits in size for generating all the theorems of A.

The only tricky thing about this proof is that it requires P to be able to know its own size in bits. And how well we are able to do this depends on the details of the particular programming language that we are using for the proof. So to get a neat result and to be able to carry out this

---

[‡] *Historical Note:* The idea of running through all possible proofs, of creativity by mechanically trying all combinations, can be traced back through Leibniz to Ramon Llull in the 1200s.

simple, elegant proof, we have to be sure to use an appropriate programming language. This is one of the key issues in algorithmic information theory, which programming language to use.[§]

6.6. Farewell to Reason: The Halting Probability Ω[¶]

So there are infinitely many elegant programs, but there are only finitely many provably elegant programs in any formal axiomatic theory A. The proof of this is rather straightforward and short. Nevertheless, this is a fundamental information-theoretic incompleteness theorem that is rather different in style from the classical incompleteness results of Gödel, Turing and others.

An even more important incompleteness result in algorithmic information theory has to do with the halting probability Ω, the numerical value of the probability that a program p whose successive bits are generated by independent tosses of a fair coin will eventually halt:

$$\Omega = \sum_{p \text{ halts}} 2^{-(\text{size in bits of } p)}.$$

To be able to define this probability Ω, it is also very important how you chose your programming language. If you are not careful, this sum will diverge instead of being ≤ 1 like a well-behaved probability should.

Turing's fundamental result is that the halting problem in unsolvable. In algorithmic information theory the fundamental result is that the halting probability Ω is algorithmically irreducible or random. It follows that the bits of Ω cannot be compressed into a theory less complicated than they are. They are irreducibly complex. It takes N bits of axioms to be able to determine N bits of the numerical value

$$\Omega = .1101011\ldots$$

of the halting probability. If your formal axiomatic theory A has $H(A) = N$, then you can determine the values and positions of at most $N + c$ bits of Ω.

In other words, the bits of Ω are logically irreducible, they cannot be proved from anything simpler than they are. Essentially the only way to determine what are the bits of Ω is to add these bits to your theory A as

---

[§]See the chapter on "The Search for the Perfect Language" in Chaitin, *Mathematics, Complexity and Philosophy*, in press.
[¶]*Farewell to Reason* is the title of a book by Paul Feyerabend, a wonderfully provocative philosopher. We borrow his title here for dramatic effect, but he does not discuss Ω in this book or any of his other works.

new axioms. But you can prove anything by adding it as a new axiom. That's not using reasoning!

So the bits of Ω refute Leibniz's principle of sufficient reason: they are true for no reason. More precisely, they are not true for any reason simpler than themselves. This is a place where mathematical truth has absolutely no structure, no pattern, for which there is no theory!

6.7. Adding New Axioms: Quasi-empirical Mathematics[‖]

So incompleteness follows immediately from fundamental information-theoretic limitations. What to do about incompleteness? Well, just add new axioms, increase the complexity $H(A)$ of your theory A! That is the only way to get around incompleteness.

In other words, do mathematics more like physics, add new axioms not because they are self-evident, but for pragmatic reasons, because they help mathematicians to organize their mathematical experience just like physical theories help physicists to organize their physical experience. After all, Maxwell's equations and the Schrödinger equation are not at all self-evident, but they work! And this is just what mathematicians have done in theoretical computer science with the hypothesis that $P \neq NP$, in mathematical cryptography with the hypothesis that factoring is hard, and in abstract axiomatic set theory with the new axiom of projective determinacy.[**]

6.8. Mathematics, Biology and Metabiology

We've discussed physical and mathematical theories; now let's turn to biology, the most exciting field of science at this time, but one where mathematics is not very helpful. Biology is very different from physics. There is no simple equation for your spouse. Biology is the domain of the complex. There are not many universal rules. There are always exceptions. Math is very important in theoretical physics, but there is no fundamental mathematical theoretical biology.

This is unacceptable. The honor of mathematics requires us to come up with a mathematical theory of evolution and either prove that Darwin

---

[‖] The term *quasi-empirical* is due to the philosopher Imre Lakatos, a friend of Feyerabend. For more on this school, including the original article by Lakatos, see the collection of quasi-empirical philosophy of math papers edited by Thomas Tymoczko, *New Directions in the Philosophy of Mathematics*.

[**] See the article on "The Brave New World of Bodacious Assumptions in Cryptography" in the March 2010 issue of the *AMS Notices*, and the article by W. Hugh Woodin on "The Continuum Hypothesis" in the June/July 2001 issue of the *AMS Notices*.

was wrong or right! We want a general, abstract theory of evolution, not an immensely complicated theory of actual biological evolution. And we want proofs, not computer simulations! So we've got to keep our model very, very simple.

That's why this proposed new field is *metabiology*, not biology.

What kind of math can we use to build such a theory? Well, it's certainly not going to be differential equations. Don't expect to find the secret of life in a differential equation; that's the wrong kind of mathematics for a fundamental theory of biology.

In fact a universal Turing machine has much more to do with biology than a differential equation does. A universal Turing machine is a very complicated new kind of object compared to what came previously, compared with the simple, elegant ideas in classical mathematics like analysis. And there are self-reproducing computer programs, which is an encouraging sign.

There are in fact three areas in our current mathematics that do have some fundamental connection with biology, that show promise for math to continue moving in a biological direction:

Computation, Information, Complexity.

DNA is essentially a programming language that computes the organism and its functioning; hence the relevance of the theory of computation for biology.

Furthermore, DNA contains biological information. Hence the relevance of information theory. There are in fact at least four different theories of information:

- Boltzmann statistical mechanics and Boltzmann entropy,
- Shannon communication theory and coding theory,
- algorithmic information theory (Solomonoff, Kolmogorov, Chaitin), which is the subject of this essay, and
- quantum information theory and qubits.

Of the four, AIT (algorithmic information theory) is closest in spirit to biology. AIT studies the size in bits of the smallest program to compute something. And the complexity of a living organism can be roughly (very roughly) measured by the number of bases in its DNA, in the biological computer program for calculating it.

Finally, let's talk about complexity. Complexity is in fact the most distinguishing feature of biological as opposed to physical science and mathematics. There are many computational definitions of complexity, usually concerned with computation times, but again AIT, which concentrates on program size or conceptual complexity, is closest in spirit to biology.

Let's emphasize what we are not interested in doing. We are certainly not trying to do systems biology: large, complex realistic simulations of biological systems. And we are not interested in anything that is at all like Fisher-Wright population genetics that uses differential equations to study the shift of gene frequencies in response to selective pressures.

We want to use a sufficiently rich mathematical space to model the space of all possible designs for biological organisms, to model biological creativity. And the only space that is sufficiently rich to do that is a software space, the space of all possible algorithms in a fixed programming language. Otherwise we have limited ourselves to a fixed set of possible genes as in population genetics, and it is hopeless to expect to model the major transitions in biological evolution such as from single-celled to multicellular organisms, which is a bit like taking a main program and making it into a subroutine that is called many times.

Recall the cover of Stephen Gould's *Wonderful Life* on the Burgess shale and the Cambrian explosion? Around 250 primitive organisms with wildly differing body plans, looking very much like the combinatorial exploration of a software space. Note that there are no intermediate forms; small changes in software produce vast changes in output.

So to simplify matters and concentrate on the essentials, let's throw away the organism and just keep the DNA. Here is our proposal:

> *Metabiology: a field parallel to biology that studies the random evolution of artificial software (computer programs) rather than natural software (DNA), and that is sufficiently simple to permit rigorous proofs or at least heuristic arguments as convincing as those that are employed in theoretical physics.*

This analogy may seem a bit far-fetched. But recall that Darwin himself was inspired by the analogy between artificial selection by plant and animal breeders and natural section imposed by malthusian limitations.

Furthermore, there are many tantalizing analogies between DNA and large, old pieces of software. Remember *bricolage*, that Nature is a cobbler, a tinkerer? In fact, a human being is just a very large piece of software, one that is 3×10^9 bases $= 6 \times 10^9$ bits $\approx$ one gigabyte of software that has been patched and modified for more than a billion years: a tremendous mess, in fact, with bits and pieces of fish and amphibian design mixed in with that for a mammal.[††] For example, at one point in gestation the human embryo has gills. As time goes by, large human software projects also turn into a tremendous mess with many old bits and pieces.

The key point is that you can't start over, you've got to make do with what you have as best you can. If we could design a human being from scratch we could do a much better job. But we can't start over. Evolution only makes small changes, incremental patches, to adapt the existing code to new environments.

So how do we model this? Well, the key ideas are:

Evolution of mutating software,

and:

Random walks in software space.

That's the general idea. And here are the specifics of our current model, which is quite tentative.

We take an organism, a single organism, and perform random mutations on it until we get a fitter organism. That replaces the original organism, and then we continue as before. The result is a random walk in software space with increasing fitness, a hill-climbing algorithm in fact.[‡‡]

Finally, a key element in our proposed model is the definition of fitness. For evolution to work, it is important to keep our organisms from stagnating. It is important to give them something challenging to do.

The simplest possible challenge to force our organisms to evolve is what is called the Busy Beaver problem, which is the problem of providing concise names for extremely large integers. Each of our organisms produces a single positive integer. The larger the integer, the fitter the organism.[*]

The Busy Beaver function of N, BB(N), that is used in AIT is defined to be the largest positive integer that is produced by a program that is less

---

[††]See Neil Shubin, *Your Inner Fish: A Journey into the 3.5-Billion-Year History of the Human Body.*

[‡‡]In order to avoid getting stuck on a local maximum, in order to keep evolution from stopping, we stipulate that there is a non-zero probability to go from any organism to any other organism, and $-\log_2$ of the probability of mutating from A to B defines an important concept, the *mutation distance*, which is measured in bits.

[*]*Alternative formulations:* The organism calculates a total function $f(n)$ of a single non-negative integer n and $f(n)$ is fitter than $g(n)$ if $f(n)/g(n) \to \infty$ as $n \to \infty$. Or the organism calculates a (constructive) Cantor ordinal number and the larger the ordinal, the fitter the organism.

than or equal to N bits in size. BB(N) grows faster than any computable function of N and is closely related to Turing's famous halting problem, because if BB(N) were computable, the halting problem would be solvable.[†]

Doing well on the Busy Beaver problem can utilize an unlimited amount of mathematical creativity. For example, we can start with addition, then invent multiplication, then exponentiation, then hyper-exponentials, and use this to concisely name large integers:

$$N + N \rightarrow N \times N \rightarrow N^N \rightarrow N^{N^N} \rightarrow \ldots$$

There are many possible choices for such an evolving software model: You can vary the computer programming language and therefore the software space, you can change the mutation model, and eventually you could also change the fitness measure. For a particular choice of language and probability distribution of mutations, and keeping the current fitness function, it is possible to show that in time of the order of 2^N the fitness will grow as BB(N), which grows faster than any computable function of N and shows that genuine creativity is taking place, for mechanically changing the organism can only yield fitness that grows as a computable function.[‡]

So with random mutations and just a single organism we actually do get evolution, unbounded evolution, which was precisely the goal of metabiology!

This theorem may seem encouraging, but it actually has a serious problem. The times involved are so large that our search process is essentially *ergodic*, which means that we are doing an exhaustive search. Real evolution is not at all ergodic, since the space of all possible designs is much too immense for exhaustive search.

It turns out that with this same model there is actually a much quicker *ideal evolutionary pathway* that achieves fitness BB(N) in time of the order of N. This path is however unstable under random mutations, plus it is much too good: Each organism adds only a single bit to the preceding organism, and immediately achieves near optimal fitness for an organism of

---

[†]Consider BB'(N) defined to be the maximum run-time of any program that halts that is less than or equal to N bits in size.

[‡]Note that to actually simulate our model an oracle for the halting problem would have to be employed to avoid organisms that have no fitness because they never calculate a positive integer. This also explains how the fitness can grow faster than any computable function. In our evolution model, implicit use is being made of an oracle for the halting problem, which answers questions whose answers cannot be computed by any algorithmic process.

its size, which doesn't seem to at all reflect the haphazard, frozen-accident nature of what actually happens in biological evolution.[§]

So that is the current state of metabiology: a field with some promise, but not much actual content at the present time. The particular details of our current model are not too important. Some kind of mutating software model should work, should exhibit some kind of basic biological features. The challenge is to identify such a model, to characterize its behavior statistically,[¶] and *to prove* that it does what is required.

Note added in proof: The mathematical structure of metabiology is starting to emerge. Please see my paper "To a mathematical theory of evolution and biological creativity" at: `http://www.cs.auckland.ac.nz/CDMTCS//researchreports/391greg.pdf`.

References

1. G. J. Chaitin, *Thinking about Gödel and Turing: Essays on Complexity, 1970–2007*, World Scientific, 2007.
2. G. J. Chaitin, *Mathematics, Complexity and Philosophy*, Midas, in press. (Draft at `http://www.cs.umaine.edu/~chaitin/midas.html`.)
3. S. Gould, *Wonderful Life*, Norton, 1989.
4. N. Koblitz and A. Menezes, "The brave new world of bodacious assumptions in cryptography," *AMS Notices* **57**, 357–365, 2010.
5. G. W. Leibniz, *Discours de métaphysique, suivi de Monadologie*, Gallimard, 1995.
6. N. Shubin, *Your Inner Fish*, Pantheon, 2008.
7. T. Tymoczko, *New Directions in the Philosophy of Mathematics*, Princeton University Press, 1998.
8. H. Weyl, *The Open World*, Yale University Press, 1932.
9. W. H. Woodin, "The continuum hypothesis, Part I," *AMS Notices* **48**, 567–576, 2001.

[§]The Nth organism in this ideal evolutionary pathway is essentially just the first N bits of the numerical value of the halting probability Ω. Can you figure out how to compute BB(N) from this?

[¶]For instance, will some kind of hierarchical structure emerge? Large human software projects are always written that way.

Chapter 7

Uncertainty in Physics and Computation

Michael A. Stay

Google and The University of Auckland

stay@google.com

.

WHAT DREW ME TO THE STUDY OF COMPUTATION AND RANDOMNESS

I've always been torn between studying physics and studying computer science, but it was the theoretical aspects of both that attracted me the most. There's clearly overlap in the computational physics arena, but I found that coming up with good numerical approximations to systems wasn't really to my taste. It was only when I began studying information theory that I came to understand that there was also an enormous overlap between the two theoretical sides of the fields. And like many others, I was intrigued with the possibility that somehow algorithmic randomness lies at the root of quantum randomness.

WHAT WE HAVE LEARNED

Szilard showed in 1929[7] that information and energy can be exchanged; Landauer[5] later showed that *temperature* is the exchange rate between the currencies of energy and information: the energy required to erase B arbitrary bits is

$$E \geq B \ln(2)kT.$$

Bennett, Gács, Li, Vitányi, and Zurek produced that wonderful paper on information distance[2] that refined Landauer's principle and connected it with computation: the energy required to turn a string x into a string y (where the energy is relative to a reversible computer M) is

$$E \geq \mathrm{KR}(x|y) \ln(2)kT,$$

where $\mathrm{KR}(x|y)$ is the length in bits of the shortest program for M that given x produces y (and since it's reversible, is also effectively a program for producing x given y). So it's clear there's a deep connection between algorithmic information theory and physics.

A real number $\omega = 0.\omega_1\omega_2\omega_3 \ldots$ is *random* if $H_U(\omega_1 \ldots \omega_n) \geq n - c_U$, where $H_U(x)$ is the length of the shortest prefix-free program producing x, U is a universal Turing machine and c_U is a natural number depending only on U. When I saw this definition for the first time, it immediately struck me that this was simply the log of an uncertainty principle. Pick a universal Turing machine U. Let Ω_U be the halting probability for U; Chaitin[3] showed that this number is algorithmically random. Finally, suppose that there are two strings s and x such that $U(sx)$ is the first n bits of Ω_U.

What is the uncertainty $\Delta\Omega_U$? Well, we have the first n bits, and a random real can be arbitrarily close to any rational, so it must be 2^{-n}.

What is the uncertainty Δx? There's no canonical way to give the distance between two strings, but there's a nice isomorphism between binary strings and natural numbers: stick a 1 on the front! Given this map, we can ask what the uncertainty is in the natural number $1x$. Well, if s is the shortest program outputting the first n bits of Ω_U, then x is the empty string and $1x = 1$. On the other hand, s might be empty or algorithmically independent from x, in which case x needs to be at least $n - c_M$ bits long; therefore $1x \geq 2^{n-c_M}$. The difference is the uncertainty: $\Delta x \geq 2^{n-c_M} - 1$.

The product of these two is an uncertainty principle formally equivalent to Heisenberg's uncertainty principle:

$$\Delta\Omega_U \cdot \Delta x \geq k_M,$$

where k_M is a real number depending only on the choice of universal machine M. Starting from this principle, Cristian Calude and I were able to show[4] that it implies Chaitin's information-theoretic incompleteness, and we also designed a quantum computer for which the halting probability and the program were complementary observables. Now, the quantum computer wasn't a universal one, and we didn't show that quantum randomness has roots in algorithmic randomness. But there's something going on there that deserves some attention.

Tadaki[8] generalized the halting probability by adding a weighting factor to the program length:

$$\Omega_U(s) = \sum_{U(p) \text{ halts}} 2^{-s|p|}.$$

I pointed out that this has the same form as the partition function in statistical mechanics:

$$\Omega_U(1/kT) = \sum_{U(p) \text{ halts}} e^{\frac{-|p|\ln(2)}{kT}}.$$

Again, the temperature is playing the role of an exchange rate; the resulting real in this summation is *partially random*, and the temperature is a measure of how well its bits can be compressed.

WHAT WE DON'T (YET) KNOW

The category of data types with α-β-η−equivalence classes of linear lambda terms between them is a symmetric monoidal closed category;[1] the same is true of the category of particle worldlines with Feynman diagrams between them and the category of Hilbert spaces and linear transformations. So programs are, in a precise way, analogous to Feynman diagrams, and a sum over programs is related to a path integral. There is something very deep to this analogy, and no one has plumbed it yet.

The zeta function of a fractal string encodes information about its dimension in its zeros;[6] the generalized halting probability is the zeta function of "the halting fractal." This brings up the question of the value of the generalized halting probability at complex inputs. What information is encoded in its zeros? Can the zeta function be analytically continued around the pole at 1? If so, how compressible is the analytical continuation of the zeta function at computable values in $\mathbb{C}$?

THE MOST IMPORTANT OPEN PROBLEMS

I don't know much about the *importance* of particular problems, but I think that the most *interesting* open problems for me are the ones dealing with what kind of algorithmic information is *actually accessible* to us: are there any finite physical systems that provably produce random bits? What computations are actually occurring when particles interact? Can we exploit them to solve problems that Turing machines can't?

THE PROSPECTS FOR PROGRESS

Very good — we're just beginning to explore, and there's a lot of low-hanging fruit waiting to be picked and appreciated.

108M. A. Stay

References

1. John Baez and Michael Stay, "Physics, Topology, Logic, and Computation: A Rosetta Stone." *New Structures in Physics*, a volume in *Lecture Notes in Physics*, to appear.
2. Charles H. Bennett, Peter Gács, Ming Li, Paul M. B. Vitányi, and Woiciech Zurek, "Thermodynamics of computation and information distance," *IEEE Transactions on Information Theory* 44 (1998), no. 4, 14071423, Preliminary version appeared in the 1993 ACM Symp. on Th. of Comp.
3. G.J. Chaitin. "A theory of program size formally identical to information theory," *J. Assoc. Comput. Mach.* 22 (1975), 329340. (Received April 1974)
4. C.S. Calude, M.A. Stay, "From Heisenberg to Gödel via Chaitin." *International Journal of Theoretical Physics*, 44 (7). 1053–1065.
5. Rolf Landauer, "Irreversibility and heat generation in the computing process," *IBM Journal of Research and Development*, vol. 5, pp. 183-191, 1961.
6. Erin Pearse, *Complex dimensions of self-similar systems*, Ph.D. thesis.
7. Szilard, L. (1929). *Zeitschrift für Physik*, 53, 840-856 [Translated and reprinted in Quantum Theory and Measurement, J. A. Wheeler and W. H. Zurek, eds., Princeton University Press, Princeton, New Jersey (1983)].
8. Kohtaro Tadaki, "A Generalization of Chaitin's Halting Probability Ω and Halting Self-Similar Sets," Hokkaido Mathematical Journal, Vol. 31, No. 1, February 2002, 219-253
9. W.K. Wootters and W.H. Zurek, "A Single Quantum Cannot be Cloned", *Nature* 299 (1982), pp. 802-803.

Chapter 8

Indeterminism and Randomness Through Physics

Karl Svozil

Institute for Theoretical Physics, Vienna University of Technology
Wiedner Hauptstraße 8-10/136, A-1040 Vienna, Austria
svozil@tuwien.ac.at
http://tph.tuwien.ac.at/~svozil

It is not totally unreasonable to speculate if and why the universe we experience with our senses and brains appears to be "(un)lawful." Indeed, the "unreasonable effectiveness of mathematics in the natural sciences"[72] seems mind-boggling and tantamount to our (non)existence. Beyond belief, there do not seem to exist *a priori* answers to such questions which would be forced upon us, say maybe by consistency constraints. But then, why should consistency and logic be considered *sacrosanct*?

In view of the opaqueness of the issues, a fierce controversy between supporters and opponents of a "clockwork universe" *versus* "cosmic chaos" has developed from antiquity onwards — cf., e.g., Aristotle's comments on the Pythagoreans in *Physics*, as well as Epicurus' *Letter to Menoeceus*. Indeed, for the sake of purported truth, many varieties of conceivable mixtures of determinism and chance have been claimed and repudiated.

The author has argued elsewhere[65] that there are many emotional reasons (not) to believe in a(n) (in)deterministic universe: does it not appear frightening to be imprisoned by remorseless, relentless predetermination; and, equally frightening, to accept one's fate as being contingent on total arbitrariness and chance? What merits and what efforts appear worthy at these extreme positions, which also unmask freedom, self-determination and human dignity as an idealistic illusion?

In order to disentangle the scientific discussion of topics such as (in)determinism, or realism *versus* idealism, from emotional overtones and possible bias, it might not be totally unreasonable to allow oneself the contemplative strategy of *evenly-suspended attention* outlined by Freud:[33] Nature is thereby treated as a "client-patient," and whatever comes up is accepted "as is," without any immediate emphasis or judgment*.

In more recent history, the European Enlightenment (illuminating also wide areas across the oceans) has brought about the belief of total causality and almost unlimited predictability, control, and manipulative capacities. Subsequently, the *principle of sufficient reason* came under pressure at two independent frontiers: Poincaré's discovery of instabilities in classical many-body motion,[30] as already envisioned by Maxwell [Ref. 19, pp. 211–212], is now considered as a precursor to *deterministic chaos*, in which the information "held" in the initial value "unfolds" through a deterministic process. Note that, with probability one, an arbitrary real number representing the initial value, which is "grabbed" (facilitated by the axiom of choice) from the "continuum urn," is provable random in the sense of algorithmic information theory;[14,23,51] i.e., in terms of algorithmic incompressibility as well as of the equivalent statistical tests. Moreover, for entirely different reasons, if one encodes universal computation into a system on n bodies, then by reduction to the halting problem of recursion theory,[4,8,28,31,53,57] (c.f. below) certain observables become provable unknowable.[66]

A second attack against determinism erupted through the development of quantum theory. Despite fierce resistance of Einstein[†], Schrödinger and De Brogli, Born expressed the new quantum *canon*, repeated by the "mainstream" ever after,[76] as follows (cf. Ref. [10, p. 866], English translation in Ref. [71, p. 54])[‡]:

---

*In Ref. 33, Freud admonishes analysts to be aware of the dangers caused by "... temptations to project, what [[the analyst]] in dull self-perception recognizes as the peculiarities of his own personality, as generally valid theory into science ..." (In German: *"Er wird leicht in die Versuchung geraten, was er in dumpfer Selbstwahrnehmung von den Eigentümlichkeiten seiner eigenen Person erkennt, als allgemeingültige Theorie in die Wissenschaft hinauszuprojizieren ... ."*)

[†]In a letter to Born, dated December 12th, 1926 Ref. [11, p. 113], Einstein expressed his conviction, "In any case I am convinced that he [[the Old One]] does not throw dice." (In German: *"Jedenfalls bin ich überzeugt, dass der [[Alte]] nicht würfelt."*)

[‡] *"Vom Standpunkt unserer Quantenmechanik gibt es keine Größe, die im Einzelfalle den Effekts eines Stoßes kausal festlegt; aber auch in der Erfahrung haben wir keinen Anhaltspunkt dafür, daß es innere Eigenschaften der Atome gibt, die einen bestimmten Stoßerfolg bedingen. Sollen wir hoffen, später solche Eigenschaften [[...]] zu entdecken und im Einzelfalle zu bestimmen? Oder sollen wir glauben, dass die Übereinstimmung von Theorie und Erfahrung in der Unfähigkeit, Bedingungen für den kausalen Ablauf anzugeben, eine prästabilisierte Harmonie ist, die auf der Nichtexistenz solcher Bedingungen beruht? Ich selber neige dazu,die Determiniertheit in der atomaren Welt aufzugeben."*

"From the standpoint of our quantum mechanics, there is no quantity which in any individual case causally fixes the consequence of the collision; but also experimentally we have so far no reason to believe that there are some inner properties of the atom which condition a definite outcome for the collision. Ought we to hope later to discover such properties [[. . .]] and determine them in individual cases? Or ought we to believe that the agreement of theory and experiment — as to the impossibility of prescribing conditions for a causal evolution — is a pre-established harmony founded on the nonexistence of such conditions? I myself am inclined to give up determinism in the world of atoms."

More specifically, Born offers a mixture of (in)determinism: while postulating a probabilistic behavior of individual particles, he accepts a deterministic evolution of the wave function (cf. Ref. [9, p. 804], English translation in Ref. [43, p. 302])[§]:

"The motion of particles conforms to the laws of probability, but the probability itself is propagated in accordance with the law of causality. [This means that knowledge of a state in all points in a given time determines the distribution of the state at all later times.]"

In addition to the indeterminism associated with outcomes of the measurements of single quanta, there appear to be at least two other types of quantum unknowables. One is complementarity, as first expressed by Pauli, Ref. [54, p. 7]. A third type of quantum indeterminism was discovered by studying quantum probabilities, in particular the consequences of Gleason's theorem:[34] whereas the classical probabilities can be constructed by the convex sum of all two-valued measures associated with classical truth tables, the structure of elementary yes–no propositions in quantum mechanics associated with projectors in three- or higher-dimensional Hilbert spaces do not allow any two-valued measure.[47,59] One of the consequences thereof is the impossibility of a consistent co-existence of the outcomes of all conceivable quantum observables (under the *noncontextuality* assumption[5] that measurement outcomes are identical if they "overlap").

Parallel to these developments in physics, Gödel[35] put an end to finitistic speculations in mathematics about possibilities to encode all math-

---

[§] *"Die Bewegung der Partikel folgt Wahrscheinlichkeitsgesetzen, die Wahrscheinlichkeit selbst aber breitet sich im Einklang mit dem Kausalgesetz aus. [Das heißt, daß die Kenntnis des Zustandes in allen Punkten in einem Augenblick die Verteilung des Zustandes zu allen späteren Zeiten festlegt.]"*

ematical truth in a finite system of rules. The recursion theoretic, formal
unknowables exhibit a novel feature: they present *provable* unknowables in
the fixed axiomatic system in which they are derived. (Note that incompleteness and undecidability exist always relative to the particular formal
system or model of universal computation.) From ancient times onwards,
individuals and societies have been confronted with a pandemonium of unpredictable behaviors and occurrences in their environments, sometimes resulting in catastrophes. Often these phenomena were interpreted as "God's
Will." In more rationalistic times, one could pretend without presenting a
formal proof that certain unpredictable behaviors are in principle deterministic, although the phenomena cannot be predicted "for various practical
purposes" ("epistemic indeterminism"). Now provable unknowables make
a difference by being immune to these kinds of speculation. The halting
problem in particular demonstrates the impossibility to predict the behavior of deterministic systems in general; it also solves the induction (rule
inference) problem to the negative.

   In order to be able to fully appreciate the impact of recursion theoretic undecidability on physics,[3,15,20,21,26,27,40,46,52,62,64,73-75] let us sketch
an algorithmic proof of the undecidability of the halting problem; i.e., the
decision problem of whether or not a program p on a given finite input finishes running (or will reach a particular halting state) or will run forever.
The proof method will use a *reductio ad absurdum*; i.e., we assume the existence of a *halting algorithm* $h(p)$ deciding the halting problem of p, as well
as some trivial manipulations; thereby deriving a complete contradiction.
The only alternative to inconsistency appears to be the nonexistence of any
such halting algorithm. For the sake of contradiction, consider an agent $q(p)$
accepting as input an arbitrary program (code) p. Suppose further that it
is able to consult a halting algorithm $h(p)$, thereby producing the *opposite*
behavior of p: whenever p halts, q "steers itself" into the halting mode;
conversely, whenever p does not halt, q forces itself to halt. A complete
contradiction results from q's behavior on itself, because whenever $q(q)$ detects (through $h(q)$) that it halts, it is supposed not to halt; conversely if
$q(q)$ detects that it does not halt, it is supposed to halt. Finally, since all
other steps in this "diagonal argument" with the exception of h are trivial,
the contradiction obtained in applying q to its own code proves that any
such program — and in particular a halting algorithm h — cannot exist.

   In physics, analogous arguments embedding a universal computer into a
physical substrate yield provable undecidable observables *via reduction to
the halting problem*. Note that this argument neither means that the sys-

tem does not evolve deterministically on a step-by-step basis, nor implies that predictions are provable impossible for all cases; that would be clearly misleading and absurd! A more quantitative picture arises if we study the potential growth of "complexity" of deterministic systems in terms of their maximal capability to "grow" before reaching a halting state through the Busy Beaver function.[12,22,29,55] Another consequence is the recursive unsolvability of the general induction (or rule inference[1,2,7,37,50]) problem for deterministic systems. As an immediate consequence of these findings it follows that no general algorithmic rule or operational method[13] exists which could "extract" some rather general law from a (coded) sequence. (Note again that it still may be possible to extract laws from "low-complex" sequences; possibly with some intuition and additional information.) Nor can there be certainty that some sequence denominated "random" is not generated by a decompression algorithm which makes it formally nonrandom; a fact well known in recursion and algorithmic information theory[14,24] but hardly absorbed by the physics community. Thereby, to quote *Shakespeare's Prospero,* any claims of absolute ("ontological") randomness decay into "thin air." Of course, one could still vastly restrict the domain of possible laws and *define* a source to be random if it "performs well" with respect to the associated, very limited collection of statistical tests, a strategy adapted by the *Swiss Federal Office of Metrology*¶.

Despite the formal findings reviewed above, which suggest that claims of absolute indeterminacy cannot be proven but represent subjective beliefs, their predominance in the physics community can be understood, or rather motivated, by the obvious inability to account for physical events, such as the outcomes of certain quantum measurements, e.g., radioactive decays,[48,49] deterministically. Why this effective incapacity to predict individual outcomes or time series of measurement data should be different from other "classical" statistical sources of randomness — even when complementarity and value indefiniteness is taken into account — remains an open question, at least from a formal point of view.

For the sake of explicit demonstration, let us consider a particular method of generation of a sequence from single quantum outcomes[18] by combination of source and beam splitter.[32,39,42,44,56,61,63,68,70] Ideally (to employ quantum complementarity as well as quantum value indefiniteness), a system allowing three or more outcomes is prepared to be in a particular pure state "contained" in a certain context (maximal observable[67] or

---

¶Cf. the *Certificate of Conformity No 151-04255*

block[38,45]), and then measured "along" a different context not containing the observable corresponding to that pure state. All outcomes except two are discarded,[14,16] and the two remaining outcomes are mapped onto the symbols "0" and "1," respectively. If independence of individual "quantum coin tosses" is assumed — a quite nontrivial assumption in view of the Hanbury Brown and Twiss effect and other statistical correlations — the concatenation and algorithmic normalization[58,69] of subsequent recordings of these encoded outcomes yield an "absolutely random sequence" relative to the unprovable axiomatic assumption of quantum randomness. Since all such operational physical sequences are finite, algorithmic information theory[14] applies to them in a limited, finite sense. Particular care should be given to the difficulties in associating an algorithmic information measure to "nontrivial" sequences of finite length.

In summary, there are two principal sources of indeterminism and randomness in physics: the first source is the deterministic chaos associated with instabilities of classical physical systems, and with the strong dependence of their future behavior on the initial value; the second source is quantum indeterminism, which can be subdivided into three subcategories: random outcomes of individual events, complementarity, and value indefiniteness.

The similarities and differences between classical and quantum randomness can be conceptualized in terms of two "black boxes", the first one of them — called the *"Poincaré box"* — containing a classical, deterministic chaotic, source of randomness; the second — called the *"Born box"* — containing a quantum source of randomness, such as a quantized system including a beam splitter. Suppose an agent is being presented with both boxes without any label on, or hint about, them; i.e., the origin of indeterminism is unknown to the agent. In a modified Turing test, the agent's task would be to find out which is the Born and which is the Poincaré box by solely observing their output.

It is an open question whether it is possible, by studying the output behavior of the *"Poincaré box"* and the *"Born box"* alone, to differentiate between them. In the absence of any criterion, there should not exist any operational method or procedure discriminating amongst these boxes. Both types of indeterminism appear to be based on metaphysical assumptions: in the classical case it is the existence of continua and the possibility to "choose" elements thereof, representing the initial values; in the quantum case it is the irreducible indeterminism of single events.

It would indeed be tempting also to compare the performance of these physical "oracles of indeterminism" with algorithmic cyclic pseudorandom generators, and with irrationals such as π. In recent studies[17] the latter, deterministic, ones seem to be doing pretty well.

In the author's conviction, the postulate of quantum randomness as well as physical randomness emerging from the continuum will be maintained by the community of physicists at large unless somebody comes up with evidence to the contrary. This pragmatic interpretation of the phenomena appears reasonable if and only if researchers are aware of its *relativity* with respect to the tests and attempts of falsification involved; and also acknowledge the tentativeness and conventionality of their assumptions.

Acknowledgments

The author gratefully acknowledges the kind hospitality of the *Centre for Discrete Mathematics and Theoretical Computer Science (CDMTCS)* of the *Department of Computer Science at The University of Auckland*. The *"Poincaré box" versus "Born box"* comparison emerged during a walk with Cristian Calude discussing differences of classical *versus* quantum randomness at the Boulevard Saint-Germain in Paris. This work was also supported by *The Department for International Relations* of the *Vienna University of Technology*.

References

1. Leonard M. Adleman and M. Blum. Inductive inference and unsolvability. *The Journal of Symbolic Logic*, 56:891–900, September 1991.
2. Dana Angluin and Carl H. Smith. A survey of inductive inference: Theory and methods. *Computing Surveys*, 15:237–269, 1983.
3. John D. Barrow. *Impossibility. The limits of science and the science of limits*. Oxford University Press, Oxford, 1998.
4. Jon Barwise. *Handbook of Mathematical Logic*. North-Holland, Amsterdam, 1978.
5. John S. Bell. On the problem of hidden variables in quantum mechanics. *Reviews of Modern Physics*, 38:447–452, 1966. Reprinted in Ref. [6, pp. 1-13].
6. John S. Bell. *Speakable and Unspeakable in Quantum Mechanics*. Cambridge University Press, Cambridge, 1987.
7. Lenore Blum and Manuel Blum. Toward a mathematical theory of inductive inference. *Information and Control*, 28(2):125–155, June 1975.
8. George S. Boolos, John P. Burgess, and Richard C. Jeffrey. *Computability and Logic*. Cambridge University Press, Cambridge, fifth edition, 2007.

9. Max Born. Quantenmechanik der Stoßvorgänge. *Zeitschrift für Physik*, 38:803–827, 1926.
10. Max Born. Zur Quantenmechanik der Stoßvorgänge. *Zeitschrift für Physik*, 37:863–867, 1926.
11. Max Born. *Physics in my generation*. Springer Verlag, New York, 2nd edition, 1969.
12. A. H. Brady. The busy beaver game and the meaning of life. In R. Herken, editor, *The Universal Turing Machine. A Half-Century Survey*, page 259. Kammerer und Unverzagt, Hamburg, 1988.
13. Percy W. Bridgman. *Reflections of a Physicist*. Philosophical Library, New York, 1950.
14. Cristian Calude. *Information and Randomness — An Algorithmic Perspective*. Springer, Berlin, 2 edition, 2002.
15. Cristian Calude, Douglas I. Campbell, Karl Svozil, and Doru Ştefănescu. Strong determinism vs. computability. In Werner DePauli Schimanovich, Eckehart Köhler, and Friedrich Stadler, editors, *The Foundational Debate. Complexity and Constructivity in Mathematics and Physics. Vienna Circle Institute Yearbook, Vol. 3*, pages 115–131, Dordrecht, Boston, London, 1995. Kluwer.
16. Cristian Calude and Ion Chiţescu. Qualitative properties of P. Martin-Löf random sequences. *Unione Matematica Italiana. Bollettino. B. Serie VII*, 3(1):229–240, 1989.
17. Cristian S. Calude, Michael J. Dinneen, Monica Dumitrescu, and Karl Svozil. How random is quantum randomness? (extended version). Report CDMTCS-372, Centre for Discrete Mathematics and Theoretical Computer Science, University of Auckland, Auckland, New Zealand, December 2009.
18. Cristian S. Calude and Karl Svozil. Quantum randomness and value indefiniteness. *Advanced Science Letters*, 1(2):165168, 2008. eprint arXiv:quant-ph/0611029.
19. Lewis Campbell and William Garnett. *The life of James Clerk Maxwell. With a selection from his correspondence and occasional writings and a sketch of his contributions to science*. MacMillan, London, 1882.
20. J. L. Casti and A. Karlquist. *Boundaries and Barriers. On the Limits to Scientific Knowledge*. Addison-Wesley, Reading, MA, 1996.
21. John L. Casti and J. F. Traub. *On Limits*. Santa Fe Institute, Santa Fe, NM, 1994. Report 94-10-056.
22. Gregory J. Chaitin. Information-theoretic limitations of formal systems. *Journal of the Association of Computing Machinery*, 21:403–424, 1974. Reprinted in Ref. 25.
23. Gregory J. Chaitin. Algorithmic information theory. *IBM Journal of Research and Development*, 21:350–359, 496, 1977. Reprinted in Ref. 25.
24. Gregory J. Chaitin. *Algorithmic Information Theory*. Cambridge University Press, Cambridge, 1987.
25. Gregory J. Chaitin. *Information, Randomness and Incompleteness*. World Scientific, Singapore, 2nd edition, 1990. This is a collection of G. Chaitin's early publications.

26. N. C. A. da Costa and F. A. Doria. Classical physics and Penrose's thesis. *Foundations of Physics Letters*, 4:363–373, August 1991.

27. N. C. A. da Costa and F. A. Doria. Undecidability and incompleteness in classical mechanics. *International Journal of Theoretical Physics*, 30:1041–1073, August 1991.

28. Martin Davis. *The Undecidable. Basic Papers on Undecidable, Unsolvable Problems and Computable Functions.* Raven Press, Hewlett, N.Y., 1965.

29. A. K. Dewdney. Computer recreations: A computer trap for the busy beaver, the hardest-working Turing machine. *Scientific American*, 251(2):19–23, August 1984.

30. Florin Diacu. The solution of the n-body problem. *The Mathematical Intelligencer*, 18(3):66–70, 1996.

31. H. Enderton. *A Mathematical Introduction to Logic.* Academic Press, San Diego, second edition, 2001.

32. M. Fiorentino, C. Santori, S. M. Spillane, R. G. Beausoleil, and W. J. Munro. Secure self-calibrating quantum random-bit generator. *Physical Review A (Atomic, Molecular, and Optical Physics)*, 75(3):032334, 2007.

33. Sigmund Freud. Ratschläge für den Arzt bei der psychoanalytischen Behandlung. In Anna Freud, E. Bibring, W. Hoffer, E. Kris, and O. Isakower, editors, *Gesammelte Werke. Chronologisch geordnet. Achter Band. Werke aus den Jahren 1909–1913*, pages 376–387, Frankfurt am Main, 1999. Fischer.

34. Andrew M. Gleason. Measures on the closed subspaces of a Hilbert space. *Journal of Mathematics and Mechanics (now Indiana University Mathematics Journal)*, 6(4):885–893, 1957.

35. Kurt Gödel. Über formal unentscheidbare Sätze der Principia Mathematica und verwandter Systeme. *Monatshefte für Mathematik und Physik*, 38:173–198, 1931. English translation in Ref. 36 and in Ref. 28.

36. Kurt Gödel. In S. Feferman, J. W. Dawson, S. C. Kleene, G. H. Moore, R. M. Solovay, and J. van Heijenoort, editors, *Collected Works. Publications 1929-1936. Volume I.* Oxford University Press, Oxford, 1986.

37. Mark E. Gold. Language identification in the limit. *Information and Control*, 10:447–474, 1967.

38. J. R. Greechie. Orthomodular lattices admitting no states. *Journal of Combinatorial Theory*, 10:119–132, 1971.

39. Ma Hai-Qiang, Wang Su-Mei, Zhang Da, Chang Jun-Tao, Ji Ling-Ling, Hou Yan-Xue, and Wu Ling-An. A random number generator based on quantum entangled photon pairs. *Chinese Physics Letters*, 21(10):1961–1964, 2004.

40. Arne Hole. Predictability in deterministic theories. *International Journal of Theoretical Physics*, 33:1085–1111, May 1994.

41. Clifford Alan Hooker. *The Logico-Algebraic Approach to Quantum Mechanics. Volume I: Historical Evolution.* Reidel, Dordrecht, 1975.

42. Vincent Jacques, E Wu, Frédéric Grosshans, Francois Treussart, Philippe Grangier, Alain Aspect, and Jean-François Roch. Experimental realization of Wheeler's delayed-choice gedanken experiment. *Science*, 315(5814):966–968, 2007.

43. Max Jammer. *The Conceptual Development of Quantum Mechanics. The History of Modern Physics, 1800-1950; v. 12.* American Institute of Physics, New York, 2 edition, 1989.

44. Thomas Jennewein, Ulrich Achleitner, Gregor Weihs, Harald Weinfurter, and Anton Zeilinger. A fast and compact quantum random number generator. *Review of Scientific Instruments,* 71:1675–1680, 2000.

45. Gudrun Kalmbach. *Orthomodular Lattices.* Academic Press, New York, 1983.

46. I. Kanter. Undecidability principle and the uncertainty principle even for classical systems. *Physical Review Letters,* 64:332–335, 1990.

47. Simon Kochen and Ernst P. Specker. The problem of hidden variables in quantum mechanics. *Journal of Mathematics and Mechanics (now Indiana University Mathematics Journal),* 17(1):59–87, 1967. Reprinted in Ref. [60, pp. 235–263].

48. Helge Kragh. The origin of radioactivity: from solvable problem to unsolved non-problem. *Archive for History of Exact Sciences,* 50:331–358, 1997.

49. Helge Kragh. Subatomic determinism and causal models of radioactive decay, 1903-1923. RePoSS: Research Publications on Science Studies 5. Department of Science Studies, University of Aarhus, November 2009.

50. M. Li and P. M. B. Vitányi. Inductive reasoning and Kolmogorov complexity. *Journal of Computer and System Science,* 44:343–384, 1992.

51. Per Martin-Löf. The definition of random sequences. *Information and Control,* 9(6):602–619, 1966.

52. Christopher D. Moore. Unpredictability and undecidability in dynamical systems. *Physical Review Letters,* 64:2354–2357, 1990. Cf. Ch. Bennett, *Nature,* **346**, 606 (1990).

53. Piergiorgio Odifreddi. *Classical Recursion Theory, Vol. 1.* North-Holland, Amsterdam, 1989.

54. Wolfgang Pauli. Die allgemeinen Prinzipien der Wellenmechanik. In S. Flügge, editor, *Handbuch der Physik. Band V, Teil 1. Prinzipien der Quantentheorie I,* pages 1–168. Springer, Berlin, Göttingen and Heidelberg, 1958.

55. Tibor Rado. On non-computable functions. *The Bell System Technical Journal,* XLI(41)(3):877–884, May 1962.

56. J. G. Rarity, M. P. C. Owens, and P. R. Tapster. Quantum random-number generation and key sharing. *Journal of Modern Optics,* 41:2435–2444, 1994.

57. Hartley Rogers, Jr. *Theory of Recursive Functions and Effective Computability.* MacGraw-Hill, New York, 1967.

58. Paul A. Samuelson. Constructing an unbiased random sequence. *Journal of the American Statistical Association,* 63(324):1526–1527, 1968.

59. Ernst Specker. Die Logik nicht gleichzeitig entscheidbarer Aussagen. *Dialectica,* 14(2-3):239–246, 1960. Reprinted in Ref. [60, pp. 175–182]; English translation: *The logic of propositions which are not simultaneously decidable,* Reprinted in Ref. [41, pp. 135-140].

60. Ernst Specker. *Selecta.* Birkhäuser Verlag, Basel, 1990.

61. André Stefanov, Nicolas Gisin, Olivier Guinnard, Laurent Guinnard, and Hugo Zbinden. Optical quantum random number generator. *Journal of Modern Optics*, 47:595–598, 2000.

62. Patrick Suppes. The transcendental character of determinism. *Midwest Studies In Philosophy*, 18(1):242–257, 1993.

63. Karl Svozil. The quantum coin toss — testing microphysical undecidability. *Physics Letters A*, 143:433–437, 1990.

64. Karl Svozil. *Randomness & Undecidability in Physics*. World Scientific, Singapore, 1993.

65. Karl Svozil. Science at the crossroad between randomness and determinism. In Cristian Calude and Karl Svozil, editors, *Millennium III*, pages 73–84. Black Sea University Foundation in colaboration with the Romanian Academy of Sciences, Bucharest, Romania, 2002.

66. Karl Svozil. Omega and the time evolution of the n-body problem. In Cristian S. Calude, editor, *Randomness and Complexity, from Leibniz to Chaitin*, pages 231–236, Singapore, 2007. World Scientific. eprint arXiv:physics/0703031.

67. Karl Svozil. Contexts in quantum, classical and partition logic. In Kurt Engesser, Dov M. Gabbay, and Daniel Lehmann, editors, *Handbook of Quantum Logic and Quantum Structures*, pages 551–586. Elsevier, Amsterdam, 2009.

68. Karl Svozil. Three criteria for quantum random-number generators based on beam splitters. *Physical Review A (Atomic, Molecular, and Optical Physics)*, 79(5):054306, 2009.

69. John von Neumann. Various techniques used in connection with random digits. *National Bureau of Standards Applied Math Series*, 12:36–38, 1951. Reprinted in *John von Neumann, Collected Works, (Vol. V)*, A. H. Traub, editor, MacMillan, New York, 1963, p. 768–770.

70. P. X. Wang, G. L. Long, and Y. S. Li. Scheme for a quantum random number generator. *Journal of Applied Physics*, 100(5):056107, 2006.

71. John Archibald Wheeler and Wojciech Hubert Zurek. *Quantum Theory and Measurement*. Princeton University Press, Princeton, 1983.

72. Eugene P. Wigner. The unreasonable effectiveness of mathematics in the natural sciences. Richard Courant Lecture delivered at New York University, May 11, 1959. *Communications on Pure and Applied Mathematics*, 13:1–14, 1960.

73. Stephen Wolfram. Undecidability and intractability in theoretical physics. *Physical Review Letters*, 54(8):735–738, Feb 1985.

74. Stephen Wolfram. Computation theory of cellular automata. *Communications in Mathematical Physics*, 96(1):15–57, 1984.

75. David H. Wolpert. Computational capabilities of physical systems. *Physical Review E*, 65(1):016128, Dec 2001.

76. Anton Zeilinger. The message of the quantum. *Nature*, 438:743, 2005.

Chapter 9

The Martin-Löf-Chaitin Thesis: The Identification by Recursion Theory of the Mathematical Notion of Random Sequence[*]

Jean-Paul Delahaye

Université des Sciences et Techniques de Lille
Laboratoire d'Informatique Fondamentale de Lille
delahaye@lifl.fr

WHAT DREW ME TO THE STUDY OF RANDOMNESS

Recursion theory initiated by Gödel, Church, Turing, Post and Kleene in the 1930's led 30 years later to a definition of randomness that seems to fulfill the main objectives stated by von Mises. The definition of random sequence by Martin-Löf in 1965 and the other works on the so-called "algorithmic theory of information" by Kolmogorov, Chaitin, Schnorr, Levin and Solomonoff (among others) may be understood as the formulation of a thesis similar to the Church-Turing Thesis about the notion of computability. Here is this new thesis we call the **Martin-Löf-Chaitin Thesis**: the intuitive informal concept of random sequences (of 0 and 1) is satisfactorily defined by the notion of Martin-Löf-Chaitin random sequences (MLC-random sequences) that is, sequences which do not belong to any recursively null set. In this paper (a short version of Ref. 23), we first recall and explain shortly the notion of MLC-random sequences; and propose a comparison between the Church-Turing Thesis and the Martin-Löf-Chaitin Thesis.

In the context of the foundation of probability, the notion of random sequences was introduced by Richard von Mises[80–82] under the name of "collectives" ("Kollectiv"). von Mises' idea was that a mathematical theory of probability should be based on a precise and absolute definition of randomness. The crucial features of collectives are the existence of limiting

---

[*]This is an updated version of a previously published chapter in "Information, complexité et hasard" by the same author which came from a LIFL report published in 1990 and in another version in Dubucs, J. (ed.) Philosophy of Probability, 145–167, Kluwer Academic Publishers, 1993.

relative frequencies within the sequence, and the invariance of the limiting relative frequencies under the operation of "admissible selection".

We claim that our theory, which serves to describe observable facts, satisfies all reasonable requirements of logical consistency and is free from contradictions and obscurities of any kind. ... I would even claim that the real meaning of the Bernoulli theorem is inaccessible to any probability theory that does not start with the frequency definition of probability. ... All axioms of Kolmogorof can be accepted within the framework of our theory as a part of it, but in no way as a substitute for the foregoing definition of probability. *von Mises, R. On the Foundations of Probability and Statistics. Ann. Math. Statist. 12. 1941. pp. 191–205.*

WHAT WE HAVE LEARNED

Unfortunately von Mises did not really arrived to a satisfactory notion of "admissible selection" and in consequence did not give a satisfactory mathematical definition of what he call "collective", that is random sequences.

The problem of giving an adequate mathematical definition of a random sequence was subjected to an intense discussion about thirty years ago. It was initiated by von Mises as early as 1919 and reached its climax in the thirties when it engaged most of the pioneers of probability theory of that time. ... von Mises urged that a mathematical theory of probability should be based on a definition of randomness, the probability of an event then being introduced as the limit of the relative frequency as the number of trials tends to infinity. ... It was objected that there is just as little need for a definition of random sequences and probabilities by means of them as there is need for a definition of points and straight lines in geometry. ... The question was not whether the theory in spe should be axiomatized or not, but what objects should be taken as primitive and what axioms should be chosen to govern them. In the axiomatization of Kolmogorov 1933 the random sequences are left outside the theory. ... von Mises wanted to define random sequences in an absolute sense, sequences that were to possess all conceivable properties of stochasticity. This program appears impossible to carry out within the measure theoretic framework of Kolmogorov 1933. ... It seems as if it were this incapability of finding an adequate mathematical definition that brought the so rapid development in the thirties to an abrupt end. ... A common feature of the experiments considered by von Mises is that they may be repeated any, or at least a very large number of times. For the sequence of the successive outcomes $x_1, x_2, ..., x_n, ...$ which is imagined to extend indefinitely, von Mises coined the term "Kollektiv". A Kollektiv has to satisfy two requirements. To formulate the first of these let n_A denote the frequency with which the event A has occurred in the first n trials, i.e., the number of points x_m with $1 \leq m \leq n$, that belong to the subset A of the sample space. For every "angebbare Punktmenge" A the limit of the relative frequencies should exist, $\lim n_A/n = pi(A)$. This limit is called the probability of the event A with respect to the given Kollektiv. ... The second axiom is more intricate. It is to express the well-known irregularity of a random sequence, the impossibility of characterizing the correspondence between the number of an experiment and its outcome by a mathematical law. In a gambler's terminology it may be called the axiom of the impossibility of a successful gambling system. Thus sequences like (0 1 0 1 0 1 0 1 0 1 0 1 0 ...), 0 denoting failure and 1 success, are excluded although the limit frequency exists, since betting at every even trial would assure us constant success. The final form of the axiom is the following. If we select a subsequence of $x_1, x_2, x_3, ..., x_n, ...$ in such a way that the decision whether x_n should be selected or not does not depend on x_n, then the limiting relative frequency of the subsequence should exist and equal that of the original sequence. ... The definition of a Kollektiv was criticised for being mathematically imprecise or even inconsistent. ... The trouble was due to the fact that the concept of effectiveness was not a rigourous mathematical one at that time. *Martin-Löf, P. The definition of Random Sequences. Information and Control, 9. 1966. pp. 602–619.*

The axiomatic construction of probability theory on the basis of measure theory[34] as a purely mathematical discipline is logically irreproachable and does not cast doubts in anybody's mind. However to be able to apply this theory rigorously in practice its **physical interpretation has to be stated clearly.** Until recently there was no satisfactory solution of this problem. Indeed, probability is usually interpreted by means of the following arguments: "If we perform many tests, then the ratio of the number of favourable outcomes to the number of tests performed will always give a number close to, and in the limit exactly equal to, the probability (or measure) of the event in question. However to say "always" here would be untrue: strictly speaking, this does not always happen, but only with probability 1 (and for finite series of tests, with probability close to 1). In this way the concept of the probability of an arbitrary event is defined through the concept of an event that has probability close to (and in the limit equal to) 1, consequently cannot be defined in this manner without an obviously circular argument. In 1919 von Mises put forward the following way of eliminating these difficulties: according to von Mises there are random and non-random sequences. From the mathematical point of view, random sequences form a set of full measure and all without exception satisfy all the laws of probability theory. It is physically possible to assume that as a result of an experiment only random sequences appear. However, the definition of random sequences proposed by von Mises and later defined more precisely by Wald (1937), Church (1940) and Kolmogorov (1963) turned out to be unsatisfactory. For example, the existence was proved of random sequences, according to von Mises (his so-called collectives) that do not satisfy the law of iterated logarithm.[79] *Zvonkin, A.K. and Levin, L.A. The Complexity of finite object and the development of the concepts of information and randomness by means of the theory of algorithms. Russ. Math. Survey, 25, 6, 1970. pp. 83–124.*

9.1. The Notion of Martin-Löf-Chaitin Random Sequence

In 1965 Kolmogorov[36] has defined the complexity $H(Y)$ of an object Y as the minimal length of the binary program which computes Y on an certain universal Turing Machine. He shows that this notion was invariant in the sense that if U and $U\prime$ are two Universal Machines then the complexity defined by the first is the same as the complexity defined by the second within a constant. Similar work was done simultaneously by Chaitin.[3,4] Kolmogorov on the basis of this definition has proposed to consider those elements of a given large finite population to be random whose complexity is maximal.

In 1966 Martin-Löf had shown that the random element as defined by Kolmogorov possess all conceivable statistical properties of randomness. He also extended the definition to infinite binary sequences and for the first time gave a precise mathematical definition of the von Mises' Kolectivs.

Several equivalent formulations are possible. We give them here. In the following we identify a real number in the interval $[0, 1]$ with its sequence of digits (hence instead of defining the notion of random infinite sequence of 0 and 1, we define the notion of random real).

The 4 following definitions are equivalent,[12] the real number $r \in [0,1]$ is random if and only if :

- **Definition 1** (random in the **Martin-Löf sense**, 1966) : *For every recursively enumerable sequence A_i of sets of intervals, every A_i with a measure less than 2^{-i} ($\mu(A_i) < 2^{-i}$): r does not belong to every A_i.*
- **Definition 2** (random in the **Solovay sense**, 1975) : *For every recursively enumerable sequence A_i of sets of intervals with a finite total measure $(\Sigma\mu(A_i) < \infty)$: r is at most in a finite number of A_i.*
- **Definition 3** (random in the **Chaitin-Levin** sense): *The complexity $H(r_n)$ of the n first digits of r satisfies* : $\exists c \forall n : H(r_n) \geq n - c$
- **Definition 4** (strongly random in the **Chaitin sense**, 1987) : *The complexity $H(r_n)$ of the n first digits of r satisfies* : $\forall k \exists N \forall n \geq N : H(r_n) \geq n + k$

In a less formal manner, the situation of the question today is that for a sequence of binary digit, **to be random** is to verify one of the equivalent properties:

- *not to fulfil any exceptional regularity effectively testable* (i.e., to pass all the sequential effective random test[57]),
- *to have an incompressible information content* (i.e., to have a maximal algorithmic complexity[6,43]),
- *to be unpredictable or impossible to win* (no gambling system can win when playing on the sequence).[67,71]

and **to be random** implies :

- *not to have any algorithmic form* (not to be definable with an algorithm as the sequence of the digits of π is).
- *to have limiting relative frequencies for every subsequence extracted by an algorithm* (effective property of von Mises-Church);
- *to be free from aftereffect*[65] (weak form of the property of von Mises-Church : the limiting relative frequencies do not change for sequences extracted by the following process : a finite sequence of 0 and 1 being fixed $x_1, x_2, \ldots, x_n$ extract the subsequence of elements which are just after each occurrence of $x_1, x_2, \ldots, x_n$.)

For a more precise history and mathematical details the reader is referred to.[23,41,51,52,60,71,77,85]

WHAT WE ARE FIGURING OUT: COMPARISON OF CHURCH-TURING'S THESIS AND MARTIN-LÖF-CHAITIN'S THESIS

It seems interesting to compare the present situation of the Church-Turing Thesis (about the notion of "algorithmically computable function") and the Martin-Löf-Chaitin Thesis (about the notion of "infinite binary random sequence").

It is necessary to precise that we only want to consider the "standard Church-Turing's Thesis" which identifies the mathematical concept of *recursive function* with the intuitive metamathematical concept of *function computable with a discrete, deterministic, finite algorithm*. Here we do not consider the physical Church-Turing's Thesis (about functions computable by machines or by physical processes) nor the mental Church-Turing's Thesis (about functions computable by a mind or brain)

The discussion about the Church-Turing Thesis is often obscured by the confusion between the standard Church-Turing's Thesis which is widely accepted and its variants which are controversial. Analogous confusion is possible about the Martin-Löf-Chaitin Thesis which is only for us the statement of the identification of a mathematical notion with a intuitive and metamathematical notion. We are not concerned in our discussion with the physical notion of chance or indeterminism, and we are not concerned with the problem of the possibility of free choice by a mind or a brain.

The Church-Turing Thesis and the Martin-Löf-Chaitin Thesis are not definitions. Really they are falsifiable and the proof of this is that the Thesis of Popper (with identifies random sequences with sequences without aftereffect) is now falsified by the results of Ville.[79]

The Church-Turing Thesis and the Martin-Löf-Chaitin Thesis are really similar: each of them is a statement of identification of a mathematical notion with an intuitive metamathematical one: the first is about the notion of algorithm, the second is about the notion of chance. It is impossible to prove these theses because they are not mathematical results, but our previous informal notions are sufficiently precise so that the possibility of refutation still exists, and also that arguments may be given in favour or against these theses.

We have made a classification of the arguments about Church-Turing's Thesis and Martin-Löf-Chaitin's Thesis and a comparison as precise as possible.

(a) Arguments by means of examples

(a1) Arguments for the Church-Turing Thesis
Usual functions as $n \to 2n$; $n \to n!$; $n \to$ n-th prime number and many others, are computable in the intuitive metamathematical sense, and it is easy to prove that they are recursive, hence the Church-Turing Thesis is not too restrictive.

(a2) Arguments for the Martin-Löf-Chaitin Thesis
Sequences as (0 0 0 0 ...) (0 1 0 1 0 1 ...) and many others are non-random in the intuitive metamathematical sense and it is easy to prove that they are not MLC-random, hence the Martin-Löf-Chaitin Thesis is not too tolerant.

(a3) Comparison and remarks
Arguments by means of examples tell us in the first case that the thesis is not too restrictive and in the second case that the thesis is not too tolerant. With a complementation (i.e., seeing the Martin-Löf-Chaitin Thesis is about non-random sequences) the two arguments are of equal strength. These arguments are arguments of **minimal adequation**: functions or se-

quences for which we have a clear obvious intuitive judgment are correctly classified by the mathematical notions.

Concerning the Church-Turing Thesis the proposed examples are often proposed by infinite families (for example the family of polynomial functions) but always denumerable families, and no non-denumerable families of examples can be given (for the entire set of recursive functions is denumerable!) About the Martin-Löf-Chaitin Thesis some non-denumerable sets of non-random sequences are easy to propose (for example the set of sequences verifying $x_{2n} = x_{2n+1}$ for every n). Hence it may be said that the Martin-Löf-Chaitin Thesis is better supported by examples than the Church-Turing Thesis.

It is clear that this type of arguments is not sufficient to reach a definitive judgment about the theses in question, but their importance is great and some examples have played an important role in the history of the definition of random sequences: the construction of sequences not satisfying the law of iterated logarithm and hence non-random (in the intuitive metamathematical sense) by Ville is what have proved that the definition of von Mises Church or of Popper had to be eliminated.

(b) Arguments by means of counterexamples

(b1) Arguments for the Church-Turing Thesis

By diagonalization we can obtain functions which are not recursive and for which we have no reason to believe that they are intuitively computable. This shows that the Church-Turing Thesis is not too tolerant (and in particular that the thesis is not empty).

(b2) Arguments for the Martin-Löf-Chaitin Thesis

Using non-constructive arguments (about null set in measure theory), or using more direct argument (definition of the number omega of Chaitin[11,12]) it can be proved that MLC-random sequences exist. The Martin-Löf-Chaitin Thesis is not too restrictive.

(b3) Comparison and remarks

As about the arguments by means of examples, within a complementation, it seems that the arguments by means of counterexamples are equivalent in strength with respect to Church-Turing's Thesis and Martin-Löf-Chaitin's Thesis. But the methods used to obtain MLC-random sequences are more sophisticated than those used to obtain non-recursive functions. Results about the incompressibility of random sequences show precisely that every

MLC-random sequence is very hard to obtain: the first n bits of a random sequence cannot be computed by a program of length less than n. So random sequences much more than non-recursive functions are somewhat unreal, and so we can say that the Martin-Löf-Chaitin Thesis is less supported by counterexamples than the Church-Turing Thesis.

(c) Arguments based on the intrinsic convincing strength of the definitions

(c1) Arguments for the Church-Turing Thesis
There are many definitions of the family of recursive functions. Each of them is based on a mathematical formalization of an informal notion of algorithmically computable function. So each of them gives more or less the feeling that the idea of what is the intuitive notion of algorithmically computable function is captured in the mathematical definition of recursive functions. Whatever mathematical formalisation you try for the notion of computable function, it is now absolutely certain that you will obtain a definition easily provable equivalent to the others.

Hence the intrinsic convincing strength of all until today tried formalisation of the notion of algorithmically computable function is summed in favour of the Church-Turing Thesis. There are several hundred such definitions, and their set is the more profound argument in favour of the Church-Turing Thesis.

These definitions are based on modelisations of machines like Turing's machines or more complicated ones (Turing, Kolmogorov-Uspinski, Markov, Gandy), on the basis of consideration about deductions of values from systems of equations (Gödel, Kreisel-Tait), on formal systems for arithmetic (Gödel, Tarski), on grammar production rules (Post, Chomsky) and many others.

(c2) Arguments for the Martin-Löf-Chaitin Thesis
About the Martin-Löf-Chaitin Thesis there are mainly three families of definitions. The first based on the notion of effective random tests (Martin-Löf). The second based on algorithmic information theory (Kolmogorov, Chaitin, Levin, Schnorr). The third based on unpredictability and impossibility of winning gambler methods (Schnorr, Chaitin).

(c3) Comparison and remarks
Each of the definitions of MLC-random sequences is interesting and gives a somewhat convincing element, but the chaotic history of the notion of

random sequences is such that none of these definitions alone is sufficient to ensure us that we have handled the good notion. It's the opposite with the Church-Turing Thesis for which one well-chosen definition can give a good definitive argument.

(d) Arguments of convergence of the definitions

(d1) Arguments for the Church-Turing Thesis

Each definition of recursive functions is in itself an interesting argument, but the fact that all the definitions are equivalent is surprising and indirectly constitutes the most convincing argument in favour of the Church-Turing Thesis. Initially there may have been several notions of algorithms, even an infinite number of notions of algorithm and no good notions. The proof that all the attempts to make definitions gives us the same class of functions is a first order argument and it is an argument which likes the mathematician for it is a non trivial one (i.e., is based on a real mathematical work).

(d2) Arguments for the Martin-Löf-Chaitin Thesis

The same thing holds for Martin-Löf-Chaitin Thesis, but with less strength. For there are less demonstrably equivalent definitions of the notion of random sequences, and also for the formulation of the good definition was preceded by a long period in which many bad thesis were proposed. But now the convergence argument is important and it is the very argument which was stressed in Ref. 41. Perhaps new definitions will be formulated which will prove equivalent to the Martin-Löf 's definition.

(d3) Comparison and remarks

Here we have an clear advantage in favour of the Church-Turing Thesis over the Martin-Löf-Chaitin Thesis for the first one is supported by several hundred equivalent definitions, meanwhile the second one has only several equivalent definitions and many non-equivalent definitions.

Please note that it is wrong to think that the convergence argument alone is a good argument. There are also many definitions of the class of primitive recursive functions (or of the class of the recursive sequences of 0 and 1), and these definitions cannot be considered as a good formalization of the notion of algorithmically computable functions (or of the notion of non-random sequences).

(e) Arguments of robustness

(e1) Arguments for the Church-Turing Thesis

The fact that it is impossible to diagonalize the class of recursive functions is what convinced Kleene in 1934 that the recently formulated proposition of Church was correct. This robustness of the class of recursive functions has many other aspects. For example all the variant notions of Turing machines (machines with several tapes, non deterministic machines, machines with a two-dimensional work-space, set of simultaneous communicating machines, etc.) give the good notion of recursive functions. This proves that the notion of recursive function is an intrinsic one, hence that it is an important one, hence that it is presumably the one expected.

(e2) Arguments for the Martin-Löf-Chaitin Thesis

The notion of random sequences resists also to small modifications in the formulations of the definitions. One of the most remarkable is based on what Chaitin calls the "complexity gap": to define random sequences with the condition $H(x_n) > n - c$, or with the condition $H(x_n) - n$ tends to infinite is equivalent. But there are many other examples of robustness of the definition of random sequences with regard to the random tests used in the Martin-Löf's definition.

(e3) Comparison and remarks

Convergence and robustness show that the mathematical notions of recursive functions and random sequences are of the same nature that the number π or the field of complex numbers: they are ubiquitous mathematical objects, and consequently are profound and important ones. Convergence and robustness are symptoms that we are right in identifying these notions with the intuitive metamathematical notions we are interested in.

Here there are anew more arguments of robustness in favour of the Church-Turing Thesis than in favour of the Martin-Löf-Chaitin Thesis.

About the question of robustness it must be said that relatively to the finite there is no robustness of the notion of recursive function or of the notion of random sequence. Even if all the finite approximations f_n of a function f (functions f_n which are equal to f for every $m \leq n$) are recursive then the limit function f is not necessarily recursive. There is a similar result for random sequences.

(f) Arguments of duration, and of resistance to concurrent propositions

(f1) Arguments for the Church-Turing Thesis

No truly concurrent theses have been formulated for identifying the notion of algorithmically computable function. No truly good argument against the (standard) Church-Turing Thesis had been proposed. These 50 years of success without real rival are a very strong argument in favour of the Church-Turing Thesis. Perhaps every work on computer science may be considered as an indirect confirmation of it. Perhaps also the mathematical use of the Church-Turing Thesis in order to shorten the proofs in recursion theory is an indirect argument, and certainly this shows the confidence of mathematicians in the Church-Turing Thesis.

(f2) Arguments for the Martin-Löf-Chaitin Thesis

The Martin-Löf-Chaitin Thesis is not attested by a similar long duration. Between 1966 and 1976, before the equivalences had been proved relating the definition in terms of effective statistical tests and the definition in terms of algorithmic information theory, the Martin-Löf-Chaitin Thesis was uncertain. Some concurrent theses and in particular the proposition of Schnorr[67,71] are not absolutely eliminated today, they are only becoming less supported than the Martin-Löf-Chaitin Thesis. The possibility of new mathematical results is always present, and a new evolution of the subject is not impossible. For example van Lambalgen says that he is convinced that a satisfactory treatment of random sequences is possible only in set theory lacking the power set axiom, in which random sequences "are not already there" and that non classical logic is the only way to obtain a definitive solution.[77]

To-day the Martin-Löf-Chaitin Thesis is supported by near all the specialists of the subject: Chaitin,[12] Kolmogorov et Uspenskii,[41] Gacs,[31] Schnorr (with a slight restriction),[71] Levin.[47]

(f3) Comparison and remarks

The Martin-Löf-Chaitin Thesis is clearly less supported by this type of arguments than the Church-Turing Thesis: it seems totally impossible that a new thesis replaces the Church-Turing one, it seems unlikely that a new thesis replaces Martin-Löf-Chaitin one, but in each case the constructive mathematics have certainly something to say.

(g) Arguments of effectivity, fruitfulness and usefulness

(g1) Arguments for the Church-Turing Thesis

Nothing can be proved true by saying that it is effective, fruitful or useful, but if a thesis is uninteresting, not applicable and does not give rise to new ideas we can imagine that the thesis is false and that we shall never see its falseness. Hence arguments of effectivity, fruitfulness and usefulness has to be considered.

The Church-Turing Thesis is effective (it enables simplification of mathematical proofs and hence allows to go further in the development of recursion theory), fruitful (it gives a profound insight in our conception of the world), useful (in computer science for example the strength of programming language is studied, and a first step is always to prove that a language is algorithmically complete, i.e., by using Church-Turing's Thesis that all recursive functions are programmable).

(g2) Arguments for the Martin-Löf-Chaitin Thesis

The Martin-Löf-Chaitin Thesis is fruitful (see Refs. 15 and 50) but it is so ineffective (in part due to strong Incompleteness Gödel's Theorem that are related) that no concrete utilization of MLC-random sequences is possible. This problem is what has motivated new researches in the theory of pseudo-random sequences.

The usefulness of the Martin-Löf-Chaitin Thesis is also attested by recent uses of the algorithmic information theory in physics[1] in biology[9] in statistics[66] and in philosophy.[44,47]

(g3) Comparison and remarks

About this type of arguments the comparison is a new in favour of Church-Turing's Thesis, but we may see in a near future new developments and new utilizations of the concept of MLC-random sequences.

9.2. Conclusion

In 1990, when I coined for the first time the term *Martin-Löf-Chaitin's Thesis*,[25,76] I drew the conclusion that the two thesis were truly similar, the Church-Turing Thesis offers a mathematical identification of the intuitive informal concept of algorithm, while the Martin-Löf-Chaitin Thesis proposes an identification of the intuitive informal concept of randomness. They are profound insights in the mathematical and philosophical understanding of our universe. I wrote back then that the first is more deeply attested and it has the advantage that it was formulated 50 years ago (now almost 70 years ago). The second one is more complicated (or seems so) and I said perhaps in 25 years ahead when turning 50 years old, it would reach to a certainty status similar to the other.

Today, twenty years later, the Martin-Löf-Chaitin Thesis has reached some maturity in its formulation.[27,76]

> Accordingly, Delahaye (1993) has proposed the Martin-Löf-Chaitin Thesis, that either of these definitions captures the intuitive notion of randomness. If this thesis is true, this undermines at least some skeptical contentions about randomness, such as the claim of Howson and Urbach (1993: 324) that it seems highly doubtful that there is anything like a unique notion of randomness there to be explicated. *Eagle, Antony, "Chance versus Randomness", The Stanford Encyclopedia of Philosophy, Edward N. Zalta (ed.), Fall 2010 Edition.* http://plato.stanford.edu/archives/fall2010/entries/chance-randomness/

Dasgupta claims that the Martin-Löf-Chaitin Thesis has been now reinforced.

> Following Delahaye [Delahaye, 1993], we use the term Martin-Löf-Chaitin Thesis for the assertion that Martin-Löf randomness and equivalently Kolmogorov-Chaitin randomness is the correct formulation of the intuitive notion of randomness for sequences. In this sense, it parallels the classic Church-Turing thesis, and is not a mathematical proposition to be proved or disproved. The Church-Turing thesis turned out to be highly successful in capturing the intuitive notion of algorithm. Delahaye has carried out a detailed comparison between the Church-Turing thesis and the Martin-Löf-Chaitin thesis, and concludes that in both cases, the resulting precise definitions provide "profound insights to the mathematical and philosophical understanding of our universe. Delahaye admits that the Church- Turing thesis is more deeply attested and that the definition of randomness of sequences is "more complicated compared to the definition of algorithm, but hopes that with time the Martin-Löf-Chaitin thesis will reach a level

of certainty similar to the Church-Turing thesis. We think that overall, Delahayes assertions still remain valid. In the past few decades, there has been a vast amount of research activity in the area of algorithmic randomness. Many definitions of randomness for sequences have been studied extensively, but none was found to be clearly superior to the Martin-Löf definition. Compared with other notions, it appears to be of optimal strength: Weaker notions turn out to be too weak, and the stronger ones too strong. In this way, the Martin-Löf-Chaitin thesis has gained strength in a slow but steady fashion. The proliferation of definitions of randomness for sequences makes the field harder for non-experts, but it should not be regarded negatively. It is an indication of the richness of the area, and the associated healthy and lively activity provides refinements and insights deep into the subject. Recall that while we consider the Church-Turing thesis as more satisfying, there was an even larger number of associated notions of computability, both stronger and weaker, that were (and still are) studied fruitfully. Perhaps the strongest evidence for the Martin-Löf-Chaitin thesis available so far is Schnorrs theorem, which establishes the equivalence between a naturally formulated typicality definition (Martin-Löf randomness) and a naturally formulated "incompressibility definition" (Kolmogorov-Chaitin randomness). Another justification of the Martin-Löf-Chaitin thesis is provided by the simplicity of the definition of Martin-Löf randomness within the arithmetical hierarchy. *Dasgupta, A. "Mathematical Foundations of Randomness". (preprint) in Bandyopadhyay, P.S. and Forster, M. (eds.), Handbook of the Philosophy of Science. Volume 7: Philosophy of Statistics, Elsevier, to appear 2010.*

References

1. Bennett, C. H. Logical Depth and Physical Complexity. In *The Universal Turing Machine: A Half-Century Survey*. Edited by R. Herken. Oxford University Press. 1988. pp. 227-257.
2. Borel, E. *Presque tous les nombres réels sont normaux*. Rend. Circ. Mat. Palermo. 27. 1909. pp. 247-271.

3. Chaitin, G. J. *On the Length of Programs for Computing Finite Binary Sequences.* J. A.C.M. 13. 1966. pp.547-569. Also in Ref. 11.

4. Chaitin, G. J. *On the Length of Programs for Computing Finite Binary Sequences, Statistical Considerations.* J.A.C.M. 16. 1969. pp.145-159. Repris dans.[11]

5. Chaitin, G. J. *Information Theoretic Limitations of Formal Systems.* J.A.C.M. 1974. pp. 403-424. Also in Ref. 11.

6. Chaitin, G. J. *A theory of program size formally identical to information theory.* J.A.C.M. 22. 1975. pp. 329-340. Also in Ref. 11.

7. Chaitin, G. J. *Randomness and Mathematical Proof.* Scientific American. 232. May 1975. pp. 47-52.

8. Chaitin, G. J. *Algorithmic Information Theory.* IBM Journal of Research and Development. 31. 1977. pp. 350-359. Also in Ref. 11.

9. Chaitin, G. J. *Toward a mathematical definition of "life".* In *The Maximum Entropy Formalism.* Levine, R. D. and Tribus M., (Eds). MIT Press. 1979. pp. 477-498. Also in Ref. 11.

10. Chaitin, G. J. *Randomness and Gödel's Theorem.* IBM Research R. RC 11582. 1985. Also in Ref. 11.

11. Chaitin, G. J. *Information, Randomness and Incompleteness: Papers on Algorithmic Information Theory.* World Scientific, Singapore, 1987.

12. Chaitin, G. J. *Algorithmic information theory.* Cambridge Tracts in Theoretical Computer Science 1. Cambridge University Press, New York, 1987.

13. Chaitin, G. J. *Incompleteness Theorems for Random Reals.* Advances in Applied Mathematics. 8. 1987. pp. 119-146. Also in Ref. 11

14. Chaitin, G. J. *Randomness in Arithmetic.* Scientific American. July 1988. pp. 80-85.

15. Chaitin, G. J. and Schwartz, J.T. *A note on Monte Carlo Primality Tests and Algorithmic Information Theory.* Communication on Pure and Applied Mathematics. 31. 1978. pp. 521-527. Also in Ref. 11.

16. Church, A. *On the concept of a random sequence.* Bulletin Amer. Math. Soc. 46. 1940. pp. 130-135.

17. Cover, T. M., Gacs, P., Gray, R. M. *Kolmogorov's contributions to information theory and algorithmic complexity.* The Annals of Probability, Vol 17, n.3, 1989, pp. 840-865.

18. Daley, R.P. *Minimal-Program Complexity of Pseudo-Recursive and Pseudo-Random Sequences.* Mathematical System Theory, vol 9, n. 1, 1975. pp. 83-94.

19. Daley, R.P. *Noncomplex Sequences: Characterizations and Examples.* J. Symbol. Logic. 41. 1976. pp. 626.

20. Delahaye, J-P. *Une Extension Spectaculaire du Théorème de Gödel.* La Recherche n. 200, juin 1988. pp. 860-862.

21. Delahaye, J-P. *Cinq Classes d'Idées.* Rapport Laboratoire d'Informatique Fondamentale de Lille. Univ. Sc. Lille, Bât M3, 59655 Villeneuve d'Ascq. Avril 1989.

22. Delahaye, J-P. *Chaitin's Equation: An Extension of Gödel's Theorem.* Notices of The American Mathematical Society. October 1989. pp. 984-987.

23. Delahaye, J-P. *Le hasard comme imprevisibilité et comme absence d'ordre.* Rapport du Laboratoire d'Informatique Fondamentale de Lille. Université de Sci. et Tech. de Lille 59655 Villeneuve d'Ascq cédex F, 1990.

24. Jean-Paul Delahaye. *Information, complexité et hasard,* Hermes Sciences Publicat. 2e éd. 1999. (First edition 1994).

25. Jean-Paul Delahaye. "Randomness, Unpredictability and Absence of Order: The Identification by the Theory of Recursivity of the Mathematical Notion of Random Sequence." in Dubucs, J (ed.) *Philosophy of Probability,* Kluwer Academic Publishers, Dordrecht, 1993, pp. 145–167.

26. Dies, J. E. *Information et Complexité.* Annales de L'institut Henry Poincaré, Section B, Calcul des Probabilités et Statistiques. Nouvelle Série Vol 12. 1976. pp.365-390 et Vol 14. 1978. pp. 113-118.

27. Dasgupta, A. "Mathematical Foundations of Randomness," in Bandyopadhyay, P.S. and Forster, M. (eds.), *Handbook of the Philosophy of Science.* Volume 7: Philosophy of Statistics, Elsevier, to appear 2010.)

28. Gacs, P. *On the symmetry of algorithmic information.* Dokl. Akad. Nauk SSSR. 218. 1974. pp. 1477-1480.

29. Gacs, P. *Exact Expressions for Some Randomness Tests.* Z. Math. Logik. Grundl. Math. 26. 1980.

30. Gacs, P. *On the relation between descriptional complexity and probability.* Theo. Comp. 22. 1983. pp.71-93.

31. Gacs, P. *Every Sequence Is Reducible to a Random One.* Information and Control. 70. 1986. pp. 186-192.

32. Gaifman, H. Snir, M. *Probabilities over rich languages, randomness and testing.* Journal of Symbolic Logic. Vol. 47. 1982. pp. 495-548.

33. Gardner, M. *Le nombre aléatoire oméga semble bien recéler les mystères de l'univers.* Pour La Science. 1980.

34. Kolmogorov, A. N. *Osnovye ponyatia teorii veroyatnostei.* ONTI. Moscow, 1936. English translation: Foundations of the theory of probability. Chelsea, New-York. 1950.

35. Kolmogorov, A. N. *On table of random numbers.* Sankhya The Indian Journal of Statistics. A25. 369. 1963. pp. 369-376.

36. Kolmogorov, A. N. *Three approaches for defining the concept of information quantity.* Information Transmission. V. 1. 1965. pp. 3-11.

37. Kolmogorov, A. N. *Logical basis for Information Theory and Probability Theory.* IEEE Transaction on Information Theory. Vol. IT14, n.5. 1968. pp. 662-664.

38. Kolmogorov, A. N. *Some Theorems on algorithmic entropy and the algorithmic quantity of information.* Uspeki Mat. Nauk, Vol. 23:2. 1968. pp. 201.

39. Kolmogorov, A. N. *Combinatorial foundations of information theory and the calculus of probabilities.* Russian Mathematical Surveys. Vol. 38.4. 1983. pp. 29-40.

40. Kolmogorov, A. N. On logical foundations of probability theory. In *Probability Theory and Mathematical Statistics.* "Lecture Notes in Mathematics." Ed. K. Ito and J. V. Prokhorov. Vol. 1021. Springer-Verlag., Berlin. 1983. pp. 1-5.

41. Kolmogorov, A.N. and Uspenskii, V. A. *Algorithms and Randomness*. SIAM Theory Probab. Appl. Vol. 32. 1987. pp. 389-412.

42. Levin, L. A. *On the notion of random sequence*. Dokl. Akad. Nauk SSSR. 212, 3. 1973.

43. Levin, L. A. *Laws of information conservation (non-growth) and aspects of the foundation of probability theory*. Problems Inform. Transmission. 10 n.3. 1974. pp. 206-210.

44. Levin, L. A. *On the principle of conservation of information in intuitionistic mathematics*. Dokl. Akad. Nauk. SSSR. Tom 227 n.6. 1976. Soviet Math. Dokl. 17 n.2. 1976. pp. 601-605.

45. Levin, L. A. *Various measures of complexity for finite objects (axiomatic descriptions)*. Soviet Math. Dokl. 17. n.2. 1976. pp. 552-526.

46. Levin, L. A. *Uniform tests of randomness*. Soviet Math. Dokl. 17, n.2. 1976. pp. 337-340.

47. Levin, L. A. *Randomness conservative inequalities: Information and independence in mathematical theories*. Inf. Contr. 61. 1984. pp. 15-37.

48. Levin, L. A. and V'Yugin, V. V. *Invariant Properties of Informational Bulks*. Lecture Notes in Computer Science n.53. Springer, Berlin. 1977. pp. 359-364.

49. Li, M., Vitanyi, P. M. B.. *A New Approach to Formal Language Theory by Kolmogorov Complexity*. Proc 16th International Colloquium on Automata Languages and Programming. 1989.

50. Li, M., Vitanyi, P. M. B. *Inductive Reasoning and Kolmogorov Complexity*. Proc. 4th Annual IEEE Structure in Complexity Theory Conference. 1989.

51. Li, M., Vitanyi, P. M. B. *Kolmogorov Complexity and Its Applications*. Handbook of Theoretical Computer Science. J. van Leeuwen Editor. North-Holland. 1990.

52. Li, M., Vitanyi, P. M. B. *An Introduction to Kolmogorov Complexity and Its Applications*. Third Edition, Springer Verlag, 2008.

53. Loveland, D. W. *A new interpretation of the von Mises' Concept of random sequence*. Zeitschr. F. Math. Logik und Grundlagen d. Math. Bd12. 1966. pp. 279-294.

54. Loveland, D. W. *The Kleene Hierarchy Classification of Recursively Random Sequences*. Trans. Amer. Math. Soc. 125. 1966. pp. 487-510.

55. Loveland, D. W. *Minimal Program Complexity Measure*. Conference Record ACM Symposium on Theory of Computing. May 1968. pp. 61-65.

56. Loveland, D. W. *A variant of the Kolmogorov concept of complexity*. Information and Control. 15. 1969. pp. 510-526.

57. Martin-Löf, P. *On the Concept of a Random Sequence*. Theory Probability Appl.. Vol. 11 1966. pp. 177-179

58. Martin-Löf, P. *The Definition of Random Sequences*. Information and Control. 9. 1966. pp. 602-619.

59. Martin-Löf, P. *Algorithms and Randomness*. Intl. Stat. Rev. 37, 265. 1969. pp. 265-272.

60. Martin-Löf, P. *The Literature on von Mises' Kollektivs Revisited*. Theoria, XXXV. 1969. pp. 12-37.

61. Martin-Löf, P. On the notion of Randomness. in *Intuitionism and Proof Theory"*. Kino, A., Myhill, J. and Vesley, R.E. (eds). North-Holland Publishing Co. Amsterdam. 1970, pp.73-78.

62. Martin-Löf, P. *Complexity Oscillations in Infinite Binary Sequences.* Zeitschrift fur Wahrscheinlichkeitstheory und Vervandte Gebiete. 19. 1971. pp.225-230.

63. Martin-Löf, P. *The notion of redundancy and its use as a quantitative measure of the discrepancy between statistical hypothesis and a set of observational data.* Scand. J. Stat. Vol 1. 1974. pp. 3-18.

64. O'Connor, M. *An Unpredictibility Approach to Finite-State Randomness.* Journal of Computer and System Sciences. 37. 1988 pp. 324-336.

65. Popper, K. R. *Logik der Forschund.* Springer. 1935. French translation: La Logique de la Découverte Scientifique. Payot, Paris. 1978.

66. Rissanen, J. *Stochastic Complexity in Statistical Inquiry.* World Scientific. Series in Computer Science. Vol 15. 1989

67. Schnorr, C. P. *A unified approach to the definition of random sequence.* Math. Systems Theory. 5. 1971. pp. 246-258.

68. Schnorr, C. P. *Zufälligkeit und Wahrscheinlichkeit.* Lecture Notes in Mathemathics. Vol 218. Berlin-Heidelberg-New York. Springer, 1971.

69. Schnorr, C. P. *The process complexity and effective random tests.* Proc. ACM Conf. on Theory of Computing. 1972. pp. 168-176.

70. Schnorr, C. P. *Process complexity and effective random tests.* J. Comput. Syst. Sci. 7. 1973. pp 376-388.

71. Schnorr, C. P. A survey of the theory of random sequences. In *Basic Problems in Methodology and Linguistics.* Ed. Butts, R. E., Hintikka, J., Reidel, D. Dordrecht. 1977. pp. 193-210.

72. Shen', A. Kh. *The concept of (a,b)-stochasticity in the Kolmogorov sense, and its properties.* Soviet Math. Dokl. Vol. 28. 1983. pp. 295-299.

73. Shen', A. Kh. *On relation between different algorithmic definitions of randomness.* Soviet Math. Dokl. Vol. 38. 1989. pp. 316-319.

74. Solomonoff, R.J. *A formal theory of inductive Inference.* Information and Control. 7. 1964. pp. 1-22.

75. Turing, A. M. *On Computable Numbers, with an application to the Entscheidungsproblem.* Proceeding of the London Mathematical Society. 2, 42, 1936-7. pp. 230-265. corrections 43. 1937. pp. 544-546.

76. Eagle, Antony, "Chance versus Randomness", The Stanford Encyclopedia of Philosophy, Edward N. Zalta (ed.), Fall 2010 Edition. http://plato.stanford.edu/archives/fall2010/entries/chance-randomness/

77. van Lambalgen, M. *von Mises' definition of random sequences reconsidered.* Journal of Symbolic Logic. Vol 52. 1987. pp. 725-755.

78. Ville, J. *Sur la notion de collectif.* C.R. Acad. Scien. Paris. 203. 1936. pp. 26-27. Sur les suites indifférentes. C.R. Acad. Scien. Paris. 202 . 1936. p.1393

79. Ville, J. *Etude critique de la notion de collectif.* Gauthier-Villars. Paris. 1939.

80. von Mises, R. *Grundlagen der Wahrscheinlichkietsrechnung.* Math. Z. 5. 100. 1919.

81. von Mises, R. *On the foundation of probability and statistics*. Am. Math. Statist. 12. 1941. pp.191-205.
82. von Mises, R. *Selected papers of Richard von Mises*. Providence Rhode Island, Amer. Math. Soc. 1964.
83. Wald, A. *Die Widerspruchsfreiheit des Kollektivbegriffes der Wahrscheinlichkeitsrechnung*. Ergebnisse eines Mathetischen Kolloquiums. 8. 1937. pp. 38-72.
84. Wald, A. *Die Widerspruchsfreiheit des Kollektivgriffes*. Actualités Sci. Indust. 735. 1938. pp. 79-99.
85. Zvonkin, A. K. and Levin, L. A. *The Complexity of finite object and the development of the concepts of information and randomness by means of the theory of algorithms*. Russ. Math. Survey. 25, 6. 1970. pp 83-124.

Chapter 10

The Road to Intrinsic Randomness

Stephen Wolfram
Wolfram Research
s.wolfram@wolfram.com

WHAT DREW ME TO THE STUDY OF COMPUTATION AND RANDOMNESS

When I was 12 years old I was interested in physics, and I was reading a college-level book about statistical physics. The cover had a picture of a simulation of idealized gas molecules getting more and more random — supposedly following the Second Law of Thermodynamics. I got really interested in that, and decided to simulate it on a computer.

Because this was 1972, and I was just a kid, I only had access to quite a primitive computer. So I had to come up with a simple idealized model to simulate. As it happens, the model I invented was a cellular automaton. But for detailed technical reasons it only showed quite simple behavior, and never generated randomness.

A few years later, I started studying cosmology, and was interested in the question of how the universe can start from seeming randomness at the Big Bang, and then generate structure in galaxies and everything we now see. I wanted to make a simple model for that process. At first I made models that I thought were direct and realistic.

But eventually I started thinking about general models that could show spontaneous organization. And in doing this, I ended up starting to do computer simulations of simple cellular automata.

At first, my main interest was starting with random initial conditions, and watching how much structure could arise.

I had noticed that some cellular automata (like rule 90) can generate elaborate nested patterns when started from simple initial conditions. But somehow I assumed that if the initial conditions were simple, a simple rule could never generate complex behavior.

141

But in 1984 I decided to do the straightforward computer experiment of running all possible cellular automata with the simplest kinds of rules, starting from the simplest initial conditions. And what I found was that rule 30 did something quite amazing. Even though its initial conditions were simple, it generated a pattern of great complexity, that in many ways seemed completely random.

Still not believing that such simple rules could generate such randomness, I tried to "crack" rule 30 using all sorts of mathematical, statistical and cryptographic techniques. But it never yielded.

And indeed the center column of the pattern it generates has proved itself a source of practical pseudorandomness — and is the only survivor of practical schemes invented when it was.

WHAT WE HAVE LEARNED

Having seen the basic phenomenon of rule 30, I began to study how widespread it is, and how it really occurs.

I started looking at randomness that occurs in all sorts of physical systems. Some, I realized, get randomness in particular measurements because of some random perturbation from another part of the system. And a few do the classic "chaos theory" thing of progressively excavating the details of random details in their initial conditions.

But what I concluded is that many systems actually operate more like rule 30: they in effect intrinsically generate randomness by what amounts to a purely algorithmic procedure.

We know of examples like this throughout mathematics. The digits of π, and the distribution of primes, are two examples. But the great significance of rule 30 is that it shows us that a system with rules we might realistically encounter in nature can exhibit this kind of intrinsic randomness.

And my guess is that intrinsic randomness is at the heart of much of the randomness that we see in the natural world.

What leads to the randomness in rule 30?

I think it is intimately associated with the Principle of Computational Equivalence. It is a reflection of the computational sophistication that I suspect exists in something like 30.

In a sense, something like rule 30 will seem non-random if we — with our various mental and computational powers — can readily predict how it will behave. But the Principle of Computational Equivalence makes the statement that almost any system not obviously simple in its behavior will show behavior that corresponds to a computation that is as sophisticated as it can be.

And that means that our mental and computational powers cannot "outrun" it. We cannot compress the behavior of the system. And so it will inevitably seem to us unpredictable and random.

This general way of thinking has many consequences.

For example, it finally gives us a clear computational way to understand the Second Law of Thermodynamics.

In effect, many systems evolve in such a way as to so "encrypt" their initial conditions that no reasonable observer — with any implementable coarse-graining scheme — can do the decryption, and recognize whether the initial conditions were simple or not.

This way of thinking also gives us a way to resolve the determinism vs. free will issue.

In effect it says that the actual behavior of a system can require an irreducible amount of computation to determine from the underlying laws. And that's why the behavior will seem "free" of deterministic underlying laws that govern it.

WHAT WE DON'T (YET) KNOW

There are certainly more experiments that could be done on seemingly random physical systems.

Intrinsic randomness has an important prediction: it suggests that the randomness we see should be repeatable.

If randomness comes from extrinsic factors — like noise in the environment or details of initial conditions — then it cannot be expected to be repeatable.

But intrinsic randomness, like the digits of π, will be repeatable.

And so, in sufficiently carefully controlled experiments we can expect that the randomness we see will indeed be repeatable.

There are all sorts of indications that this is exactly what's seen in lots of experiments. But the experiments haven't been done with an eye to this issue, so we don't know for sure. It would be good if we did.

THE MOST IMPORTANT OPEN PROBLEMS

It'd be great to understand more about the Principle of Computational Equivalence. We have some excellent empirical evidence for it in cellular automata and Turing machines. But it would be nice to get more. It'd be great, for example, to prove that rule 30 is computation universal.

In fact, it'd be great to prove almost anything about rule 30. It'll probably be difficult to show much about the global randomness of the center column, just as it's difficult to prove much about the randomness of the digits of π.

But perhaps we'll be able to prove something.

And then there are other things to prove. For example, is the problem of inverting rule 30 NP complete? In the case of an arbitrary cellular automaton rule, it is. But what about the specific case of rule 30? Does the kind of cryptographic randomness that this would imply occur already for a rule as simple as rule 30?

There's more to know about randomness in our universe and in physics, too.

Can we find a deterministic underlying theory for all of physics? If so, does it generate intrinsic randomness? Is that intrinsic randomness what leads to the apparent randomness that's been identified in quantum mechanics?

THE PROSPECTS FOR PROGRESS

As more basic science is done exploring the computational universe, we will see more and more examples of computational systems that generate randomness.

More and more empirical information will build up on just how randomness generation can occur. And that's what we need to be able to abstract general principles.

What I expect is that the Principle of Computational Equivalence will gradually emerge with more and more evidence.

But there will be many details — many specific conditions for its validity in particular cases, many detailed predictions about what's possible and what's not.

Right now we know about periodicity, and we know about nesting. We don't really know any other general classes of regularities — general features that demonstrate non-randomness.

I'm sure there'll be many hierarchies of such features that can be identified, and some of them will probably occur in decent numbers of systems.

On the other side, each of these regularities is associated with some kind of computational test that can be applied to systems to reveal non-randomness.

And I also expect that a much more robust theory can be developed of hierarchies of computational (or statistical) tests for randomness.

All of this in effect feeds into a more sophisticated model for the "observer". And that's important not only at a formal level, but also in understanding what we'll perceive about the natural world.

Indeed, I have a suspicion that a better understanding of the observer is what's at the heart of some of the issues around understanding how the fundamental theory of physics can be built.

In a completely different direction, I think more and more of our technology will be "mined" from the computational universe. And as such, it will often show what appears to be randomness.

Right now one can usually tell artifacts from natural systems by the fact that they look simpler — less random.

The reason is typically that they are engineered in an incremental way that forces them to show that simplicity.

But when they're mined from the computational universe, they won't have to have that simplicity.

Just like in nature, they'll be able to show all sorts of complexity and randomness.

People will think very differently about randomness in the future.

Because so much of our technology — and no doubt also our art — will be mined from the computational universe. Where there are rules, which we can identify and perceive. But where the overall behavior that seems random is ubiquitous.

PART III

Algorithmic Inference and Artificial Intelligence

Chapter 11

Algorithmic Probability — Its Discovery — Its Properties and Application to Strong AI

Ray J. Solomonoff

*Computer Learning Research Centre
Royal Holloway, University of London**
and
IDSIA, Galleria 2, CH–6928 Manno–Lugano, Switzerland
http://world.std.com/~rjs/pubs.html*

INTRODUCTION

We will first describe the discovery of Algorithmic Probability — its motivation, just how it was discovered, and some of its properties. Section two discusses its Completeness — its consummate ability to discover regularities in data and why its Incomputability does not hinder to its use for practical prediction. Sections three and four are on its Subjectivity and Diversity — how these features play a critical role in training a system for strong AI. Sections five and six are on the practical aspects of constructing and training such a system. The final Section, seven, discusses present progress and open problems.

WHAT DREW ME TO THE STUDY OF COMPUTATION AND RANDOMNESS

My earliest interest in this area arose from my fascination with science and mathematics. However, in first studying geometry, my interest was more in how proofs were discovered than in the theorems themselves. Again, in science, my interest was more in how things were discovered than in the contents of the discoveries. The Golden Egg was not as exciting as the goose that laid it.

These ideas were formalized into two goals: one goal was to find a general method to solve all mathematical problems. The other goal was to find a general method to discover all scientific truths. I felt the first

*Visiting Professor.

problem to be easier because mathematics was deterministic and scientific truths were probabilistic. Later, it became clear that the solutions to both problems were identical![†]

WHAT WE HAVE LEARNED

Some important heuristic ideas:

First — From Rudolph Carnap: That the state of the universe could be represented by a long binary string, and that the major problem of science was the prediction of future bits of that string, based on its past.

Second — From Marvin Minsky and John McCarthy: the idea of a universal Turing machine. That any such machine could simulate any describable function or any other Turing machine (universal or not). That it had a necessary "Halting Problem" — that there had to be inputs to the machine such that one could never be sure what the output would be.

Third — Noam Chomsky's description of Generative and Nongenerative grammars. To use them for prediction, in 1958 I invented Probabilistic Grammars (Described in the appendix of Ref. 6).

The final discovery occurred in 1960[7] when I began investigating the most general deterministic Grammar — based on a universal Turing machine. It's probabilistic version had some striking properties and suggested to me that a probabilistic grammar based on a universal Turing machine would be the most general type of grammar possible — and would perhaps be the best possible way to do prediction.

This was the birth of Algorithmic Probability (ALP). In the initial version, we take a universal Turing machine with an input tape and an output tape. Whenever the machine asks for an input bit, we give it a zero or a one with probability one-half. The probability that (if and when the machine stops) the output tape is a certain binary string, x, is the universal probability of x. This was a universal distribution on finite strings.

I was much more interested in sequential prediction (as in Carnap's problem), so I generalized it in the following way: We use a special kind of universal Turing machine. It has three tapes — unidirectional input and output tapes, and an infinite bidirectional work tape. We populate the input tape with zeros and ones, each with probability one-half. The

---

[†]The subject of the beginning of this essay has been treated in some detail in "The Discovery of Algorithmic Probability".[11] Here, we will summarize some ideas in that paper and deal with important subsequent developments.

probability of the string x is the probability that the output tape will be a string that begins with x.

This second universal distribution can be used for sequential prediction in the following way: suppose $P(x1)$ is the probability assigned by the distribution to the string, $x1$. Let $P(x0)$ be probability assigned to $x0$. Then the probability that x will be followed by 1 is

$$P(x1)/((P(x0) + P(x1)).$$ (11.1)

I will be usually be referring to this second model when I discuss ALP. It is notable that ALP doesn't need Turing machines to work properly. Almost all of its properties carry over if we use any computer or computer language that is "universal" — i.e., that it can express all computable functions in an efficient way. Just about all general purpose computers are "universal" in this sense, as are general programming languages such as Fortran, LISP, C, C++, Basic, APL, Mathematica, Maple, ...

11.1. Completeness and Incomputability

Does ALP have any advantages over other probability evaluation methods? For one, it's the only method known to be *complete*. The completeness property of ALP means that if there is any regularity in a body of data, our system is guaranteed to discover it using a relatively small sample of that data. More exactly, say we had some data that was generated by an *unknown* probabilistic source, P. Not knowing P, we use instead, P_M, the Algorithmic Probabilities of the symbols in the data. How much do the symbol probabilities computed by P_M differ from their true probabilities, P?

The Expected value with respect to P of the total square error between P and P_M is bounded by $-1/2 \ln P_0$.

$$E_P \Big[\sum_{m=1}^{n} (P_M(a_{m+1} = 1|a_1, a_2 \cdots a_m) - P(a_{m+1} = 1|a_1, a_2 \cdots a_m))^2 \Big] \leq -\frac{1}{2} \ln P_0 \,,$$

$$\ln P_0 \approx k \ln 2 \,.$$ (11.2)

P_0 is the a priori probability of P. It is the probability we would assign to P if we knew P.

k is the *Kolmogorov complexity* of the data generator, P. It's the shortest binary program that can describe P, the generator of the data.

This is an extremely small error rate. The error in probability approaches zero more rapidly than $1/n$. Rapid convergence to correct probabilities is a most important feature of ALP. The convergence holds for any P that is describable by a computer program and includes many functions that are formally *incomputable*. The convergence proof is in Ref. 9. It was discovered in 1968, but since there was little general interest in ALP at that time I didn't publish until 1975 (Sol 75) and it wasn't until 1978[9] that a proof was published. The original proof was for a mean square loss function and a normalized universal distribution. — but the proof itself showed it to be also true for the more general KL loss function. Later, Peter Gács[1] showed it would work for a universal distribution that was not normalized and Marcus Hutter[2] showed it would work for arbitrary (non-binary) alphabets, and for a variety of loss functions.

While ALP would seem to be the best way to predict, the scientific and mathematical communities were disturbed by another property of algorithmic probability: — it was incomputable! This incomputability is attributable to "the halting problem" — that there will be certain inputs to the Turing machine for which we can never be certain as to what the output will be.

It is, however, possible to get a sequence of approximations to ALP that converge to the final value, but at no point can we make a useful estimate as to how close we are to that value.

Fortunately, for practical prediction, we very rarely need to know "the final value". What we really need to know is "How good will the present approximate predictions be in the future (out of sample) data"? This problem occurs in all prediction methods and algorithmic probability is often able to give us insight on how to solve it.

It is notable that completeness and incomputability are complementary properties: It is easy to prove that any *complete* prediction method must be incomputable. Moreover, any computable prediction method cannot be *complete* — there will always be a large space of regularities for which its predictions are catastrophically poor.

Since incomputability is no barrier to practical prediction, and computable prediction methods necessarily have large areas of ineptitude, it would seem that ALP would be preferred over any computable prediction methods.

There is, however another aspect of algorithmic probability that people find disturbing — it would seem to take too much time and memory to find a good prediction. In Section 5 we will discuss this at greater length. There is a technique for implementing ALP that seems to take as little time as possible to find regularities in data.

11.2. Subjectivity

Subjectivity in science has usually been regarded as Evil. — that it is something that does not occur in "true science" — that if it does occur, the results are not "science" at all. The great statistician, R. A. Fisher, was of this opinion. He wanted to make statistics "a true science" free of the subjectivity that had been so much a part of its history.

I feel that Fisher was seriously wrong in this matter, and that his work in this area has profoundly damaged the understanding of statistics in the scientific community — damage from which it is recovering all too slowly.

Two important sources of error in statistics are finite sample size and model selection error. The finite sample part has been recognized for some time. That model selection error is a necessary part of statistical estimation is an idea that is relatively new, but our understanding of it has been made quite clear by ALP. Furthermore, this kind of error is very subjective, and can depend strongly on the lifelong experience of a scientist.

In ALP, this subjectivity occurs in the choice of "reference" — a universal computer or universal computer language. In the very beginning, (from the "invariance theorem") it was known that this choice could only influence probability estimates by a finite factor — since any universal device can simulate any other universal device with a finite program. However, this "finite factor" can be enormous — switching between very similar computer languages will often give a change of much more than 2^{1000} in probability estimates!

To understand the role of subjectivity in the life of a human or an intelligent machine, let us consider the human infant. It is born with certain capabilities that assume certain a priori characteristics of its environment-to-be. It expects to breathe air, its immune system is designed for certain kinds of challenges, it is usually able to learn to walk and converse in whatever human language it finds in its early environment. As it matures, its a priori information is modified and augmented by its experience.

The AI system we are working on is of this sort. Each time it solves a problem or is unsuccessful in solving a problem, it updates the part of its a priori information that is relevant to problem solving techniques. In a manner very similar to that of a maturing human being, its a priori information grows as the life experience of the system grows.

From the foregoing, it is clear that the subjectivity of algorithmic probability is a necessary feature that enables an intelligent system to incorporate experience of the past into techniques for solving problems of the future.

11.3. Diversity

In Section 1 we described ALP with respect to a universal Turing machine with random input. An equivalent model considers all prediction methods, and makes a prediction based on the weighted sum of all of these predictors. The weight of each predictor is the product of two factors: the first is the a priori probability of each predictor — It is the probability that this predictor would be described by a universal Turing machine with random input. If the predictor is described by a small number of bits, it will be given high a priori probability. The second factor is the probability assigned by the predictor to the data of the past that is being used for prediction. We may regard each prediction method as a kind of model or explanation of the data. Many people would use only the best model or explanation and throw away the rest. Minimum Description Length,[3] and Minimum Message Length[13] are two commonly used approximations to ALP that use only the best model of this sort. When one model is much better than any of the others, then Minimum Description Length and Minimum Message Length and ALP give about the same predictions. If many of the best models have about the same weight, then ALP gives better results.

However, that's not the main advantage of ALP's use of a diversity of explanations. If we are making a single kind of prediction, then discarding the non-optimum models usually has a small penalty associated with it. However if we are working on a sequence of prediction problems, we will often find that the model that worked best in the past, is inadequate for the new problems. When this occurs in science we have to revise our old theories. A good scientist will remember many theories that worked in the past but were discarded — either because they didn't agree with the new data, or because they were a priori "unlikely". New theories are character- istically devised by using failed models of the past, taking them apart, and using the parts to create new candidate theories. By having a large diverse set of (non-optimum) models on hand to create new trial models, ALP is in the best possible position to create new, effective models for prediction.

When ALP is used in Genetic Programming, its rich diversity of mod- els can be expected to lead to very good, very fast solutions with little likelihood of "premature convergence".

11.4. Computation Costs

If Algorithmic Probability is indeed so very effective, it is natural to ask about its computation costs — it would seem that evaluating a very large number of prediction models would take an enormous amount of time. We have, however, found that by using a search technique similar to one used by Levin for somewhat different kinds of problems, that it is possible to perform the search for good models in something approaching optimum speed. It may occasionally take a long time to find a very good solution — but no other search technique could have found that solution any faster.

A first approximation of how the procedure works: Suppose we have a universal machine with input and output tapes and a very big internal memory. We have a binary string, x, that we want to extrapolate — to find the probability of various possible continuations of x. We could simply feed many random strings into the machine and watch for inputs that gave outputs that started with x. This would take a lot of time, however. There is a much more efficient way:

We select a small time limit, T, and we test all input strings such that

$$t_k < T2^{-l_k} . \qquad (11.3)$$

Here l_k is the length of the kth input string being tested, and t_k is the time required to test it. The test itself is to see if the output starts with string x. If we find no input that gives output of the desired kind, we double T and go through the same test procedure. We repeat this routine until we find input strings with output strings that start with x. If we give each such output a weight of 2^{-l_k} (l_k being the length of its input), the weighted output strings will get a probability distribution over the possible continuations of the string, x.

In the example given, all input strings of a given length were assumed to have the same probability. As the system continues to predict a long binary sequence, certain regularities will occur in input sequences that generate the output. These regularities are used to impose a nonuniform probability distribution on input strings of a given length. In the future this "adaptation" of the input distribution enables us to find much more rapidly, continuations of the data string that we want to predict.

11.5. Training Sequences

It is clear that the sequence of problems presented to this system will be an important factor in determining whether the mature system will be very much more capable than the infant system. Designing "training sequences" of this sort is a crucial and challenging problem in the development of strong intelligence.

In most ways, designing a training sequence for an intelligent machine, is very similar to designing one for a human student. In the early part of the sequence, however, there is a marked difference between the two. In the early training sequence for a machine, we know exactly how the machine will react to any input problem. We can calculate a precise upper bound on how long it will take the machine to solve early problems. It is just

$$T_i/P_i \, . \qquad\qquad (11.4)$$

P_i is the probability that the machine assigns to the solution that is known by the trainer. T_i is the time needed to test that solution. I call this upper bound the "conceptual jump size" (CJS). It tells us how hard a problem is for a particular AI. I say "upper bound" because the system may discover a better, faster, solution than that known by the trainer.

This CJS estimate makes it easy to determine if a problem is feasible for a system at a particular point in its education. The P_i for a particular problem will vary during the life of the system. For a properly constructed training sequence, the P_i associated with a particular problem should increase as the system matures.

Eventually in any training sequence for a very intelligent machine, the trainer will not be able to understand the system in enough detail to compute CJS values. The trainer will then treat the machine as a human student. By noting which problems are easy and which are difficult for the machine the trainer will make a very approximate model of the machine and design training problems using that model.

Learning to train very intelligent machines should give very useful insights on how to train human students as well.

THE PROSPECTS FOR PROGRESS

A system incorporating some of the features we have discussed has been programmed by Schmidhuber.[4] It was able to discover a recursive solution to the "Tower of Hanoi" problem, after finding a recursive solution to one of its earlier, easier problems.

For further progress, we need larger, more detailed training sequences — Writing sequences of this sort is a continuing "Open Problem".[10]

The process of updating the system in view of its past experience is another important area of ongoing research. We have considered PPM (Prediction by Partial Matching[12]), APPM (an augmented version of PPM) and SVR (Support Vector Regression[5]) as possible updating systems. The improvement of the updating algorithm remains another continuing "Open Problem".

References

1. Gács. P. Theorem 5.2.1 in *An Introduction to Kolmogorov Complexity and Its Applications*, Springer–Verlag, N.Y., 2nd edition, pp. 328-331, 1997.
2. Hutter, M.,"Optimality of Universal Bayesian Sequence Prediction for General Loss and Alphabet," http://www.idsia.ch/ marcus/ai/
3. Rissanen, J. "Modeling by the Shortest Data Description," *Automatica*, 14:465–471, 1978.
4. Schmidhuber, J., "Optimal Ordered Problem Solver," TR IDSIA-12-02, 31 July 2002. http://www.idsia.ch/ juergen/oops.html
5. Sapankevych, N. and Sankar, R., "Time Series Prediction Using Support Vector Machines: A Survey," *IEEE Computational Intelligence* Vol. 4, No. 2, pp 24–38, May 2009.
6. Solomonoff, R.J. "A Progress Report on Machines to Learn to Translate Languages and Retrieve Information," *Advances in Documentation and Library Science*, Vol. III, pt. 2, pp. 941–953. (Proceedings of a conference in September 1959.)
7. Solomonoff, R.J. "A Preliminary Report on a General Theory of Inductive Inference." (Revision of Report V–131, Feb. 1960), Contract AF 49(639)–376, Report ZTB–138, Zator Co., Cambridge, Mass., Nov, 1960.
8. Solomonoff, R.J. "Inductive Inference Theory – a Unified Approach to Problems in Pattern Recognition and Artificial Intelligence," *Proceedings of the 4th International Conference on Artificial Intelligence*, pp 274–280, Tbilisi, Georgia, USSR, September 1975.
9. Solomonoff, R.J. "Complexity–Based Induction Systems: Comparisons and Convergence Theorems," *IEEE Trans. on Information Theory*, Vol IT—24, No. 4, pp. 422–432, July 1978.
10. Solomonoff, R.J. "A System for Incremental Learning Based on Algorithmic Probability," *Proceedings of the Sixth Israeli Conference on Artificial Intelligence, Computer Vision and Pattern Recognition*, pp. 515–527, Dec. 1989.
11. Solomonoff, R.J. "The Discovery of Algorithmic Probability," *Journal of Computer and System Sciences*, Vol. 55, No. 1, pp. 73–88, August 1997.
12. Teahan, W.J. "Probability Estimation for PPM," *Proc. of the New Zealand Computer Science Research Students' Conference*, University of Waikato, Hamilton, New Zealand, 1995. http://cotty.16x16.com/compress/peppm.htm
13. Wallace, C.S and Boulton, D.M. "An Information Measure for Classification," *Computer Journal*, 11:185–194, 1968.

Chapter 12

Algorithmic Randomness as Foundation of Inductive Reasoning and Artificial Intelligence

Marcus Hutter

SoCS, RSISE, IAS, CECS
Australian National University
Canberra, ACT, 0200, Australia
www.hutter1.net

INTRODUCTION

This chapter is a brief personal account of the past, present, and future of algorithmic randomness, emphasizing its role in inductive inference and artificial intelligence. It is written for a general audience interested in science and philosophy. Intuitively, randomness is a lack of order or predictability. If randomness is the opposite of determinism, then algorithmic randomness is the opposite of computability. Besides many other things, these concepts have been used to quantify Ockham's razor, solve the induction problem, and define intelligence.

WHAT DREW ME TO THE STUDY OF COMPUTATION AND RANDOMNESS

Some sequences of events follow a long causal "computable" path, while others are so "random" that the coherent causal path is quite short. I am able to trace back quite far my personal causal chain of events that eventually led me to computability and randomness (C&R), although the path looks warped and random.

At about 8 years of age, I got increasingly annoyed at always having to tidy up my room. It took me more than 20 years to see that computation and randomness was the solution to my problem (well — sort of). Here's a summary of the relevant events:

First, my science fiction education came in handy. I was well-aware that robots were perfectly suitable for all kinds of boring jobs, so they should be good for tidying up my room too. Within a couple of years I had built a series of five increasingly sophisticated robots. The "5th generation" one was humanoid-shaped, about 40cm high, had two arms and hands, and one broad roller leg. The frame was metal and the guts cannibalized from my remote controlled car and other toys.

With enough patience I was able to maneuver Robbie5 with the remote control to the messy regions of my room, have it pick up some Lego pieces and place them in the box they belonged to. It worked! And it was lots of fun. But it didn't really solve my problem. Picking up a block with the robot took at least 10 times longer than doing it by hand, and even if the throughput was the same, I felt I hadn't gained much.

Robbie5 was born abound a year before my father brought home one of the first programmable pocket calculators in 1978, a HP-19C. With its 200 bytes of RAM or so it was not quite on par with Colossus (a super computer which develops a mind of its own in the homonymous movie), but HP-19C was enough for me to realize that a computer allows programming of a robot to perform a sequence of steps autonomously. Over the following 15 years, I went through a sequence of calculators and computers, wrote increasingly sophisticated software, and studied computer science with a Masters degree in Artificial Intelligence (AI). My motivation in AI of course changed many times over the years, from the dream of a robot tidying up my room to more intellectual, philosophical, economic, and social motivations.

Around 1992 I lost confidence in any of the existing approaches towards AI, and despite considerable effort for years, didn't have a ground-breaking idea myself.

While working in a start-up company on a difficult image interpolation problem, I realized one day in 1997 that simplicity and compression are key, not only for solving my problem at hand, but also for the grand AI problem.

It took me quite a number of weekends to work out the details. Relatively soon I concluded that the theory I had developed was too beautiful to be novel. I had rediscovered aspects of Kolmogorov complexity and Solomonoff induction. Indeed, I had done more. My system generalized Solomonoff induction to a universally optimal general-purpose reinforcement learning agent.

In order to prove some of my claims it was necessary to become more deeply and broadly acquainted with Algorithmic Information Theory (AIT).

AIT combines information theory and computation theory to an objective and absolute notion of information in an individual object, and gives rise to an objective and robust notion of randomness of individual objects. Its major sub-disciplines are Algorithmic "Kolmogorov" Complexity (AC), Algorithmic "Solomonoff" Probability (AP), Algorithmic "Martin-Löf" Randomness (AR), and Universal "Levin" Search (UL).[17] This C&R book, mainly refers to AR.

This concludes my 25(!) year journey to C&R. In the last 10 years, I have contributed to all the 4 subfields. My primary driving force when doing research in C&R is still AI, so I've most to say about AP, and my answers to the following questions are biased towards my own personal interests. Indeed, a more objective and balanced presentation would cause a lot of redundancy in this book, which is likely not desirable.

WHAT WE HAVE LEARNED

Let me begin with what *I* have learned: The most important scientific insight I have had is the following: Many scientists have a bias towards elegant or beautiful theories, which usually aligns with some abstract notion of simplicity. Others have a bias towards simple theories in the concrete sense of being analytically or computationally tractable. With theories I essentially mean mathematical models of some aspect of the real world, e.g., of a physical system like an engine or the weather or stock market.

Way too late in my life, at age 30 or so, I realized that the most important reason for preferring simple theories is a quite different one: Simple theories tend to be better for what they are developed for in the first place, namely predicting in related but different situations and using these predictions to improve decisions and actions.

Indeed, the principle to prefer simpler theories has been popularized by William of Ockham (1285–1349) ("Entities should not be multiplied unnecessarily") but dates back at least to Aristotle.[9]

Kolmogorov complexity[25] is a universal objective measure of complexity and allows simplicity and hence Ockham's "razor" principle to be to quantified. Solomonoff[43] developed a formal theory of universal prediction along this line, actually a few years before Kolmogorov introduced his closely related complexity measure. My contribution in the 200X[15,19] was to generalize Solomonoff induction to a universally intelligent learning agent.[39]

This shows that Ockham's razor, inductive reasoning, intelligence, and the scientific inquiry itself are intimately related. I would even go so far as to say that science *is* the application of Ockham's razor: *Simple* explanations of observed real-world phenomena have a higher chance of leading to correct predictions.[21]

What does all this have to do with C&R? We cannot be certain about *anything* in our world. It might even end or be completely different tomorrow. Even if some proclaimed omniscient entity told us the future, there is no scientific way to verify the premise of its omniscience. So induction has to deal with uncertainty. Making worst-case assumptions is not a generic solution; the generic worst-case is "anything can happen". Considering restricted model classes begs the question about the validity of the model class itself, so is also not a solution. More powerful is to model uncertainty by probabilities and the latter is obviously related to randomness.

There have been many attempts to formalize probability and randomness: Kolmogorov's axioms of probability theory[24] are the default characterization. Problems with this notion are discussed in item (d) of Question 4. Early attempts to define the notion of randomness of *individual* objects/sequences by von Mises,[36] Wald,[45] and Church[4] failed, but finally Martin-Löf[37] succeeded. A sequence is Martin-Löf random if and only if it passes all effective randomness tests or, as it turns out, if and only if it is incompressible.

WHAT WE DON'T YET KNOW

Lots of things, so I will restrict myself to open problems in the intersection of AIT and AI. See Ref. 20 for details.

(i) Universal Induction: The induction problem is a fundamental problem in philosophy[7,14] and statistics[22] and science in general. The most important fundamental philosophical and statistical problems around induction are discussed in:[18] Among others, they include the problems of old evidence, ad-hoc hypotheses, updating, zero prior, and invariance. The arguments in Ref. 18 that Solomonoff's universal theory M overcomes these problems are forceful but need to be elaborated on further to convince the (scientific) world that the induction problem is essentially solved. Besides these general induction problems, universal induction raises many additional questions: for instance, it is unclear whether M can predict all computable *sub*sequences of a sequence that is itself not computable, how to formally identify "natural" Turing machines,[38] Martin-Löf convergence of M, and whether AIXI (see below) reduces to M for prediction.

(ii) Universal Artificial Intelligence (UAI): The AIXI model integrates Solomonoff induction with sequential decision theory. As a unification of two optimal theories in their own domains, it is plausible that AIXI is optimal in the "union" of their domains. This has been affirmed by positive pareto-optimality and self-optimizingness results.[16] These results support the claim that AIXI is a universally optimal generic reinforcement learning agent, but unlike the induction case, the results so far are not yet strong enough to allay all doubts. Indeed, the major problem is not to *prove* optimality but to *come up* with sufficiently strong but still satisfiable optimality notions in the reinforcement learning case. A more modest goal than proving optimality of AIXI is to ask for additional reasonable convergence properties, like posterior convergence for unbounded horizon. The probably most important fundamental and hardest problem in game theory is the grain-of-truth problem.[23] In our context, the question is what happens if AIXI is used in a multi-agent setup[42] interacting with other instantiations of AIXI.

(iii) Defining Intelligence: A fundamental and long standing difficultly in the field of AI is that intelligence itself is not well defined. Usually, formalizing and rigorously defining a previously vague concept constitutes a quantum leap forward in the corresponding field, and AI should be no exception. AIT again suggested an extremely general, objective, fundamental, and formal measure of machine intelligence,[8,11,15,27] but the theory surrounding it has yet to be adequately explored. A comprehensive collection, discussion and comparison of verbal and formal intelligence tests, definitions, and measures can be found in Ref. 30.

THE MOST IMPORTANT OPEN PROBLEMS

There are many important open technical problems in AIT. I have discussed some of those that are related to AI in Ref. 20 and the previous answer. Here I concentrate on the most important open problems in C&R which I am able to describe in non-technical terms.

(a) The development of notions of complexity and individual randomness didn't end with Kolmogorov and Martin-Löf. Many variants of "plain" Kolmogorov complexity C[25] have been developed: prefix complexity K,[3,10,29] process complexity,[41] monotone complexity Km,[28] uniform complexity,[32,33] Chaitin complexity Kc,[3] Solomonoff's universal prior

$M = 2^{-KM}$,[43,44] extension semimeasure Mc,[5] and some others.[35] They often differ only by $O(\log K)$, but this can lead to important differences. Variants of Martin-Löf randomness are: Schnorr randomness,[40] Kurtz randomness,[26] Kolmogorov-Loveland randomness,[31] and others.[2,6,46] All these complexity and randomness classes can further be relativized to some oracle, e.g., the halting oracle, leading to an arithmetic hierarchy of classes. Invoking resource-bounds moves in the other direction and leads to the well-known complexity zoo[1] and pseudo-randomness.[34] Many of these classes should be discussed in the other contributions to this book. Which definition is the "right" or "best" one, and in which sense? Current research on algorithmic randomness is more concerned about abstract properties and convenience, rather than practical usefulness. This is in marked contrast to complexity theory, in which the classes also sprout like mushrooms,[1] but the majority of classes delineate important practical problems.

(b) The historically oldest, non-flawed, most popular, and default notion of individual randomness is that of Martin-Löf. Let us assume that it is or turns out to be the "best" or single "right" definition of randomness. This would uniquely determine which individual infinite sequences are random and which are not. This unfortunately does not hold for finite sequences. This non-uniqueness problem is equivalent to the problem that Kolmogorov complexity depends on the choice of universal Turing machine. While the choice is asymptotically, and hence for large-scale practical applications, irrelevant, it seriously hinders applications to "small" problems. One can argue the problem away,[16] but finding a unique "best" universal Martin-Löf test or universal Turing machine would be more satisfactory and convincing. Besides other things, it would make inductive inference absolutely objective.

(c) Maybe randomness can, in principle, only be relative: What looks random to me might be order to you. So randomness depends on the power of the "observer". In this case, it is important to study problem-specific randomness notions, and clearly describe and separate the application domains of the different randomness notions, like classical sufficient statistics depends on the model class. Algorithmic randomness usually includes all computable tests and goes up the arithmetic hierarchy. For practical applications, limited classes, like all efficiently computable tests, are more relevant. This is the important domain of pseudo-random number generation. Could every practically useful complexity class correspond to a randomness class with practically relevant properties?

(d) It is also unclear whether algorithmic randomness or classical probability theory has a more fundamental status. While measure theory is mathematically, and statistics is practically very successful, Kolmogorov's probability axioms are philosophically crippled and, strictly speaking, induce a purely formal but meaningless measure theory exercise. The easygoing frequentist interpretation is circular: The probability of head is p, if the long-run relative frequency tends to p almost surely (with probability one). But what does 'almost surely' mean? Applied statistics implicitly invokes Cournot's somewhat forgotten principle: An event with very small probability, singled out in advance, will not happen. That is, a probability 1 event will happen for sure in the real world. Another problem is that it is not even possible to ask the question of whether a particular single sequence of observations is random (w.r.t. some measure). Algorithmic randomness makes this question meaningful and answerable. A downside of algorithmic randomness is that not every set of measure 1 will do, but only constructive ones, which can be much harder to find and sometimes do not exist.[13]

(e) Finally, to complete the circle, let's return to my original motivation for entering this field: Ockham's razor (1) is the key philosophical ingredient for solving the induction problem and crucial in defining science and intelligence, and (2) can be quantified in terms of algorithmic complexity which itself is closely related to algorithmic randomness. The formal theory of universal induction[18,44] is already well-developed and the foundations of universal AI have been laid.[16] Besides solving specific problems like (i)–(iii) and (a)–(d) above, it is also important to "translate" the results and make them accessible to researchers in other disciplines: present the philosophical insights in a less-mathematical way; stress that sound mathematical foundations are crucial for advances in most field, and induction and AI should be no exception; etc.

THE PROSPECTS FOR PROGRESS

The prospects for the open problems (a)–(e) of Question 4, I believe are as follows:

(a) I am not sure about the fate of the multitude of different randomness notions. I can't see any practical relevance for those in the arithmetic hierarchy. Possibly the acquired scientific knowledge from studying the different classes and their relationship can be used in a different field in an unexpected way. For instance, the ones in the arithmetic hierarchy may be useful in the endeavor of unifying probability and logic.[12] Possibly the whole idea of *objectively* identifying individually which strings shall be regarded as random will be given up.

(b) All scientists, except some logicians studying logic, essentially use the same classical logic and axioms, namely ZFC, to do *deductive* reasoning. Why do not all scientists use the same definition of probability to do *inductive* reasoning? Bayesian statistics and Martin-Löf randomness are the most promising candidates for becoming universally accepted for inductive reasoning.

Maybe they will become universally accepted some time in the future, for pragmatic reasons, or simply as a generally agreed upon convention, since no one is interested in arguing over it anymore. While Martin-Löf uniquely determines infinite random sequences, the randomness for finite sequences depends on the choice of universal Turing machine. Finding a unique "best" one (if possible) is, in my opinion, the most important open problem in algorithmic randomness. A conceptual breakthrough would be needed to make progress on this hard front. See[38] for a remarkable but failed recent attempt.

(c) Maybe pursuing a single definition of randomness is illusory. Noise might simply be that aspect of the data that is not useful for the particular task or method at hand. For instance, sufficient statistics and pseudo-random numbers have this task-dependence. Even with a single fundamental notion of randomness (see b) there will be many different practical approximations. I expect steady progress on this front.

(d) Bayesian statistics based on classical probability theory is incomplete, since it does not tell you how to choose the prior. Solomonoff fixes the prior to a negative exponential in the model complexity. Time and further research will convince classical statisticians to accept this (for them now) exotic choice as a kind of "Gold standard" (as Solomonoff put it). All this is still within the classical measure theoretic framework, which may be combined with Cournot or with Martin-Löf.

(e) Finally, convincing AI researchers and philosophers about the importance of Ockham's razor, that algorithmic complexity is a suitable quantification, and that this led to a formal (albeit non-computable) conceptual solution to the induction and the AI problem should be a matter of a decade or so.

References

1. S. Aaronson, G. Kuperberg, and C. Granade. Complexity zoo, 2005. http://www.complexityzoo.com/.
2. C. S. Calude. *Information and Randomness: An Algorithmic Perspective.* Springer, Berlin, 2nd edition, 2002.
3. G. J. Chaitin. A theory of program size formally identical to information theory. *Journal of the ACM*, 22(3):329–340, 1975.
4. A. Church. On the concept of a random sequence. *Bulletin of the American Mathematical Society*, 46:130–135, 1940.
5. T. M. Cover. Universal gambling schemes and the complexity measures of Kolmogorov and Chaitin. Technical Report 12, Statistics Department, Stanford University, Stanford, CA, 1974.
6. R. Downey and D. R. Hirschfeldt. *Algorithmic Randomness and Complexity.* Springer, Berlin, 2007.
7. J. Earman. *Bayes or Bust? A Critical Examination of Bayesian Confirmation Theory.* MIT Press, Cambridge, MA, 1993.
8. C. Fiévet. Mesurer l'intelligence d'une machine. In *Le Monde de l'intelligence*, volume 1, pages 42–45, Paris, November 2005. Mondeo publishing.
9. J. Franklin. *The Science of Conjecture: Evidence and Probability before Pascal.* Johns Hopkins University Press, 2002.
10. P. Gács. On the symmetry of algorithmic information. *Soviet Mathematics Doklady*, 15:1477–1480, 1974.
11. D. Graham-Rowe. Spotting the bots with brains. In *New Scientist magazine*, volume 2512, page 27, 13 August 2005.
12. H. Gaifman and M. Snir. Probabilities over rich languages, testing and randomness. *Journal of Symbolic Logic*, 47:495–548, 1982.
13. M. Hutter and An. A. Muchnik. On semimeasures predicting Martin-Löf random sequences. *Theoretical Computer Science*, 382(3):247–261, 2007.
14. D. Hume. *A Treatise of Human Nature, Book I.* [Edited version by L. A. Selby-Bigge and P. H. Nidditch, Oxford University Press, 1978], 1739.
15. M. Hutter. A theory of universal artificial intelligence based on algorithmic complexity. Technical Report cs.AI/0004001, München, 62 pages, 2000. http://arxiv.org/abs/cs.AI/0004001.
16. M. Hutter. *Universal Artificial Intelligence: Sequential Decisions based on Algorithmic Probability.* Springer, Berlin, 2005. 300 pages, http://www.hutter1.net/ai/uaibook.htm.
17. M. Hutter. Algorithmic information theory: a brief non-technical guide to the field. *Scholarpedia*, 2(3):2519, 2007.
18. M. Hutter. On universal prediction and Bayesian confirmation. *Theoretical Computer Science*, 384(1):33–48, 2007.
19. M. Hutter. Universal algorithmic intelligence: A mathematical top→down approach. In *Artificial General Intelligence*, pages 227–290. Springer, Berlin, 2007.
20. M. Hutter. Open problems in universal induction & intelligence. *Algorithms*, 3(2):879–906, 2009.

21. M. Hutter. A complete theory of everything (will be subjective). 2009. arXiv:0912.5434.
22. E. T. Jaynes. *Probability Theory: The Logic of Science*. Cambridge University Press, Cambridge, MA, 2003.
23. E. Kalai and E. Lehrer. Rational learning leads to Nash equilibrium. *Econometrica*, 61(5):1019–1045, 1993.
24. A. N. Kolmogorov. *Grundlagen der Wahrscheinlichkeitsrechnung*. Springer, Berlin, 1933. [English translation: *Foundations of the Theory of Probability*. Chelsea, New York, 2nd edition, 1956].
25. A. N. Kolmogorov. Three approaches to the quantitative definition of information. *Problems of Information and Transmission*, 1(1):1–7, 1965.
26. S. A. Kurtz. *Randomness and Genericity in the Degrees of Unsolvability*. PhD thesis, University of Illinois, 1981.
27. S. Legg. *Machine Super Intelligence*. PhD thesis, IDSIA, Lugano, 2008.
28. L. A. Levin. On the notion of a random sequence. *Soviet Mathematics Doklady*, 14(5):1413–1416, 1973.
29. L. A. Levin. Laws of information conservation (non-growth) and aspects of the foundation of probability theory. *Problems of Information Transmission*, 10(3):206–210, 1974.
30. S. Legg and M. Hutter. Universal intelligence: A definition of machine intelligence. *Minds & Machines*, 17(4):391–444, 2007.
31. D. E. Loveland. A new interpretation of von Mises' concept of a random sequence. *Zeitschrift für Mathematische Logik und Grundlagen der Mathematik*, 12:279–294, 1966.
32. D. W. Loveland. On minimal-program complexity measures. In *Proc. 1st ACM Symposium on Theory of Computing*, pages 61–78. ACM Press, New York, 1969.
33. D. W. Loveland. A variant of the Kolmogorov concept of complexity. *Information and Control*, 15(6):510–526, 1969.
34. M. Luby. *Pseudorandomness and Cryptographic Applications*. Princeton University Press, 1996.
35. M. Li and P. M. B. Vitányi. *An Introduction to Kolmogorov Complexity and its Applications*. Springer, Berlin, 3rd edition, 2008.
36. R. von Mises. Grundlagen der Wahrscheinlichkeitsrechnung. *Mathematische Zeitschrift*, 5:52–99, 1919. Correction, *Ibid.*, volume 6, 1920, [English translation in: *Probability, Statistics, and Truth*, Macmillan, 1939].
37. P. Martin-Löf. The definition of random sequences. *Information and Control*, 9(6):602–619, 1966.
38. M. Müller. Stationary algorithmic probability. *Theoretical Computer Science*, 411(1):113–130, 2010.
39. T. Oates and W. Chong. Book review: Marcus Hutter, universal artificial intelligence, Springer (2004). *Artificial Intelligence*, 170(18):1222–1226, 2006.
40. C. P. Schnorr. *Zufälligkeit und Wahrscheinlichkeit*, volume 218 of *Lecture Notes in Mathematics*. Springer, Berlin, 1971.
41. C. P. Schnorr. Process complexity and effective random tests. *Journal of Computer and System Sciences*, 7(4):376–388, 1973.

42. Y. Shoham and K. Leyton-Brown. *Multiagent Systems: Algorithmic, Game-Theoretic, and Logical Foundations.* Cambridge University Press, 2008.

43. R. J. Solomonoff. A formal theory of inductive inference: Parts 1 and 2. *Information and Control*, 7:1–22 and 224–254, 1964.

44. R. J. Solomonoff. Complexity-based induction systems: Comparisons and convergence theorems. *IEEE Transactions on Information Theory*, IT-24:422–432, 1978.

45. A. Wald. Die Widerspruchsfreiheit des Kollektivbegriffs in der Wahrscheinlichkeitsrechnung. In *Ergebnisse eines Mathematischen Kolloquiums*, volume 8, pages 38–72, 1937.

46. Y. Wang. *Randomness and Complexity.* PhD thesis, Universität Heidelberg, 1996.

Chapter 13

Randomness, Occam's Razor, AI, Creativity and Digital Physics

Jürgen Schmidhuber

IDSIA, Galleria 2, 6928 Manno-Lugano, Switzerland
University of Lugano
Switzerland, TU München, Germany
juergen@idsia.ch
http://www.idsia.ch/~juergen

WHAT DREW ME TO THE STUDY OF COMPUTATION AND RANDOMNESS

The topic is so all-encompassing and sexy. It helps to formalize the notions of Occam's razor and inductive inference,[9,10,12,19,36,37] which are at the heart of all inductive sciences. It is relevant not only for Artificial Intelligence[7,28,30,33] and computer science but also for physics and philosophy.[14,18,20] Every scientist and philosopher should know about it. Even artists should, as there are complexity-based explanations of essential aspects of aesthetics and art.[15,16,32]

WHAT WE HAVE LEARNED

In the new millennium the study of computation and randomness, pioneered in the 1930s,[5,8,9,12,36,39] has brought substantial progress in the field of theoretically optimal algorithms for prediction, search, inductive inference based on Occam's razor, problem solving, decision making, and reinforcement learning in environments of a very general type.[7,21,23,25,26,28,30,33] It led to asymptotically optimal universal program search techniques[6,22,33] for extremely broad classes of problems. Some of the results even provoke nontraditional predictions regarding the future of the universe[14,18,20,29] based on Zuse's thesis[40,41] of computable physics.[14,18,19,27] The field also is relevant for art, and for clarifying what science and art have in common.[15–17,24,31,32]

WHAT WE DON'T (YET) KNOW

A lot. It is hard to write it all down, for two reasons: (1) lack of space. (2) We don't know what we don't know, otherwise we'd know, that is, we wouldn't not know.

THE MOST IMPORTANT OPEN PROBLEMS

13.1. Constant Resource Bounds for Optimal Decision Makers

The recent results on universal problem solvers living in unknown environments show how to solve arbitrary well-defined tasks in ways that are theoretically optimal in various senses, e.g., Refs. 7 and 33. But present universal approaches sweep under the carpet certain problem-independent constant slowdowns, burying them in the asymptotic notation of theoretical computer science. They leave open an essential remaining question: If an agent or decision maker can execute only a fixed number of computational instructions per unit time interval (say, 10 trillion elementary operations per second), what is the best way of using them to get as close as possible to the recent theoretical limits of universal AIs? Once we have settled this question there won't be much left to do for human scientists.

13.2. Digital Physics

Another deep question: If our universe is computable,[40,41] and there is no evidence that it isn't,[27] then which is the shortest algorithm that computes the entire history of our particular universe, without computing any other computable objects?[14,18,29] This can be viewed as the ultimate question of physics.

13.3. Coding Theorems

Less essential open problems include the following. A previous paper[18,19] introduced various generalizations of traditional computability, Solomonoff's algorithmic probability, Kolmogorov complexity, and Super-Omegas more random than Chaitin's Omega,[1,2,35,38] extending previous work on enumerable semimeasures by Levin, Gács, and others.[3,4,11,12,42] Under which conditions do such generalizations yield *coding theorems* stating that the probability of guessing any (possibly non-halting) program computing some object in the limit (according to various degrees of limit-computability[19]) is essentially the probability of guessing its shortest program?[13,19]

13.4. Art and Science

Recent work[24,31,32] pointed out that a surprisingly simple algorithmic principle based on the notions of data compression and data compression *progress* informally explains fundamental aspects of attention, novelty, surprise, interestingness, curiosity, creativity, subjective beauty, jokes, and science & art in general. The crucial ingredients of the corresponding *formal* framework are (1) a continually improving predictor or compressor of the continually growing sensory data history of the action-executing, learning agent, (2) a computable measure of the compressor's progress (to calculate intrinsic *curiosity* rewards), (3) a reward optimizer or reinforcement learner translating rewards into action sequences expected to maximize future reward. In this framework any observed data becomes temporarily interesting by itself to the self-improving, but computationally limited, subjective observer once he learns to predict or compress the data in a better way, thus making it subjectively simpler and more *beautiful*. Curiosity is the desire to create or discover more non-random, non-arbitrary, regular data that is novel and *surprising* not in the traditional sense of Boltzmann and Shannon[34] but in the sense that it allows for compression progress because its regularity was not yet known. This drive maximizes *interestingness*, the first derivative of subjective beauty or compressibility, that is, the steepness of the learning curve. From the perspective of this framework, scientists are very much like artists. Both actively select experiments in search for simple but new ways of compressing the resulting observation history. Both try to create new but non-random, non-arbitrary data with surprising, previously unknown regularities. For example, many physicists invent experiments to create data governed by previously unknown laws allowing to further compress the data. On the other hand, many artists combine well-known objects in a subjectively novel way such that the observer's subjective description of the result is shorter than the sum of the lengths of the descriptions of the parts, due to some previously unnoticed regularity shared by the parts (art as an eye-opener). Open question: which are practically feasible, reasonable choices for implementing (1–3) in curious robotic artists and scientists?

THE PROSPECTS FOR PROGRESS

Bright. Sure, the origins of the field date back to a human lifetime ago,[5,8,9,36,39] and its development was not always rapid. But if the new millennium's progress bursts[30] are an indication of things to come, we should expect substantial achievements along the lines above in the near future.

References

1. C. S. Calude. Chaitin Ω numbers, Solovay machines and Gödel incompleteness. *Theoretical Computer Science*, 2001.
2. G. J. Chaitin. *Algorithmic Information Theory*. Cambridge University Press, Cambridge, 1987.
3. P. Gács. On the symmetry of algorithmic information. *Soviet Math. Dokl.*, 15:1477–1480, 1974.
4. P. Gács. On the relation between descriptional complexity and algorithmic probability. *Theoretical Computer Science*, 22:71–93, 1983.
5. K. Gödel. Über formal unentscheidbare Sätze der Principia Mathematica und verwandter Systeme I. *Monatshefte für Mathematik und Physik*, 38:173–198, 1931.
6. M. Hutter. The fastest and shortest algorithm for all well-defined problems. *International Journal of Foundations of Computer Science*, 13(3):431–443, 2002.
7. M. Hutter. *Universal Artificial Intelligence: Sequential Decisions based on Algorithmic Probability*. Springer, Berlin, 2004. (On J. Schmidhuber's SNF grant 20-61847).
8. A. N. Kolmogorov. *Grundbegriffe der Wahrscheinlichkeitsrechnung*. Springer, Berlin, 1933.
9. A. N. Kolmogorov. Three approaches to the quantitative definition of information. *Problems of Information Transmission*, 1:1–11, 1965.
10. L. A. Levin. Universal sequential search problems. *Problems of Information Transmission*, 9(3):265–266, 1973.
11. L. A. Levin. Laws of information (nongrowth) and aspects of the foundation of probability theory. *Problems of Information Transmission*, 10(3):206–210, 1974.
12. M. Li and P. M. B. Vitányi. *An Introduction to Kolmogorov Complexity and its Applications (2nd edition)*. Springer, 1997.
13. J. Poland. A coding theorem for enumerable output machines. *Information Processing Letters*, 91(4):157–161, 2004.
14. J. Schmidhuber. A computer scientist's view of life, the universe, and everything. In C. Freksa, M. Jantzen, and R. Valk, editors, *Foundations of Computer Science: Potential - Theory - Cognition*, volume 1337, pages 201–208. Lecture Notes in Computer Science, Springer, Berlin, 1997.
15. J. Schmidhuber. Femmes fractales, 1997.
16. J. Schmidhuber. Low-complexity art. *Leonardo, Journal of the International Society for the Arts, Sciences, and Technology*, 30(2):97–103, 1997.

17. J. Schmidhuber. Facial beauty and fractal geometry. Technical Report TR IDSIA-28-98, IDSIA, 1998. Published in the Cogprint Archive: http://cogprints.soton.ac.uk.

18. J. Schmidhuber. Algorithmic theories of everything. Technical Report IDSIA-20-00, quant-ph/0011122, IDSIA, Manno (Lugano), Switzerland, 2000. Sections 1-5: see Ref. 19; Section 6: see Ref. 20.

19. J. Schmidhuber. Hierarchies of generalized Kolmogorov complexities and nonenumerable universal measures computable in the limit. *International Journal of Foundations of Computer Science*, 13(4):587–612, 2002.

20. J. Schmidhuber. The Speed Prior: a new simplicity measure yielding near-optimal computable predictions. In J. Kivinen and R. H. Sloan, editors, *Proceedings of the 15th Annual Conference on Computational Learning Theory (COLT 2002)*, Lecture Notes in Artificial Intelligence, pages 216–228. Springer, Sydney, Australia, 2002.

21. J. Schmidhuber. Towards solving the grand problem of AI. In P. Quaresma, A. Dourado, E. Costa, and J. F. Costa, editors, *Soft Computing and complex systems*, pages 77–97. Centro Internacional de Mathematica, Coimbra, Portugal, 2003. Based on Ref. 26.

22. J. Schmidhuber. Optimal ordered problem solver. *Machine Learning*, 54:211–254, 2004.

23. J. Schmidhuber. Completely self-referential optimal reinforcement learners. In W. Duch, J. Kacprzyk, E. Oja, and S. Zadrozny, editors, *Artificial Neural Networks: Biological Inspirations - ICANN 2005, LNCS 3697*, pages 223–233. Springer-Verlag Berlin Heidelberg, 2005. Plenary talk.

24. J. Schmidhuber. Developmental robotics, optimal artificial curiosity, creativity, music, and the fine arts. *Connection Science*, 18(2):173–187, 2006.

25. J. Schmidhuber. Gödel machines: Fully self-referential optimal universal self-improvers. In B. Goertzel and C. Pennachin, editors, *Artificial General Intelligence*, pages 199–226. Springer Verlag, 2006. Variant available as arXiv:cs.LO/0309048.

26. J. Schmidhuber. The new AI: General & sound & relevant for physics. In B. Goertzel and C. Pennachin, editors, *Artificial General Intelligence*, pages 175–198. Springer, 2006. Also available as TR IDSIA-04-03, arXiv:cs.AI/0302012.

27. J. Schmidhuber. Randomness in physics. *Nature*, 439(3):392, 2006. Correspondence.

28. J. Schmidhuber. 2006: Celebrating 75 years of AI - history and outlook: the next 25 years. In M. Lungarella, F. Iida, J. Bongard, and R. Pfeifer, editors, *50 Years of Artificial Intelligence*, volume LNAI 4850, pages 29–41. Springer Berlin / Heidelberg, 2007. Preprint available as arXiv:0708.4311.

29. J. Schmidhuber. Alle berechenbaren Universen (All computable universes). *Spektrum der Wissenschaft Spezial (German edition of Scientific American)*, (3):75–79, 2007.

30. J. Schmidhuber. New millennium AI and the convergence of history. In W. Duch and J. Mandziuk, editors, *Challenges to Computational Intelligence*,

volume 63, pages 15–36. Studies in Computational Intelligence, Springer, 2007. Also available as arXiv:cs.AI/0606081.

31. J. Schmidhuber. Simple algorithmic principles of discovery, subjective beauty, selective attention, curiosity & creativity. In *Proc. 10th Intl. Conf. on Discovery Science (DS 2007), LNAI 4755*, pages 26–38. Springer, 2007. Joint invited lecture for *ALT 2007 and DS 2007*, Sendai, Japan, 2007.

32. J. Schmidhuber. Simple algorithmic theory of subjective beauty, novelty, surprise, interestingness, attention, curiosity, creativity, art, science, music, jokes. *Journal of SICE*, 48(1):21–32, 2009.

33. J. Schmidhuber. Ultimate cognition à la Gödel. *Cognitive Computation*, 2009, in press.

34. C. E. Shannon. A mathematical theory of communication (parts I and II). *Bell System Technical Journal*, XXVII:379–423, 1948.

35. T. Slaman. Randomness and recursive enumerability. Technical report, Univ. of California, Berkeley, 1999. Preprint, http://www.math.berkeley.edu/~slaman.

36. R. J. Solomonoff. A formal theory of inductive inference. Part I. *Information and Control*, 7:1–22, 1964.

37. R. J. Solomonoff. Complexity-based induction systems. *IEEE Transactions on Information Theory*, IT-24(5):422–432, 1978.

38. R. M. Solovay. A version of Ω for which ZFC can not predict a single bit. In C. S. Calude and G. Păun, editors, *Finite Versus Infinite. Contributions to an Eternal Dilemma*, pages 323–334. Springer, London, 2000.

39. A. M. Turing. On computable numbers, with an application to the Entscheidungsproblem. *Proceedings of the London Mathematical Society, Series 2*, 41:230–267, 1936.

40. K. Zuse. Rechnender Raum. *Elektronische Datenverarbeitung*, 8:336–344, 1967.

41. K. Zuse. *Rechnender Raum*. Friedrich Vieweg & Sohn, Braunschweig, 1969. English translation: *Calculating Space*, MIT Technical Translation AZT-70-164-GEMIT, Massachusetts Institute of Technology (Proj. MAC), Cambridge, Mass. 02139, Feb. 1970.

42. A. K. Zvonkin and L. A. Levin. The complexity of finite objects and the algorithmic concepts of information and randomness. *Russian Math. Surveys*, 25(6):83–124, 1970.

PART IV

Randomness, Information and Computability

Chapter 14

Randomness Everywhere: My Path to Algorithmic Information Theory

Cristian S. Calude

Department of Computer Science, 38 Princes Street,
University of Auckland, New Zealand
cristian@cs.auckland.ac.nz
www.cs.auckland.ac.nz/~cristian

WHAT DREW ME TO THE STUDY OF COMPUTATION AND RANDOMNESS

I initially worked in computability theory and Blum's abstract complexity theory. In the late 1970's I started a weekly research seminar on AIT at the University of Bucharest which lasted till my departure in 1992. The main participants were Şerban Buzeţeanu, Cezar Câmpeanu, Ion Chiţescu, Nelu Dima, Nicolae Duţă, Gabriel Istrate, Ion Macarie, Ion Măndoiu, Mihaela Maliţa, Ana-Maria Sălăjean (Mandache), Ileana Streinu, Monica Tătărăm, and Marius Zimand.

The initial motivation of my interest in AIT came from two papers, one by Per Martin-Löf and one by Martin Davis. Together with my colleague Ion Chiţescu I read Martin-Löf's famous 1966 paper on the definition of random sequences. We thought a lot about the last section of the paper where the theory is extended from the uniform probability distribution to an arbitrary computable probability distribution and we found examples of primitive recursive infinite sequences which are Martin-Löf random according to simple computable probability distributions. If one wishes to preserve the intuition of randomness — incomputability seems to be a condition any general definition of randomness has to satisfy — we should exclude computable probability distributions that allow these pathological examples. The paper was accepted to the *Workshop on Recursion Theoretical Aspects of Computer Science, Purdue University* (1981), but we couldn't attend the meeting due to restrictions imposed on academic travel by the communist government of Romania of that time. We published it in

C. Calude, I. Chiţescu (On Per Martin-Löf random sequences, *Bull. Math. Soc. Sci. Math. R. S. Roumanie* (N.S.) 26 (74) (1982), 217–221).

In the AIT seminar we read Davis' paper "What is a computation?" (published in L. A. Steen, *Mathematics Today: Twelve Informal Essays*, Springer-Verlag, New York, 1978, 241–267. Reprinted in C. S. Calude (ed.) *Randomness & Complexity, from Leibniz to Chaitin*, World Scientific, Singapore, 2007, 89–122) which gives a beautiful presentation of one of Chaitin's information-theoretic forms of incompleteness. The new approach to incompleteness and Chaitin's Omega number have been a source of inspiration for my own work.

In January 1993 Greg Chaitin was my first visitor in Auckland. Since then, I have been privileged to meet him regularly, to understand some of his results not from papers, not from books, but from stimulating discussions, and to cooperate on different projects (including two joint papers, Randomness everywhere, *Nature* 400, 22 July (1999), 319–320)* and What is ... a halting probability? *Notices of the AMS* 57, 2 (2010), 236–237.)

2002: with G. Chaitin.

I have started to work in AIT when only a handful of people were active in the field. Today, AIT is vastly more popular as one can see by the wealth of results presented in dedicated workshops and conferences, some collected in two recent books: A. Nies. *Computability and Randomness*, Oxford University Press, 2009 and R. Downey, D. Hirschfeldt. *Algorithmic Randomness and Complexity*, Springer, 2010.

---

*This note inspired a poem R. M. Chute (Reading a note in the journal Nature I learn, *Beloit Poetry Journal* 50 (3), Spring 2000, 8; also in the book by the same author, *Reading Nature*, JWB, Topsham, 2006).

Working in AIT offered me the chance not only to learn a good piece of mathematics, but also a bit of programming (in a register machine language and Mathematica[†]) and quantum physics (is quantum randomness algorithmic random?), and, more importantly, the privilege to co-operate with eminent colleagues and friends: Alastair Abbott, Azat Arslanov, Gregory Chaitin, Cezar Câmpeanu, Ion Chiţescu, Richard J. Coles, Michael J. Dinneen, Monica Dumitrescu, Cristian Grozea, Nicholas Hay, Peter Hertling, Hajimi Ishihara, Gabriel Istrate, Helmut Jürgensen, Bakhadyr Khoussainov, Eva Kurta, Shane Legg, J. P. Lewis, Giuseppe Longo, Solomon Marcus, F. Walter Meyerstein, André Nies, Thierry Paul, Gheorghe Păun, Tania K. Roblot, Arto Salomaa, Kai Salomaa, Chi-Kou K. Shu, Ludwig Staiger, Mike A. Stay, Frank Stephan, Karl Svozil, Monica Tătărâm, Sebastiaan A. Terwijn, Ioan Tomescu, Yongee Wang, Klaus Weihrauch, Takeshi Yamaguchi, Tudor Zamfirescu, Marius Zimand.

1992: with G. Păun, A. Salomaa and S. Marcus.

[†]The computation of exact bits of a natural Omega number in C. S. Calude, M. J. Dinneen and C.-K. Shu. Computing a glimpse of randomness, *Experimental Mathematics* 11, 2 (2002), 369–378 and C. S. Calude, M. J. Dinneen. Exact approximations of omega numbers, *Int. Journal of Bifurcation & Chaos* 17, 6 (2007), 1937–1954 has attracted some attention, sometimes in an unexpected form: the musical piece entitled "For Gregory Chaitin," by Michael Winter, a composer living in Los Angeles, "realizes the binary expansion of Omega" from our first paper. He explains in the score that "the sequence of digits are read linearly, '0' and '1' representing two distinct [sonic] events."

C. S. Calude

2007: with L. Staiger.

2006: with N. Hay, S. Legg and M. Zimand.

2007: with G. Chaitin and K. Svozil.

I was very lucky to meet some early contributors to AIT: in chronological order of our first meetings, Vladimir Uspenskii (Varna, 1986), Greg Chaitin (Auckland 1993), Per Matin-Löf (Vienna, 1994), Ray Solomonoff (Istanbul, 2006), Claus-Peter Schnorr (Daghstul, 2006).

2002: with V. Uspensky and son.

2006: with R. Solomonoff.

Unfortunately I also learnt that AIT politics can be as ruthless as party politics: for one's sanity one should not take it too seriously.

WHAT WE HAVE LEARNED

On April 19, 1988 *The New York Times* published the article "The Quest for True Randomness Finally Appears Successful" by James Gleick. It started with the statement:

> One of the strangest quests of modern computer science seems to be reaching its goal: mathematicians believe they have found a process for making perfectly random strings of numbers.

Is this claim true? The claim is too vague to be technically refuted. Pseudo-randomness, that recently is extremely good (see for example Mathematica 7 six pseudo-random generators described at http://reference.wolfram.com/mathematica/tutorial/RandomNumberGeneration.html) is Turing computable, so not random in AIT sense. Some forms of pseudo-randomness can be described by versions of AIT using weaker machines. For example, if we replace Turing machines with finite transducers we can obtain a nice characterisation of "finite-state random sequences": they are exactly the Borel normal sequences (V. N. Agafonov. Normal sequences and finite automata, *Soviet Mathematics Doklad* 9 (1968), 324–325; a finite-state theory of finite random strings was proposed in C. S. Calude, K. Salomaa, Tania K. Roblot. Finite-State Complexity and Randomness, *CDMTCS Research Report* 374, 2009, 24 pp).

Randomness is a fascinating but elusive concept, appearing with various meanings in different fields, from natural sciences to finance and politics, and from mathematics to religion.

> *Randomness is the very stuff of life, looming large in our everyday experience*

wrote E. Beltrami in his book *What Is Random?* (Copernicus, NY, 1999, p. xi).

Randomness has appeared since pre-historic times. We can find it, for example, in Homer's *Iliad*, in writings by Aristotle and Epicurus and in the Old Testament, in spite of the fact that gods were believed to control every single detail of life (a precursor of what later became the cosmological hypothesis of determinism). Chance was part of God's power as explained by Aristotle in *Physics Book II*, 4:

> *Others there are who believe that chance is a cause but that it is inscrutable to human intelligence as being a divine thing and full of mystery.*

Many popular perceptions of randomness are based on logical errors. The typical argument about "due numbers" goes by saying that because all numbers (from a finite set) have to appear in a random sequence, a number which has not appeared yet has more chances to come up — it's due — than a number that has already appeared, even no number is removed from the sequence after its occurrence. The same type of error appears in identifying the opposite effect: "cursed numbers" are numbers which will appear less often because they have come up less in the past.

In physics randomness has appeared in statistical mechanics, and, more importantly, it's an essential part of the foundations of quantum mechanics. Probability theory formulates mathematical laws for chance events: randomness is assumed, but not defined. AIT proposed definitions for (algorithmic) random finite strings and infinite sequences which are compatible with the requirements used in probability theory and statistics (see C. S. Calude. Who is afraid of randomness?, in E. von Collani (ed.). *Defining the Science of Stochastics*, Heldermann Verlag, 2004, 95–116). Initially, AIT was a subject at the intersection of computability, probability and information theories with applications in mathematical logic. It evolved in various directions, particularly interesting for me being its intersections with experimental mathematics, quantum physics and formal proofs.

Is algorithmic randomness a faithful model of randomness? Certainly, algorithmic randomness is a natural, adequate model of randomness, which captures many aspects one naturally associates with randomness. As any model, it has limits. Understanding randomness will never be complete, it's an unending process to which AIT is contributing in a significant way.

For a long time randomness was perceived as a bad phenomenon, to be fooled by[‡] or to fear. The cosmological hypothesis of determinism excludes randomness from the universe; only unpredictability is allowed. The opposite hypothesis, the universe is random, proposed and studied in C. Calude, A. Salomaa. Algorithmically coding the universe, in G. Rozenberg, A. Salomaa (eds.). *Developments in Language Theory*, World Scientific, Singapore, 1994, 472–492, and C. S. Calude, F. W. Meyerstein. Is the universe lawful? *Chaos, Solitons & Fractals* 10, 6 (1999), 1075–1084, has interesting consequences, many positive.

Actually, the more we understand randomness the more we learn how to use it because of its intrinsic "fairness." This trend can be seen also in AIT, where most initial applications were negative and only recently positive applications emerged. My favourite mathematical application refers to probabilistic algorithms. In the first and simplest form the result applies to probabilistic checking of primality. To probabilistically test whether an integer $n > 1$ is prime, we take a sample of k natural numbers uniformly distributed between 1 and $n - 1$, inclusive, and for each chosen i we check whether a simple predicate $W(i, n)$ holds. If $W(i, n)$ holds for at least one i then n is composite. Otherwise, the test is inconclusive; in this case if one declares n to be prime then the probability to be wrong is smaller than 2^{-k}. Predicates W have been proposed by Miller-Rabin and Solovay-Strassen. G. Chaitin and J. Schwartz (A note on Monte Carlo primality tests and algorithmic information theory, *Comm. Pure and Applied Math.* 31, 4 (1978), 521–527) proved that if the sample is not only uniformly distributed, but algorithmically random, then for almost all n the algorithm returns the correct answer.[§] The result extends to most "decent" probabilistic algorithms as shown in C. Calude, M. Zimand. A relation between correctness and randomness in the computation of probabilistic algorithms, *Internat. J. Comput. Math.* 16 (1984), 47–53.

[‡]A book like N. N. Taleb. *Fooled by Randomness: The Hidden Role of Chance in Life and in the Markets* (Texere, NY, 2001) sells very well in time of global recession.
[§]Why do we need probabilistic algorithms when it is known that primality testing is in P? Because no known deterministic algorithm is "fast enough" for practical use.

From Pythagoras and Euclid to Hilbert and Bourbaki, mathematical proofs were essentially based on axiomatic-deductive reasoning. A *formal proof* is written in a formal language consisting of certain strings of symbols from a fixed alphabet. Formal proofs are precisely specified without any ambiguity because all notions are explicitly defined, no steps are omitted, no appeal to any kind of intuition is made. They satisfy Hilbert's criterion of mechanical testing:

> *The rules should be so clear, that if somebody gives you what they claim is a proof, there is a mechanical procedure that will check whether the proof is correct or not, whether it obeys the rules or not.*

An *informal proof* is a rigorous argument expressed in a mixture of natural language and formulae that is intended to convince a knowledgeable mathematician of the truth of a statement, the theorem.

In theory, each informal proof can be converted into a formal proof. However, this is rarely, almost never, done in practice. For some mathematicians Gödel's incompleteness theorem killed Hilbert's dream to base mathematics on formal proofs. This may be true "in theory," but not in the current practice of mathematics. Due to the advent of computer technology, formal proofs are becoming increasingly used in mathematics: for a comprehensive overview see T. C. Hales. Formal proof, *Notices of the AMS*, 11 (2008),1370–1380. The first formal proofs in AIT were proposed in C. S. Calude, N. J. Hay. Every computably enumerable random real is provably computably enumerable random, *Logic Jnl. IGPL* 17 (2009), 325–350. The motivation came from a problem suggested by R. Solovay regarding the provability of randomness in Peano Arithmetic, a problem that "begged" a formal proof. Using the proof-assistant Isabelle formal proofs for the Kraft-Chaitin theorem (see http://www.cs.auckland.ac.nz/~nickjhay/KraftChaitin.thy) and the theorem that *left-computable random reals are provable random in Peano Arithmetic* have been obtained. In fact, Isabelle was used not only as a system for writing formal proofs, but also as an assistant in the process of discovering those proofs. A more general result, namely, *a real is left-computable ε–random iff it is the halting probability of an ε–universal prefix-free Turing machine*, which has as consequence that *left-computable ε–random reals are provable ε–random in Peano Arithmetic* was proved by "pen-on-paper" in C. S. Calude, N. J. Hay, F. C. Stephan. Representation of Left-Computable ε–Random Reals, *CDMTCS Research Report* 365, 2009, 11 pp, and formally by N. Hay, http://www.cs.auckland.ac.nz/~nickjhay/SolovayRepresentation.thy.

WHAT WE DON'T (YET) KNOW

We don't have a theory of complexity of computably enumerable sets (see G. Chaitin. Algorithmic entropy of sets, *Computers & Mathematics with Applications* 2 (1976), 233–245, R. M. Solovay. On random r.e. sets, in A. Arruda , N. Da Costa, R. Chuaqui (eds.). *Non-Classical Logics, Model Theory, and Computability, Proceedings of the Third Latin-American Symposium on Mathematical Logic,* Campinas, July 11–17, 1976, Studies in Logic and the Foundations of Mathematics, Vol. 89, North-Holland, Amsterdam, 1977, 283–307). At the opposite pole, the finite-state theory of randomness is still in its early days.

We don't know so many other interesting things! An excellent collection of mathematical open problems is discussed in J. Miller and A. Nies. Randomness and computability: Open questions, *Bull. Symb. Logic.* 12, 3 (2006) 390–410; an updated version is available from http://www.cs.auckland.ac.nz/~nies.

THE MOST IMPORTANT OPEN PROBLEMS IN THE FIELD

Each researcher has a set of favourite open problems. I am mainly interested in

(1) *The relations between AIT and Gödel's incompleteness.*
(2) *The degree of randomness of quantum randomness.* As quantum randomness certified by value-indefiniteness[¶] is incomputable, cf. C. S. Calude, K. Svozil. Quantum randomness and value indefiniteness, *Advanced Science Letters* 1 (2008), 165–168, it is important to study

  • *finite tests "certifying" a specific randomness behaviour of a string of quantum random bits* (cf. C. S. Calude, M. Dinneen, M. Dumitrescu, K. Svozil. How Random Is Quantum Randomness? (Extended Version), *CDMTCS Research Report 372,* 2009, 70 pp.), and

  • *the quality of simulation of computationally intensive processes using quantum random bits.*

(3) *AIT results that can be formally proved in Peano Arithmetic or weaker formal theories.*

[¶]There cannot exist any consistent, context independent, omniscience for quantum systems with three or more mutually exclusive outcomes, cf. N. D. Mermin. Hidden variables and the two theorems of John Bell, *Reviews of Modern Physics* 65 (1993), 803–815.

THE PROSPECTS FOR PROGRESS

Any theory of randomness has to prove, sooner or later, its relevance to various areas which naturally operate with the informal concept of randomness, e.g., biology (see, for example, G. Longo. Randomness and determination, from physics and computing towards biology, in M. Nielsen et al. (eds.). *SOFSEM 2009*, LNCS 5404, Springer, Berlin 2009, 49–61), physics (see K. Svozil. *Randomness & Undecidability in Physics*, World Scientific, Singapore, 1993), game theory, random sampling, random noise, etc. It is expected that AIT will continue to flowerish from its own mathematical questions, but also in connection with applications in other subjects.

Chapter 15

The Impact of Algorithmic Information Theory on Our Current Views on Complexity, Randomness, Information and Prediction

Peter Gács

Computer Science Department
Boston University
gacs@bu.edu

WHAT DREW ME TO THE STUDY OF COMPUTATION AND RANDOMNESS

I learned about algorithmic information theory first as a student when I happened on Kolmogorov's 1965 and 1968 articles on the open shelves of the library of the Hungarian Mathematical Institute. I taught myself computability theory in order to understand what Kolmogorov was writing about. It seems like all the main questions that algorithmic information theory was trying to answer were interesting to me at the time:

- What is individual information?
- What is individual randomness?
- What prediction methods are worth considering?

I must have had some philosophical interests originally (maybe just from an inherited obligation, at the time, of understanding Marxism...). It had cost me much work to realize how hopelessly muddled is a lot of what passes for philosophy, but recipes of sterile pedantry to repair it did not appeal to me either. Mathematical logic, probability theory, statistics, information theory, and then later algorithmic information theory introduce mathematical models that, when intelligently applied, bring clarity and new understanding in areas that seemed earlier to belong "only" to philosophy.

Within mathematics, I have always felt that my strength is a kind of stubbornness, not letting go until complete understanding. Algorithmic information theory probably attracted me since it promised some personal advantages in an area that requires both conceptual work and a feel for quantitative relations: for example logic and probability theory.

Looking back, personally it seems that almost each of my (very few) papers in this area is just the solution of some technical problem posed by others, or is based on the observation of a rather technical connection. I did not contribute much to creating new concepts or new problems. The philosophical attraction of algorithmic information theory just helped feeling good while working on technical questions (say, distinguishing between two slightly different versions of complexity) or doing expository work.

WHAT WE HAVE LEARNED

I will interpret this question as "what have we learned about issues of some significance going beyond internal technical matters of the field"? This question invites a lot of critical winnowing of a by now considerable mass of mathematical results: strictly speaking, each new theorem provides new information.

Limiting to unlimited computational resources   If the original questions I quoted above are taken to define algorithmic information theory then it becomes unmanageably broad: some very important answers to the original questions behind algorithmic information theory: what is randomness? what is information? have to do with polynomial computability, pseudorandomness, extraction of randomness, and so on. Probably more exciting developments occurred in these areas of computer science. However, in what we generally understand under algorithmic information theory, computation time is not limited.

Asymptotic point of view   By now, the asymptotic point of view pervades theoretical computer science. This has not always been so: when the invariance theorem of description complexity has been proved by Kolmogorov and Solomonoff, the additive constant in it (and in almost all the inequalities of the theory) roused controversy, and is still causing it among many critics of the field. It is fair to say that algorithmic information theory played a pioneering role teaching the fruitfulness of an asymptotic point of view going beyond the mutual relations of some functions, and relating to a whole web of relations among finite objects.

Individual objects versus probability distributions   Before the development of algorithmic information theory, information-theoretical questions could be rigorously formulated only for probability distributions, not for individual objects. Concerning randomness, there were earlier attempts to formulate randomness for individual (infinite) sequences, but without enough connection to other properties to make them interesting. Even though we will never know the description complexity of a large individual sequence, the rigorous definability of this and related properties boosts our intuition and informs us in coming up with new concepts and algorithms.

Size of description   One of the fundamental discoveries of algorithmic information theory is that in many applications, the appropriate simplicity/complexity scale is along the dimension that is the *size of description*. This has not always been evident, even in the theory of randomness. For example Laplace came close to understanding individual randomness when he wrote that the number of "regular" (that is nonrandom) sequences is small. But he defined "regular" as "regular, or in which we see an order holding sway that is easy to grasp". Our current understanding is that the crucial notion of "regularity" in the context of nonrandomness includes shortness of description, but not necessarily easiness to grasp. When the latter criterion is also introduced, we step into the realm of pseudorandomness theory.

One of the original applications of the theory is, and continues to be, inductive inference. Whether it is justified to call "Occam's Razor" or not, there is a principle saying that in science, among candidate laws fitting a given body of experimental data, the simplest one should be chosen. Setting aside for a moment the issue of justification of the principle, a rigorous thinker will immediately raise the question of its meaning, interpretation. For example, what is "simplest"? Ray Solomonoff not only settled on the meaning "shortest description", but also gave what is probably still the most attractive inference formula based on this idea. There are many problems with this uncomputable (and not even semicomputable) formula, and since then many other, more practical inference methods were created, of course also more restricted in their area of applicability. But the issue of simplicity, frequently in the form of control against overfitting, has remained paramount. In view of the great variety of these methods, let me return to the question of the original justification and risk a general statement. As a minimum, when searching for the real law behind an ever increasing body of experimental data, we must be sure that sooner or later we will find it. If we keep jumping to more and more complex candidates, then we may overlook forever the real, but relatively simple law. In order to be sure, we must list *all* possible laws in some order. This modest requirement still allows for many different definitions of simplicity used to define the order, but is not vacuous. I suspect that it will be hard to find general agreement on anything more specific.

Monotonic computations, universal objects Computably enumerable sets, lower or upper semicomputable functions, play an important role in many of the central papers of algorithmic information theory. Part of the reason is that these computability concepts frequently allow some kind of universal or dominating object. Description complexity is the first example: complexity with respect to the optimal machine is smallest within an additive constant. Then there are universal tests, universal semimeasures. These features of the theory were not so evident in the original ideas leading to algorithmic information theory, and by far not in all results of algorithmic information theory are they prominent. But they are in its most attractive results, so by now it is fair to say that we "learned" the value of building structures in which monotonicity and universality can be found, and that these features are characteristic of the theory. Where semicomputability is not applicable, say in the concept of information (at least between infinite sequences), algorithmic information theory has been less successful at coming up with compelling definitions and results.

WHAT WE DON'T (YET) KNOW

By now the theory is quite mature. It has, of course, many internal, technical problems and will generate many new ones. But I interpret the question in a more basic way: what are some of the questions that have motivated the origins of the theory and are still not solved? Let me list a couple.

I am interested in understanding uniform tests of randomness, the relation between the role of universal measures, neutral measures and their connection to information better than it is understood now.

Much of algorithmic information theory is confined to the study of sequences, but I feel that sequences are only an initial and simplest model for a situation in which information is revealed gradually about a (finite or infinite) object. Even in case of sequences, one is forced to consider this issue in connection with randomness tests that depend on the measure in a semicomputable way. Relation between description complexity and probability is so much at the core of the original questions of algorithmic information theory, that I consider important to clear up the question of how far this relation is conserved in these more complex situations.

THE MOST IMPORTANT OPEN PROBLEMS AND THE PROSPECTS FOR PROGRESS

Even if maybe these are the questions that most interest the readers of the present book, I do not feel qualified to offer authoritative answers to them. In my own research, I either just worked on problems of others that I felt on the one hand challenging and interesting, and on the other hand able to solve, or worked on issues which I felt needed conceptual clarification.

By now the most active group of researchers pursuing algorithmic information theory handle it as an area that just helps enrich the arsenal of methods and questions of recursion theory. Despite some of the impressive results already achieved, these questions are not what would have attracted me to the field in the first case. Though I avoided naming names until now, the issues concerning complicated information and reducibility relations between finite objects, often seen in works of the Russian school around the late Andrei Muchnik, are closer to my heart.

Mostly, instead of predicting specific developments, I would declare the field still having the potential for surprise, for the kind of result that could not be predicted since we have not even known what questions to ask. For example, thinking about reversible computation in the early 1990-ies (triggered by earlier concern with thermodynamics, a field in much need of conceptual help from algorithmic information theory), led to the observation that information distance is the maximum of two mutual complexities instead of their sum. The result surprised even those who obtained it.

Chapter 16

Randomness, Computability and Information

Joseph S. Miller

Department of Mathematics
University of Wisconsin
jmiller@math.wisc.edu

WHAT DREW ME TO THE STUDY OF COMPUTATION AND RANDOMNESS

It is easy to go back and find the signposts that seem to point to the present. By only telling the right parts of my story, I can make it sound like I was destined to become a computability theorist studying randomness. The truth is that the outcome was never clear to me, but there are certainly signposts worth mentioning.

My interest in computation goes back to the Texas Instruments TI-99/4A that my parents bought when I was ten. I learned to program (in BASIC) because that's what you did with home computers back then. I think that anybody with years of programing experience has an innate understanding of what it means for something to be *computable*; I certainly don't think I ever gave the notion — subtle in Alan Turing's time — a moment's thought. A function is computable if you can write a computer program to implement it. That's all there is to it.

I got bachelor's degrees in computer science and math; the former mostly as a precaution, but I enjoyed both programming (especially as a pastime) and theory. When I was in college I happened to read large parts of two of Gregory Chaitin's books: "Algorithmic Information Theory" and "Information-theoretic Incompleteness". As I recall, I stumbled across them in the library; I have no idea what I was looking for at the time. The books were quite interesting. The key insight is to define the complexity (or *information content*) of a binary string as the length of the shortest program that outputs the string. Simple strings can be compressed and complicated stings cannot. It's an elegant idea. There are various forms of *Kolmogorov complexity*, but they are all are defined in this way, with slightly different interpretations of "shortest program". Chaitin developed the theory of (algorithmic) information and randomness, and connected it to Gödel's incompleteness theorem, a fundamental result in the foundations of mathematics. As much as I enjoyed these themes, I never expected that I would do technical research in the field, nor even that there was an active field in which to work.

Like many in my age group, I had already been exposed to Gödel's theorem by Douglas Hofstadter's delightful "Gödel, Esher, Bach". This book had a large impact on me and it's unlikely that I would have taken a two semester graduate *Introduction to Logic* when I was in college were it not for Hofstadter. I enjoyed the course, but went to grad school certain that logic was not my subject. Despite that, I attended the logic seminar (partly because I liked the logic graduate students). Soon I was speaking in the seminar and had chosen a logician (Anil Nerode) as my advisor. Even then I meandered through a variety of possible thesis topics, finally settling on computable analysis, a hybrid of computability theory and classical mathematics. This suited me because I was still somewhat in denial about having become a mathematical logician.

I got my degree just as many computability theorists were developing a serious interest in randomness. Rod Downey invited me to spend a year with him as a postdoc in New Zealand and I was enthusiastic because he was writing "Algorithmic Randomness and Complexity" with one of my friends from graduate school, Denis Hirschfelt. Randomness offered a rich mix of computability theory and mathematics, not entirely unlike the subject of my dissertation. I welcomed the chance to jump into what had become an active, exciting field.

WHAT WE HAVE LEARNED

Those of us who study randomness use the word *information* is a way that is at odds with the colloquial meaning; one thing we have learned is that the two meanings are not just different, in some ways they are negatively correlated. First, we must understand the difference. As an analogy, say you tuned your radio to static. After ten minutes of listening to white noise, it is highly unlikely that you would have learned anything about current events, why dolphins don't drown when they sleep,* or any other subject. If

---

*This is apparently because only half of a dolphin's brain is asleep at any given time.

you want to learn something, you would do better listening to ten minutes of NPR. Colloquially, we would say that NPR contains information and radio static doesn't. To distinguish it from the other notion, let's call this kind of information "useful information".

Now assume that you have recorded the two ten minute segments as compressed audio files.[†] The radio static should be much *less* compressible than the ten minutes of NPR. Why? Because it is random noise and compression is only possible when there are patterns to exploit. So, in the sense of Kolmogorov complexity, the white noise contains *more* information.

Let's turn to the mathematical context, drawing on our analogy to help understand the distinction between the two notions of information. Consider an infinite binary sequence X. For example, we might have $X = 00100100001111110\ldots$, continuing without end and perhaps with no discernible pattern. In computability theory, our measure of "useful information" is the *computational power* of X, in other words, what can be computed *from* X. Say that we have the binary sequence X written on an (infinite) hard drive. With this drive plugged into our computer, we can write programs that have access to X. If X is computable, these programs can't do anything that we couldn't do without the information on the hard drive.[‡] But if X is not computable, we can write programs that do new things. As a trivial example, there is a program that computes X by simply copying it off the drive. More generally, it is possible that, using X, we can compute another sequence Y (which may not, itself, be computable). In that case we say that Y is *computable from* X or that Y is *Turing reducible to* X. It is also possible, for example, that X computes a function that grows faster than every computable function. Such sequences form an important class in computability. There are plenty of other ways that X could be computationally useful, in each case allowing us to do something that we couldn't do without access to the hard drive containing X.

[†]The most commonly used forms of audio compression do not perfectly preserve the uncompressed audio. By making changes that are (ideally) not detectable by the listener, they allow for much more efficient compression. For our example it is best to assume we are using *lossless* compression. The fact that white noise contains so little information in the colloquial sense is closely related to the fact that we cannot perceptually distinguish two similar white noise samples. This in turn allows for efficient compression of white noise if all we want to do is maintain the perceptual experience.

[‡]It is entirely possible that access to a computable sequence X would allow us to perform some computations much faster than would otherwise be possible, but we are not interested in the efficiency of computation here; a program that won't halt until well after the sun has burnt itself out is still a perfectly good program from our perspective.

Next we want to capture the randomness-theoretic information contained in X. For that we use the Kolmogorov complexity of its (finite) initial segments. As was already mentioned, the Kolmogorov complexity of a string is the length of the most compressed form of the string. Think of it as the length of the irreducible part; what's left when all redundancy and pattern is removed. Hopefully, you see that it makes sense to call this *information*, even though it is quite different from what we meant by "information" in the previous paragraph.

If none of the initial segments of X can be compressed (beyond some fixed amount and using an appropriate variation of Kolmogorov complexity), we say that X is *Martin-Löf random*.[§] By this definition, a random sequence has a lot of information, in the sense that its initial segments cannot be generated by programs shorter than themselves. This is an elegant definition, and useful if we want to talk about the two meanings given to "information". It may not be immediately clear what it has to do with our intuitive notion of randomness, which has more to do with flipping coins and rolling dice. However, a Martin-Löf random sequence is guaranteed to "look like" a sequence you would get by flipping a coin and assigning 1 to heads and 0 to tails. It is worth a digression to understand the connection. If you flip a coin 1,000 times, you would expect roughly 500 heads. How can this be if every possible sequence of heads and tails is equally likely? The reason is that the vast majority of sequences of 1,000 heads and tails have roughly 500 heads, so the probability that the sequence you generate will have this property is very high. Just to put numbers to this observation, if you tried this experiment one trillion times, you would expect to see fewer than 400 or more than 600 heads in only about 180 of those trials! What does this have to do with compression? Since there are comparatively few sequences of length 1,000 with fewer than 400 or more than 600 ones, each of these can be assigned as the output to a relatively short program. Hence, these pathological sequences have Kolmogorov complexity less than their lengths. So a Martin-Löf random sequence will, if you keep a running count, have roughly the same number of ones as zeros. Other properties that you would expect a randomly generated sequence — one generated by flipping a fair coin — to have can be analyzed in the same way. The point is that Martin-Löf random sequences look random in exactly the way you would hope.

[§] This is not actually how Per Martin-Löf defined his notion of randomness in 1966. It is an approach to the same notion of randomness that was developed by Leonid Levin, Gregory Chaitin and Claus-Peter Schnorr in the seventies.

Now that they have both been described, it should be clear that randomness-theoretic information and computability-theoretic information are very different notions. But remember our goal is to show that the two meanings are not just different, but at odds with each other. Let us return to our radio analogy. Taking Martin-Löf random sequences as our analog of white noise and using the computational power of X as a measure of its useful-information, how well does our analogy fit the formal situation? Not as well as we might hope; Péter Gács and Antonín Kučera independently proved that there is no limit on the computational power of Martin-Löf random sequences. In other words, whatever you want to compute, there is a random sequences that computes it, so Martin-Löf randoms *can* contain useful information. One weakness in the analogy is that we are allowing ourselves to work much harder to extract this information than one is willing to work when listening to the radio. The bigger weakness — one that will let us rescue our analogy when it is resolved — is that a Martin-Löf random sequence X is not "absolutely random", whatever that might mean, but only random to the extent that computer programs cannot exploit patterns in X to compress its initial segments. What we will argue is that the more random a sequence is, the less computationally useful, *and conversely* the less computationally useful a random sequence is, the more random. But for this we have to understand what we might mean by "more random".

One way to strengthen the notion of Martin-Löf randomness is to draw on our earlier discussion of computability theory. Say that our programs have access to a sequence Z written on an (infinite) hard drive. If we cannot compress the initial segments of a sequence X *using* Z, then we say that X is Martin-Löf random *relative to Z*. This gives us a stronger notion of randomness, depending on the computational power of Z. This is not the only way that we strengthen Martin-Löf randomness in practice, but it will be enough to make our point.

We can now give examples illustrating the negative correlation between the two types of information. First, assume that X computes Y and that both sequences are Martin-Löf random. This assumption places an *upper* bound on the computability-theoretic information in Y: it cannot be more computationally useful than X because anything that Y computes, X also computes. On the other hand, it turns out that Y inherits X's randomness in the sense that if X is random relative to Z, then so is Y. This can be seen as a *lower* bound on the degree of randomness of Y: any Z that compresses Y must also compress X. So we have evidence supporting the claim that the less computationally useful a random sequence is, the more random.

For the other direction, assume that the Kolmogorov complexity of the initial segments of X always exceeds that of Y, and again that both are random. In this case we have placed an *upper* bound on the randomness-theoretic information in Y. Do we get a corresponding *lower* bound on the computability-theoretic information? We do; it turns out that Y compresses every sequence that X compresses. In other words, Y is at least as useful in exploiting patterns in other sequences as X is. Thus we have evidence that the less random a sequence is, the more computationally useful. There are many other technical results that flesh out our claim that randomness-theoretic and computability-theoretic information are not just very different notions, but for sufficiently random sequences, inversely related.

Something very interesting happens at the other extreme. The sequences that have the lowest possible randomness-theoretic information turn out to be exactly the sequences that are computationally useless in a specific sense. In other words, if we look all the way at the low information end of the spectrum, the two types of information seem to coincide. This should make more sense if we return to our radio analogy. Consider a ten minute long pure tone. Like white noise, this is bad programming and contains no *useful* information. On the other hand, if you were to record this segment as a compressed audio file, you would find that it takes up very little space, much less even than the ten minute segment recorded from NPR. So it contains very little information, no matter which meaning we use.

Let's turn to the mathematical context. We say that X is *K-trivial* if each initial segments of X has the lowest possible Kolmogorov complexity of any finite string of that length (up to a fixed tolerance and using an appropriate variation of Kolmogorov complexity). These are the sequences with no randomness-theoretic information. We have already identified the sequences with absolutely no computability computability-theoretic information; these are just the computable sequences. It isn't hard to see that every computable sequence is K-trivial. However, Robert Solovay constructed a noncomputable K-trivial sequence. Once again, we have to be careful to rescue our analogy. We need to restrict our attention to a specific kind of computation. We say that X is *low-for-random* if every Martin-Löf random sequence is Martin-Löf random *relative to X*. In other words, having X doesn't let you compress anything that isn't already compressible.

While a low-for-random sequence might be useful for certain types of computations, they are assumed to be useless in this one specific sense.¶ A beautiful and deep result of André Nies (building on the work of others) states that the K-trivial and the low-for-random sequences are the same. Being computationally useless in the sense of not aiding compression is the same as being maximally compressible. On the low end of the spectrum, the two notions of information align nicely.

WHAT WE DON'T (YET) KNOW

There is a great deal that we don't know. Some of our ignorance can be stated in the form of precise technical questions (see Question 4). Answers to those questions may impact our intuitive understanding of randomness and reveal unexpected patterns. On the other hand, these high level patterns and revised intuitions direct our technical research and lead to precise questions. I want to focus on the difficulty of this latter process, on the patterns and intuitions that we don't yet know how to formalize. Some of the more intriguing areas of our ignorance are those that we haven't been able to translate into explicit mathematical questions, let alone answer.

There are a number of interesting high level patterns that have emerged in our technical results for which we haven't found a unifying explanation. For example, there are several other characterizations of the *low-for-random* sequences, beyond the two that have been mentioned. This is a common feature of natural mathematical notions, that they have many, often very different looking, characterizations. All of the different ways we have found to characterize the low-for-random sequences have something in common: they refer to Kolmogorov complexity, to Martin-Löf randomness, or to convergent series (which can be viewed as a hidden reference to Kolmogorov complexity). Unlike properties that have similar definitions, low-for-random seems to have no purely computability-theoretic characterization. However, it is not clear how you could formulate a result to capture this observation, or even what should be meant by "purely computability-theoretic". The difficulty inherent in satisfactorily formalizing intuitive notions like "purely computability-theoretic" — especially when it's not clear that experts would agree on the intuition — is an obstacle to translating our ignorance into precise questions.

¶Although a low-for-random sequence need not be computable, there are very strict limitations on its computational usefulness. For example, it cannot compute a function that grows faster than every computable function.

Another example, one that we have already come across, is that it is not clear what it should mean for one sequence to be *more random* than another. In the answer to Question 2, we saw evidence supporting the claim that there is a negative correlation between the two types of information, that the more random a sequence is, the less computationally useful, and the less computationally useful a random sequence is, the more random. This is a higher level pattern with a lot of technical evidence to support it (and a compelling analogy!), but we don't know how to formulate a result that ties all of this evidence together into a coherent picture. Part of the reason is that we don't have an agreed upon precise formulation of "more random". If the Kolmogorov complexity of the initial segments of X always exceeds that of Y, then it seems reasonable to say that X is more random than Y. Similarly, if X is random relative to Z whenever Y is random relative to Z (i.e., any Z than can compress X can also compress Y), then it is reasonable to say that X is more random than Y. These are the two ways we interpreted "more random" in the answer to Question 2, and it is worth pointing out that for Martin-Löf random sequences, the first implies the second. But there are reasons to believe that neither relationship captures the intuition that X is more random than Y. Even the second condition seems to be too demanding.

It may be the case that intuitive notions like "purely computability-theoretic" or "more random" are too vague to support formalization, and that the higher level patterns we would like to capture are only partial truths, allowing no unifying explanation. The point is that a large part of our ignorance lies outside of the long lists of technical question that we haven't yet answered.

THE MOST IMPORTANT OPEN PROBLEMS

I will discuss just one problem. There is no shortage of open problems in the intersection of computability theory and randomness,[‖] but this one has survived focused investigation since it was formulated in 1998 by Muchnik, Semenov and Uspensky, which gives it seniority in a young and fast-moving field. Age alone does not make it important, of course.

Up until now we have focused exclusively on Martin-Löf randomness. There are good reasons why it is the most widely studied notion of its

---

[‖] André Nies and I published a list of open problems, many of which remain open (Joseph S. Miller and André Nies. Randomness and computability: open questions. *Bulletin of Symbolic Logic*, 12(3):390–410, 2006).

type, but it is not the only attempt to define randomness for a sequence using the idea that computer programs are too weak to find any pattern in the sequence, nor is it the easiest to explain. In 1971, Claus-Peter Schnorr used programs to bet against sequences, rooting his definition in the unpredictability of randomness. A program betting on a sequence X starts with $1 and bets on one bit at a time (remembering the outcomes of previous bets but obviously not being allowed to look ahead). If there is no upper bound on the amount of money the program reaches while betting against X, we say it *wins against* X. If no betting program wins against X, then X is called *computably random*. As a simple example, assume that every other bit of X is zero. We can easily write a program that bets nothing on the odd bits and bets all of its capital that each even bit is a zero. This program wins against X, so X is not computably random. This notion of randomness is weaker than Martin-Löf's in the sense that every Martin-Löf random sequence is computably random but the reverse is false.**

Schnorr also gave a characterization of Martin-Löf randomness in terms of betting strategies, but these strategies went somewhat beyond what is possible for betting programs. For this reason, Schnorr criticized Martin-Löf's definition of randomness as too restrictive. However, it is possible that with a slight change in the kind of bets we let our programs make, Martin-Löf randomness might be characterized in a way more in line with computable randomness. This is the context in which we should understand Muchnik, Semenov and Uspensky's question.

Say that we allow a program to bet on the bits of X in *any order*. At each step in the betting game, the program is allowed to choose an amount of its current capital *and* a position of X that hasn't been chosen before (hence hasn't been revealed). It could begin by betting $0.50 that the 17th position is a zero. Its second bet is conditional on the outcome of the first. For example, if it wins, it might bet $1.25 that the 3rd position is a one, but if it loses, bet $0.25 that the 51st position is a zero. We call this a *non-monotonic* betting strategy. Once again, we say that the program *wins* if there is no upper bound on the amount of money the program reaches while betting against X. If no non-monotonic betting program wins against X, then X is called *non-monotonically random*. The open question is simple:

**It has been shown that any sequence that is computably random but not Martin-Löf random computes a function that grows faster than every computable function. In other words, assuming an *upper* limit on the randomness of a somewhat random sequence puts a *lower* limit on its computational power (i.e., useful information). This gives us another example of the negative correlation between the two notions of information that we discussed in Question 2.

is this randomness notion exactly the same as Martin-Löf randomness? It is known that it is quite a bit stronger than computable randomness and that every Martin-Löf random sequence is non-monotonically random. If they turn out to be the same it would be a striking response to Schnorr's criticism.

THE PROSPECTS FOR PROGRESS

I would like to see greater integration of tools from probability and measure theory, the branches of mathematics that classically explore the notion of randomness. I think it is entirely possible that some of our open problems, such as the one discussed in Question 4, will be answered using more sophisticated ideas from classical mathematics. These ideas can also spawn new research directions. For example, Bjørn Kjos-Hanssen and Willem Fouché have studied the analog of Martin-Löf randomness for continuous functions using ideas that probability theorists developed to model and study Brownian motion. I would also like to see tools from the study of computability-theoretic randomness applied to measure theory. There are few examples of this as of yet, but one was recently given by Jan Reimann, who used these tools to give a new proof of Frostman's Lemma, a result from fractal geometry. Overall, I see the integration of classical mathematics into the study of Kolmogorov complexity and randomness to be an important part of the long term development of the field.

Chapter 17

Studying Randomness Through Computation

André Nies*

*Department of Computer Science
University of Auckland*
andre@cs.auckland.ac.nz

WHAT DREW ME TO THE STUDY OF COMPUTATION AND RANDOMNESS

My first contact with the area was in 1996 when I still worked at the University of Chicago. Back then, my main interest was in structures from computability theory, such as the Turing degrees of computably enumerable sets. I analyzed them via coding with first-order formulas. During a visit to New Zealand, Cris Calude in Auckland introduced me to algorithmic information theory, a subject on which he had just finished a book.[3] We wrote a paper[4] showing that a set truth-table above the halting problem is not Martin-Löf random (in fact the proof showed that it is not even weakly random [33, 4.3.9]). I also learned about Solovay reducibility, which is a way to gauge the relative randomness of real numbers with a computably enumerable left cut. These topics, and many more, were studied either in Chaitin's work[6] or in Solovay's visionary, but never published, manuscript,[35] of which Cris possessed a copy.

In April 2000 I returned to New Zealand. I worked with Rod Downey and Denis Hirschfeldt on the Solovay degrees of real numbers with computably enumerable left cut. We proved that this degree structure is dense, and that the top degree, the degree of Chaitin's Ω, cannot be split into two lesser degrees.[9] During this visit I learned about K-triviality, a notion formalizing the intuitive idea of a set of natural numbers that is far from random.

---

*Partially supported by the Marsden Fund of New Zealand, Grant No. 08-UOA-187.

To understand K-triviality, we first need a bit of background. Sets of natural numbers (simply called *sets* below) are a main topic of study in computability theory. Sets can be "identified" with infinite sequences of bits. Given a set A, the bit in position n has value 1 if n is in A, otherwise its value is 0. A *string* is a finite sequence of bits, such as 11001110110. Let $K(x)$ denote the length of a shortest prefix-free description of a string x (sometimes called the prefix-free Kolmogorov complexity of x even though Kolmogorov didn't introduce it). We say that $K(x)$ is the *prefix-free complexity* of x. Chaitin[6] defined a set $A \subseteq \mathbb{N}$ to be K-*trivial* if each initial segment of A has prefix-free complexity no greater than the prefix-free complexity of its length. That is, there is $b \in \mathbb{N}$ such that, for each n,

$$K(A \restriction n) \leq K(n) + b.$$

(Here $A \restriction n$ is the string consisting of the first n bits of A. On the right hand side the number n is represented in base 2 by a string.)

Martin-Löf[22] introduced a mathematical notion of randomness that is nowadays regarded as central. It is commonly referred to as Martin-Löf (ML-) randomness. We will discuss it in detail in the next section. K-triviality of sets is the opposite of ML-randomness: K-trivial sets are "antirandom". For, the Levin-Schnorr Theorem says that Z is Martin-Löf random if and only if there is a constant d such that for each n, we have $K(Z \restriction n) \geq n - d$; on the other hand, Z is K-trivial if the values $K(Z \restriction n)$ are within a constant of their lower bound $K(n)$, which is at most $2 \log n$.

Chaitin showed that each K-trivial set is Δ_2^0, that is, the set is Turing below the halting problem. Downey, Hirschfeldt and I worked our way through Chaitin's proof, the construction in Solovay's manuscript of an incomputable K-trivial, and Calude and Coles' improvement where the set constructed is also computably enumerable (c.e.), which means that one can effectively list its elements in some order. Downey realized that there is a connection between K-triviality of A and a notion introduced by Zambella:[37] a set $A \subseteq \mathbb{N}$ is *low for ML-randomness* if each ML-random set is already ML-random relative to A.

The phrase "relative to A" means that we can include queries to A in the computations determining, for instance, a ML-test. In this context A is called an "oracle set". Kučera and Terwijn[20] proved that such set can be c.e. but incomputable. Thus, the notion of relative ML-randomness does not always distinguish the oracles with some computational power.

A *lowness property* of a set expresses that the set is, in some specific sense, close to being computable. Unlike K-triviality, which expresses being far from random, Zambella's property is a lowness property defined in terms of relativized randomness.

At first these topics seemed exceedingly strange to me. Starting from April 2000, it took me almost exactly two years to understand the notions of K-triviality, and being low for ML-randomness.

Downey realized, during his visit to the University of Chicago in February 2001, that the dynamics of the Kučera-Terwijn construction of a set that is low for ML-randomness can be adapted for an easy construction of an incomputable, but c.e. K-trivial set. From 2002 on, the language of cost functions was developed as an abstract framework for such constructions; see [33, Section 5.3].

End of May 2001 I left the University of Chicago. I had more than a year ahead of me for pure research, knowing that I would safely start at the University of Auckland in mid-2002. In July 2001, Denis Hirschfeldt and I together travelled to Italy. After that, we met with Rod Downey at the Vienna summer conference of the Association for Symbolic Logic. Progress was slow but steady. For a while, we believed that a K-trivial set can be Turing complete! However, in discussions of Downey and Hirschfeldt, obstructions to building a Turing complete K-trivial emerged.

In August 2001 a group of researchers including Denis Hirschfeldt, Frank Stephan and Jan Reimann met at the University of Heidelberg. Denis had the crucial idea how to turn these obstructions into a proof that each K-trivial set is Turing incomplete. Eventually we published these findings in Ref. 10. The mechanism in Hirschfeldt's construction has been described by a stack of decanters holding precious wine.[11] The height of the stack is essentially given by the constant b in the definition of K-triviality of the set A. Wine is first poured into the top decanter (in smaller and smaller quantities). A decanter that is not at the bottom can be emptied into the next lower decanter. The purpose is to fill the bottom decanter up to a certain amount, while spilling as little as possible; this yields a contradiction to the Turing completeness of A. An elaborate argument one could call a "garbage lemma" shows that the amount one spills is indeed bounded. Such garbage lemmas recur in several related results.

After the stay in Heidelberg I went to Novosibirsk for a month, and worked with Andrei Morozov on questions related to algebra and its interaction with logic. I returned to the topics discussed above during an epic trip to Lake Baikal, Mongolia and then China on the Trans-Siberian Railway. I remember working on this in Goryachinsk, a resort for Soviet war veterans on the remote Eastern side of the lake 3 hours by minibus

from Ulan-Ude, the capital of the Russian province of Buryatia. I also remember a hotel room in Southern China where I was trying to write up a
proof that in the c.e. Turing degrees, each proper Σ^0_3 ideal has a low$_2$ upper
bound.[1] This is the mix of old and new I was immersed in. As the class
of K-trivials is closed under effective join $\oplus$ and every K-trivial is Turing
incomplete,[10] the c.e. K-trivials seemed to be a natural candidate for such
an ideal, except that we didn't know yet whether the K-trivials are closed
downward under Turing reducibility $\leq_T$.

WHAT WE HAVE LEARNED

A set of natural numbers can be studied under two aspects, its *randomness* and its *computational complexity*. We now understand both aspects.
We also know that they are closely related. There are strong interactions
from computability to randomness, and conversely, from randomness to
computability.

17.1. The Randomness Aspect of a Set

For infinite sequences of bits, there is no single formal notion corresponding
to our intuition of randomness. Our intuition is simply too vague for that.
Instead, there is a hierarchy of formal randomness notions, determined by
the strength of the algorithmic methods that are allowed for defining a test
concept. This can be traced back to the admissible selection rules of von
Mises.[36]

The infinite sequences of bits form the points of a topological space
called *Cantor Space*. Martin-Löf[22] defined a set to be random in a formal
sense if it passes each test in a certain collection of effective tests: a *ML-
test* is a sequence $(G_m)_{m\in\mathbb{N}}$ of uniformly c.e. open sets in Cantor space of
"size" at most 2^{-m} (formally, the size is the product measure λG_m). Z
passes this test if Z is not in all G_m. Z is *Martin-Löf random* if it passes
all ML-tests.

Many notions in the hierarchy of formal randomness notions can be defined via modifying Martin-Löf's test notion. If passing effective statistical
tests such as the law of large numbers is all you want, then your notion
might be Schnorr randomness, which is weaker than ML-randomness. A
test $(G_m)_{m\in\mathbb{N}}$ has to satisfy the additional condition that the size of G_m is
a computable real number uniformly in m.

If computability theoretic criteria matter to you, then Schnorr randomness is not enough because there is a Schnorr random set Z where the

sequence of bits in the even positions is Turing equivalent to the bits in the odd positions [33, 3.5.22]. This cannot happen any longer for a ML-random set. But maybe you also think that a real with computably enumerable left cut, such as Ω, should not be called random (when viewed in its binary representation). In that case, try weak 2-randomness. A Π_2^0 class in Cantor Space has the form $\{Z \colon Z \restriction_n \in R$ for infinitely many $n\}$, where R is some computable set of strings. Z is *weakly 2-random* if Z is in no Π_2^0 class of measure 0. Next is 2-randomness, namely ML-randomness relative to the halting problem. This notion was already studied in 1981 by Kurtz.[21] He showed that each 2-random Z is c.e. in some set $Y <_T Z$. Similar to ML-randomness, it has a characterization via incompressibility of initial segments: Z is 2-random $\Leftrightarrow$ for infinitely many n the initial segment $Z \restriction_n$ is incompressible in the sense of plain Kolmogorov complexity C (see [33, 3.6.20]).

So far, all tests were definable in arithmetic. If such tests are not sufficient, your notion might be Δ_1^1 randomness, surprisingly proposed by Martin-Löf in Ref. 23 as "the" formal randomness notion. A Δ_1^1 class is a sort of effective Borel class, and a set is Δ_1^1 random if it is in no Δ_1^1 class of measure 0. Martin-Löf's main result in that short paper states that there is no universal test in this sense (also see [33, after 9.3.5]). The strongest effective notion is Π_1^1 randomness, studied in Ref. 15. All null Π_1^1 classes are now tests. There is a largest one (Ref. 16, or see Ref. 15 for a direct proof). Interestingly, this notion is relevant in effective model theory: if a countable structure has with probability 1 a presentation computable in an oracle, then it already has a presentation computable in each Π_1^1 random oracle. (This is due to Kalimullin and Nies, slightly extending work of Greenberg, Montalban, and Slaman. See the the March 2010 entry in the Logic Blog on my web site.)

To summarize, the intuitive idea of randomness for sets corresponds to a whole hierarchy of formal notions. We mentioned most of the main notions:

Π_1^1-random $\Rightarrow \Delta_1^1$-random $\Rightarrow$
2-random $\Rightarrow$ weak 2-random $\Rightarrow$ ML-random $\Rightarrow$ Schnorr random.

For three notions, there is a universal test. Do you know which ones they are?

17.2. The Computational Complexity Aspect of a Set

In contrast to randomness, we have a clear intuition of what a computable set (or function) is. The Church-Turing thesis states that this intuitive notion has a clear-cut formal counterpart, the sets computable by a Turing machine. If we search our mind for an intuition on the complexity of *in*computable sets, things become less clear. Perhaps there is an intuition what it means to be "close to computable". However, the formal notions that have been proposed, the so-called lowness properties already mentioned in Section 13.4, are rather disparate. They can even exclude each other outside the computable sets. For instance, a set A is called *computably dominated* if each function that can be computed with A as an oracle is dominated by a computable function. The only computably dominated sets that are Turing below the halting problem are the computable sets. A diagram of 17 lowness properties is given on page 361 of Ref. 33.

17.3. Using Computability to Understand Randomness

In the beginning of this section I explained how computability theoretic tools are used to introduce formal randomness notions. Once defined formally, one can also study randomness, or its absence, via computability. Let A be a set of natural numbers. There are theorems supporting each of the following principles.

(1) A is far from random $\Leftrightarrow$ A is close to computable.
(2) Suppose A already has a randomness property. Then
    A is more random $\Leftrightarrow$ A closer to computable.

I will now give mathematical evidence for both principles. At present the main evidence for (1) is the following.

Theorem 17.1 *A is K-trivial $\Leftrightarrow$ A is low for Martin-Löf randomness.*

After the China visit at the end of 2001, I went on to Thailand, and then took a plane to the US to work with Richard Shore at Cornell. Now, at the beginning of 2002, I found myself travelling by bus, starting from the South of Mexico, through all the countries of Central America. I ended up on a small island called Isla Grande off the North coast of Panama, where I began working on the question posed in Ref. 20 whether each set that is low for ML-randomness is Δ_2^0. Eventually, I was able to obtain an affirmative answer. Having left Isla Grande, I wrote up a 7 page paper with this result in the internet cafés of Panama City and submitted it to the 2002 FOCS conference, where it was promptly rejected. See Ref. 28 for this proof. Later on, I improved the methods to obtain the implication "$\Leftarrow$" of Theorem 17.1. In 2003 I obtained an even stronger result involving computable randomness, which concludes this line of argument (if each set that is ML-random is still computably random in A, then A is K-trivial). The result appeared in Ref. 29; see the unpopular Theorem 8.3.10 in my book.[33]

I hitched a ride on a yacht from Ciudád Colon in Panama to Isla Mujeres in Mexico. I spent two weeks in the beautiful ocean off the Eastern Coast of Central America. My company consisted of two elderly gentlemen who hated each other, one paranoid cat, and an adventurer from the US who had the project to hide his savings in a bank on the Caribbean Island of St. Thomas because of a paternity lawsuit that awaited him at home. These were the conditions under which I started thinking about the implication "$\Rightarrow$" of Theorem 17.1. After a stop at Playa Tulum I went on to visit friends in Jalapa, the capital of the Mexican state of Veracruz. Staying there for a month, I got closer and closer to proving that remaining implication, without believing it was true at that time. I started from the decanter proof in Ref. 10 that each K-trivial is Turing incomplete. As an intermediate result, I proved that each c.e. K trivial has a lowness property called c.e. traceability, which for c.e. sets is equivalent to being array recursive. Next, I showed that the K-trivial are closed downward under Turing reducibility (which is not at all clear from the definition). Given a K-trivial set A and a set B that is Turing below A, I built a prefix-free machine showing that B is K-trivial. Unlike the previous decanter proof, this construction has a tree of runs of decanters. There must be a "golden run", namely a run that does not return while all the runs it calls do return. At the golden run node the required object is constructed, in this case, a prefix-free machine showing that B is K-trivial.

From Jalapa I went to Chicago to meet Denis Hirschfeldt. He realized that the golden run method shows the stronger result that each K-trivial set A is low for K: using A as an oracle does not yield shorter prefix-free descriptions of strings. This property, introduced by Muchnik Jr. in 1999, easily implies being low for ML-randomness. To define it formally, A is low for K if there is a constant d such that $K^A(y) \geq K(y) - d$ for each string y. Interestingly, we cannot find the golden run node in this construction, we can only prove that it exists. This is necessarily so: there is no effective way to obtain, from an index for the c.e. set A and the constant b for K-triviality of A, the constant d via which A is low for K [33, 5.5.5]. (In the construction, only the double jump $\emptyset''$ can find this golden run.)

Next I will discuss the principle (2) above: if A already has a randomness property, then

$$A \text{ is more random} \Leftrightarrow A \text{ is closer to computable.}$$

This almost seems to contradict the principle (1), but note that (1) is about sets that are far from random, while (2) is about sets that are already (somewhat) random. The implication "$\Leftarrow$" has been called *randomness enhancement*: satisfying a lowness property enhances the degree of randomness of A.[27] There are numerous instances of the principle (2).

- Randomness properties stronger than ML-randomness are usually closed downwards under $\leq_T$ within the ML-random sets, so they are given by an "abstract" lowness property. Further, if A is ML-random, then A is weakly 2-random $\Leftrightarrow$ every Δ_2^0 set below A is computable (Hirschfeldt and Miller; see [33, p. 135]), and A is 2-random $\Leftrightarrow \Omega$ is ML-random in A.[32]
- Let Z be Schnorr random set. If Z it is not high, then it is already ML-random. If Z is even computably dominated, then in fact Z is weakly 2-random.[32]
- If A is Δ_1^1 random, then A is Π_1^1 random $\Leftrightarrow$ each function f that is hyperarithmetical in A is dominated by a hyperarithmetical function (Kjos-Hanssen, Nies, Stephan and Yu;[17] also see [33, 9.4.6]).

17.4. Using Randomness to Understand Computability

Computability theory is all about the computational complexity of sets of natural numbers. One can gauge the complexity of a set A by locating A in classes of sets that all have a similar complexity. Examples of such classes are the computable sets, the high sets (i.e., $\emptyset'' \leq_T A'$), the Δ_2^0 sets (i.e., A is Turing below the halting problem), and the ω-c.e. sets (A is Turing below the halting problem, and, in addition, the reduction has a computably bounded use). The Limit Lemma of Shoenfield says that a set A is $\Delta_2^0 \Leftrightarrow$ the bit values $A(x)$ can be computably approximated with a finite number of mind changes; A is ω-c.e. $\Leftrightarrow$ in addition, the number of mind changes is computably bounded. Randomness-related concepts can be used both to introduce, and to study classes of similar complexity.

Mostly the classes of similar complexity are lowness properties. A common paradigm for lowness is the *weak-as-an-oracle* paradigm: A is weak in a specific sense when used as an oracle set in a Turing machine computation. Via randomness-related concepts, two new lowness paradigms have emerged.[13,27,33]

The *Turing-below-many* paradigm says that A is close to computable because A is easy to obtain from an oracle set, in the sense that the class of oracles computing A is large. Here, a class of oracles is considered large if it contains random sets of a certain kind. So far, all sets satisfying an instance of the Turing-below-many paradigm are Δ_2^0 sets.

The *inertness* paradigm says that a set A is close to computable because it is computably approximable with a small number of mind changes. In particular, such a set is Δ_2^0 by the Limit Lemma. For a mathematical formulation of the inertness paradigm, one can use so-called *cost functions*. A cost function $c(x, s)$ is a computable function defined on pairs of natural numbers x, s. The values $c(x, s)$ are non-negative rationals. Cost functions are used to bound the total quantity of changes of a Δ_2^0 set, and especially that of a computably enumerable set. At a stage s, if x is least such that I change my guess at $A(x)$, then I have to pay $c(x, s)$. To achieve lowness, my goal is to build a set A that *obeys* c in the sense that the total cost of changes is finite.

K-triviality has been characterized via the inertness paradigm.[29] Let $c_\Omega(x, s)$ be the measure of descriptions entering the domain of the universal prefix-free machine between stages x and s; thus, $c_\Omega(x, s) = \Omega_s - \Omega_x$. The single cost function c_Ω does the job: In Ref. 30 it is shown that A is K-trivial $\Leftrightarrow$ some computable approximation of A obeys c_Ω. Just like c_Ω, most of the other examples of cost functions are based on randomness-related concepts.

A further lowness property of a set is *strong jump traceability*. It was discovered when Santiago Figueira visited Auckland for 3 months in 2003.[12] Cholak, Downey and Greenberg[7] showed that the c.e. strongly jump traceable sets form a proper subclass of the c.e. K-trivials. The class has now been characterized via all three lowness paradigms. The original definition is by the weak-as-an-oracle paradigm, expressing that the jump $J^A(x)$ has very few possible values (it is equivalent to require[12] that the relative Kolmogorov complexity $C^A(y)$ of a string y is not far below $C(y)$, which makes the notion an analog of being low for K). The characterization via the Turing-below-many paradigm says that A is Turing below each ML-random set Z that is ω-c.e. If A itself is ω-c.e., this also expresses being far from random. For, there are many ω-c.e. ML-random sets: besides Ω, we have the ones obtained via the Low Basis Theorem. They can even be ML-random relative to one other. If an ω-c.e. A is "known" to all of them, it must be far from random itself. (Are you convinced by this argument?) The characterization via the inertness paradigm says that a c.e. set A is strongly jump traceable $\Leftrightarrow$ A obeys all so-called benign cost functions (for Δ_2^0 sets in general, at present only "$\Leftarrow$" is known).

The Turing-below-many paradigm seems to be more powerful than the weak-as-an-oracle paradigm, because it allows us to get closer to being computable. It even brings to light proper subclasses of the strongly jump traceable sets, for instance the sets Turing below all ω^2-c.e. ML-random sets.[31]

WHAT WE DON'T (YET) KNOW

We hope that new developments add to the present body of knowledge, and that, in the worst case, they supersede known results. However, new developments may also render previous results irrelevant.

The field of computability and randomness has now reached a state of "early maturity". Some notions, and results involving them, are generally agreed to be fundamental, for instance Martin-Löf randomness and K-triviality. For other notions, time will show. Let me assess the present knowledge critically.

17.5. Are We Studying the Right Randomness Notions?

There are various criteria for a good notion. The criteria (b)–(d) below work for all of mathematics, while criterion (a) only applies to some of it.

(a) *The notion corresponds to some intuitive idea.* This is true for most of the randomness notions of Subsection 17.1, and also for computable randomness, which is between ML-randomness and Schnorr randomness. Usually these notions formalize at least one of the three intuitive "randomness paradigms" introduced in Ref. 8 (typicalness, unpredictability, incompressibility of initial segments).

(b) *There are natural examples, or constructions, of instances of the notion.* This is true for ML-randomness and 2-randomness, where the examples are Ω, and Ω relative to $\emptyset'$, respectively.

(c) *The notion interacts richly with other sub-areas.* Again, this is true for several randomness notions of Subsection 17.1, for instance because of their interaction with computability.

(d) *The notion is a "sink".* That is, one reaches the same notion from different directions. To support this, there are coincidence results for Schnorr randomness, ML-randomness, and 2-randomness. Originally defined via tests, these three notions can be characterized via incompressibility of initial segments. Computable randomness can also be characterized in different ways: by definition, Z is computably random if no computable betting strategy wins on Z. However, recent research of Brattka, Miller and myself[2] shows that it is equivalent to require that each non-decreasing computable function defined on the unit interval is differentiable at (the real corresponding to) Z.

Of course, there may be undiscovered randomness notions that perform just as well with these criteria.

17.6. Do the Randomness Notions Really Form a Hierarchy?

For the notions presently known, the answer is "yes, mostly". One notable exception is Demuth randomness (see Section 3.6 of Ref. 33), which is between 2-randomness and ML-randomness, but is incomparable with weak 2-randomness. This notion is interesting because, unlike weak 2-randomness, it is compatible with being Δ_2^0. It interacts strongly with lowness properties. For instance, each c.e. set A below a Demuth random set Y is strongly jump traceable.[18] If Y is Δ_2^0 then such a set A can be incomputable. Polynomial randomness is incomparable with Schnorr randomness, because the proof of Ref. 32 that some Schnorr random is not computably random actually produces a Schnorr random that is not polynomially random (also see [33, Thm. 7.5.10]).

If the answer to the hierarchy question eventually becomes "no", it would be harder to claim that these notions have anything to do with our intuition of randomness.

17.7. Are We Studying the Right Lowness Properties?

Let us apply the criteria for good notions (a)–(d) in Subsection 17.5. Many lowness properties perform quite well in (b)–(d). As for (a), each of them catches a bit of our intuition of being "close to computable", but none of them formalizes an intuitive idea just by itself. I will check the criteria (b)–(d) for two lowness properties: lowness for ML-randomness, and superlowness.

Lowness for ML-randomness = K-triviality (b) has a natural construction: a c.e. set obeying the cost function c_Ω (see Subsection 17.4). It (c) interacts well with randomness, and (d) coincides with heaps of other classes, such as being low for K, being a base for ML-randomness, and being low for weak 2-randomness (see [33, Section 5]).

We say that a set A is superlow if A' is truth-table below the halting problem $\emptyset'$. Superlow c.e. sets can be built via finite injury. Non-c.e. superlow sets with interesting properties, such as being ML-random or PA-complete, can be built via the low basis theorem. So (b) is satisfied. For c.e. sets, superlowness is equivalent to a property called jump traceability (see [33, 8.4.23]), which gives us (d).

17.8. So, Again, are our Notions Intrinsic, or Accidental?

They are the former, hopefully. The randomness and lowness notions would seem less accidental if they were introduced in different ways. For instance, they could be specializations of more general formal notions. To obtain these more general notions one could try to formalize the randomness and the lowness paradigms discussed earlier on, which so far are informal meta-notions.

To be even more heretical, we could ask whether the whole distinction between the randomness and the computational complexity aspect of sets is more than a historical accident, caused by the fact that people with different backgrounds were working at different times and in different places. Currently we develop these two aspects separately and then find interactions. Perhaps one day they will be unified into a single theory. Perhaps one day there will only be a general theory of access to the information of a set of natural numbers.

THE MOST IMPORTANT OPEN PROBLEMS IN THE FIELD

I expect there will be many interesting new problems in areas that are just being developed, in particular the interaction of algorithmic randomness with computable analysis,[2] and ergodic theory. Instead, I will discuss two major problems on randomness or its interaction with computability that have been around for a while.

17.9. Covering a K-trivial by an Incomplete Random

Kučera[19] built an incomputable but computably enumerable set A below any given Δ_2^0 ML-random set Z. If Z does not compute the halting problem (for instance, Z is the bits of Ω in even positions), this yields an injury-free solution to Post's problem[34] whether some incomputable c.e. set is Turing incomplete.

The cost function construction of a K-trivial set yields a further injury-free solution to Post's problem (Ref. 10, or see [33, Section 5.3]). If Z is a Martin-Löf random set that does not compute the halting problem, then every c.e. set A Turing below Z is K-trivial.[14] Thus, in a sense, Kučera's solution to Post's problem is a special case of the solution via building a K-trivial. It is open whether the converse holds: given A, build Z. Essentially we are asking whether the two injury-free solutions to Post's problem are equivalent. Since every K-trivial set is below a c.e. K-trivial,[29] we may omit the hypothesis that A is computably enumerable.

Question 17.1 [8,14,25] *Is every K-trivial set Turing below an incomplete Martin-Löf random set?*

Countless people have worked on this. There is no consensus which way the answer will go.

There are several variants of Question 17.1. For instance, we say that A is *ML-noncuppable* if $A \oplus Z \geq_T \emptyset'$ implies $Z \geq_T \emptyset'$ for ML-random Z. Every c.e. ML-noncuppable is K-trivial (see [33, 8.5.15]).

Question 17.2 [25] *Is every K-trivial set ML-noncuppable?*

For background on the next variant of Question 17.1, see Subsection 17.6.

Question 17.3 [18] *Is every strongly jump traceable (c.e.) set Turing below a Demuth random?*

17.10. Kolmogorov-Loveland Randomness

An infinite sequence of bits (i.e., a set) is computably random if no computable betting strategies succeeds on it. Such a strategy places a bet on the next bit position in the usual ascending fashion. We say that a set Z is *Kolmogorov-Loveland random* (KL-random) if no computable betting strategy succeeds even when it is allowed to choose the next bit position on which it places a bet. The implications are

$$\text{Martin-Löf random} \Rightarrow \text{KL-random} \Rightarrow \text{computably rd.} \Rightarrow \text{Schnorr random.}$$

All implications except the leftmost one are known to be strict. The strictness of that implication is a major open question.

Question 17.4 [25,26] *Does KL-randomness differ from Martin-Löf randomness?*

A negative answer would defeat Schnorr's critique of ML-randomness, because KL-randomness is defined using a computable test concept. In Ref. 24 we obtained various results showing that KL-randomness is, at the very least, much closer to Martin-Löf-randomness than the other notions. For instance, the computable dimension of a KL-random set is 1.

Lowness for KL-randomness implies K-triviality.[29] Separating the two lowness properties would basically give an affirmative answer to Question 17.4. The following would interesting to begin with.

Question 17.5 *Is some incomputable set low for KL-randomness?*

Recall that co-infinite sets can be identified with reals in the unit interval via the representation in base 2. When viewed as notions about reals, computable randomness and KL-randomness appear to rely on the representation of the real in base 2. However, strategies that bet on rational intervals can be used to show the base invariance for computable randomness. We do not know whether KL-randomness of a real actually depends on the choice of the base 2.

Question 17.6 *Is KL-randomness of a real number base-invariant?*

This is probably just as hard to answer as Question 17.4. Most likely, for each base, KL-randomness induces a distinct class of reals; these notions are all incomparable, and therefore all different from Martin-Löf randomness on reals.

THE PROSPECTS FOR PROGRESS

Back in the 16th century, a prospect was an extensive view of a landscape. Imagine you stand on a mountain top and see the immense area of computability and randomness stretched out before you. Where do you want to go?

Let's say you are a young researcher with some knowledge of the books,[5,8,33] but not too much respect for the results in there. Then you might make progress on the new directions suggested in Section 17.4.

We would make progress on the open problems in Section 17.8 by convening a group of researchers, young and old, on that mountain top.

Acknowledgments

I thank Christopher Porter for comments on earlier drafts of this paper, and the editor Hector Zenil for getting me to write this.

References

1. G. Barmpalias and A. Nies. Upper bounds for ideals in the computably enumerable degrees. Submitted.
2. V. Brattka, J. Miller, and A. Nies. Computable randomness and differentiability. To appear.
3. C. Calude. *Information and randomness*. Monographs in Theoretical Computer Science. An EATCS Series. Springer-Verlag, Berlin, 1994. With forewords by Gregory J. Chaitin and Arto Salomaa.
4. C. Calude and A. Nies. Chaitin Ω numbers and strong reducibilities. *J.UCS*, 3(11):1162–1166, 1997.
5. Cristian S. Calude. *Information and randomness*. Texts in Theoretical Computer Science. An EATCS Series. Springer-Verlag, Berlin, second edition, 2002. With forewords by Gregory J. Chaitin and Arto Salomaa.
6. G. Chaitin. A theory of program size formally identical to information theory. *J. Assoc. Comput. Mach.*, 22:329–340, 1975.
7. P. Cholak, R. Downey, and N. Greenberg. Strongly jump-traceability I: the computably enumerable case. *Adv. in Math.*, 217:2045–2074, 2008.
8. R. Downey and D. Hirschfeldt. *Algorithmic randomness and complexity*. Springer-Verlag, Berlin. To appear.
9. R. Downey, D. Hirschfeldt, and A. Nies. Randomness, computability and density. *SIAM J. Computing*, 31:1169–1183, 2002.
10. R. Downey, D. Hirschfeldt, A. Nies, and F. Stephan. Trivial reals. In *Proceedings of the 7th and 8th Asian Logic Conferences*, pages 103–131, Singapore, 2003. Singapore University Press.
11. R. Downey, D. Hirschfeldt, A. Nies, and S. Terwijn. Calibrating randomness. *Bull. Symbolic Logic*, 12(3):411–491, 2006.
12. S. Figueira, A. Nies, and F. Stephan. Lowness properties and approximations of the jump. *Ann. Pure Appl. Logic*, 152:51–66, 2008.
13. N. Greenberg, D. Hirschfeldt, and A. Nies. Characterizing the strongly jump traceable sets via randomness. To appear.
14. D. Hirschfeldt, A. Nies, and F. Stephan. Using random sets as oracles. *J. Lond. Math. Soc. (2)*, 75(3):610–622, 2007.
15. G. Hjorth and A. Nies. Randomness via effective descriptive set theory. *J. London Math. Soc.*, 75(2):495–508, 2007.
16. A. Kechris. The theory of countable analytical sets. *Trans. Amer. Math. Soc.*, 202:259–297, 1975.
17. B. Kjos-Hanssen, A. Nies, F. Stephan, and A. Nies. Higher kurtz randomness. To appear.
18. A. Kučera and A. Nies. Demuth randomness and computational complexity. To appear.

19. A. Kučera. An alternative, priority-free, solution to Post's problem. In *Mathematical foundations of computer science, 1986 (Bratislava, 1986)*, volume 233 of *Lecture Notes in Comput. Sci.*, pages 493–500. Springer, Berlin, 1986.

20. A. Kučera and S. Terwijn. Lowness for the class of random sets. *J. Symbolic Logic*, 64:1396–1402, 1999.

21. S. Kurtz. *Randomness and genericity in the degrees of unsolvability*. Ph.D. Dissertation, University of Illinois, Urbana, 1981.

22. P. Martin-Löf. The definition of random sequences. *Inform. and Control*, 9:602–619, 1966.

23. Per Martin-Löf. On the notion of randomness. In *Intuitionism and Proof Theory (Proc. Conf., Buffalo, N.Y., 1968)*, pages 73–78. North-Holland, Amsterdam, 1970.

24. W. Merkle, J. Miller, A. Nies, J. Reimann, and F. Stephan. Kolmogorov-Loveland randomness and stochasticity. *Ann. Pure Appl. Logic*, 138(1-3):183–210, 2006.

25. J. Miller and A. Nies. Randomness and computability: Open questions. *Bull. Symbolic Logic*, 12(3):390–410, 2006.

26. Andrei A. Muchnik, A. Semenov, and V. Uspensky. Mathematical metaphysics of randomness. *Theoret. Comput. Sci.*, 207(2):263–317, 1998.

27. A. Nies. Applying randomness to computability. Series of three lectures at the ASL summer meeting, Sofia, 2009.

28. A. Nies. Low for random sets: the story. Preprint, available at http://www.cs.auckland.ac.nz/nies/papers/, 2005.

29. A. Nies. Lowness properties and randomness. *Adv. in Math.*, 197:274–305, 2005.

30. A. Nies. Calculus of cost functions. To appear.

31. A. Nies. Subclasses of the c.e. strongly jump traceable sets. To appear.

32. A. Nies, F. Stephan, and S. Terwijn. Randomness, relativization and Turing degrees. *J. Symbolic Logic*, 70(2):515–535, 2005.

33. André Nies. *Computability and randomness*, volume 51 of *Oxford Logic Guides*. Oxford University Press, Oxford, 2009.

34. E. Post. Recursively enumerable sets of positive integers and their decision problems. *Bull. Amer. Math. Soc.*, 50:284–316, 1944.

35. R. Solovay. Handwritten manuscript related to Chaitin's work. IBM Thomas J. Watson Research Center, Yorktown Heights, NY, 215 pages, 1975.

36. R. von Mises. Grundlagen der Wahrscheinlichkeitsrechnung. *Math. Zeitschrift*, 5:52–99, 1919.

37. D. Zambella. On sequences with simple initial segments. ILLC technical report ML 1990-05, Univ. Amsterdam, 1990.

Chapter 18

Computability, Algorithmic Randomness and Complexity

Rodney G. Downey

School of Mathematics, Statistics and Operations Research
Victoria University of Wellington, New Zealand
rod.downey@vuw.ac.nz

WHAT DREW ME TO THE STUDY OF COMPUTATION AND RANDOMNESS

I think mathematical ability manifests itself in many different ways*. In particular, mathematicians can be drawn to space and geometry, can have strong analytic intuition, can be drawn to formalism, they can be drawn to counting arguments, etc. There is definitely no unique type of mathematician. Maybe those mathematicians who are drawn to algorithmic thinking have found a home in computer science. For myself, I have always found myself drawn to thinking algorithmically. As a student, I recall studying algebra. Naturally, we would be prescribed problems to solve and sit exams. Usually, I would find that instead of some short elegant proof I would grind out some longer, but more basic often algorithmic version. Likely this reflected lack of study, as I was pretty lazy as an undergraduate, so I had not read the notes mostly! Even when studying analysis, I saw this as an algorithmic game in that given ϵ, somehow I would try to compute δ, viewing this as a game of us versus an opponent.

In my honours year[†], at Queensland University, I recall studying Szmielew's decision procedure for the elementary theory of abelian groups and the word problem for groups. After this I moved to Monash to work on effective algebra which was very fashionable at the time. In effective or

---

*In the following I have sort of combined the questions into a long essay. I have also made several comments which would have references, but I refer to the Downey-Hirschfeldt monograph as a suitable source for references.

[†]The British system has three years of undergraduate study for a degree and then one more year called the honours year, where if you achieve first class or second upper honours, you can then move directly on to do a Ph.D.

computable algebra, one tries to understand the effective content of mathematics. This is the extent to which mathematics can be made algorithmic. One imagines the data as being *presented* in some computable fashion, and then asks for the extent to which aspects or processes of the data can be made computable.

A nice illustrative example from combinatorics is Dilworth's decomposition theorem. The classical version says that if the size of the largest antichain in a partially ordered set is k, then the partially ordered set can be expressed as the union of k linearly ordered chains. The computable version asks whether this can be done computably. To wit, given a computable partially ordered set, $(P, \leq)$ (meaning that the domain is computable and the relation $\leq$ is computable), can we decompose this into k *computable* chains? Kierstead showed that the answer is no, *but* we can always decompose into $f(w) = \frac{5^w - 1}{4}$ many computable chains, using an *online* algorithm. For those of you looking for a nice problem, figure out what the correct bound is. We have this exponential upper bound, but don't even have polynomial lower bounds for $f(w)$.

WHAT WE HAVE LEARNED

Computability theory allows one to calibrate how difficult computational tasks are. To do this structures like the *degrees of unsolvability* (and various other hierarchies) were invented by Post, Turing and others. Here A and B have the same degree of unsolvability if using A as read only memory we can compute membership of B with an algorithm, (so that $A \leq_T B$) and conversely (so that $B \leq_T A$). These equivalence classes partition the world into collections of *equicomputability*. There is a basic spine generated by the *halting problem* which, in data n, m, asks if there is an algorithm that decides if the n'th algorithm halts on input m. A fundamental result is that the halting problem is algorithmically unsolvable, and this gives rise to a hierarchy, the degree of the halting problem, called $\mathbf{0'}$, the degree of the halting problem *if* I give you the halting problem as read only memory, called $\mathbf{0''}$, and so on.

As an illustration, the problem of deciding if a Diophantine equation in many variables with integer coefficients has a positive rational zero is famously known to be of degree $\mathbf{0'}$. Similarly, the word problem for finitely presented groups asks : what is the algorithmic complexity the complexity of determining whether two words are equal in a given finitely presented group. This is also of degree $\mathbf{0'}$. My favourite example is Conway's generalization of Collatz functions. Recall that the Collatz function is $f(x) = \frac{x}{2}$ if x is even and $f(x) = 3x + 1$ if x is odd. Then a famous conjecture is that if you look at the iterates $f(x), f(f(x)), \ldots$ then the sequence returns to 1 eventually. Conway asked what about if we choose a set of rational numbers $\langle q_1, \ldots, q_n \rangle$ and looked at a similar problem but with congruences chosen via these rationals. That is, define $f_{\langle q_1, \ldots, q_n \rangle}(x) = q_m \cdot x$ if $x \equiv m$ mod n, and choose the rationals so that the result is a integer whenever x is an integer. Then what can be predicted about the sequence for this f and a given x? Conway showed that we can choose a finite set of rationals so that this simulates precisely the action of a given algorithm. Each of these results is proven by a similar kind of simulation and hence we see that these disparate areas of mathematics can simulate computation.

The reader might note that I was a bit sloppy in that I defined *the* halting problem when the definition seems to depend on the coding of the programmes. This turns out to be of no consequence, since we can show that all reasonable versions of the problem are the same up to a very strong simulation, called an m-reduction.

Problems can be very much harder than the halting problem. At a far extreme, there is a notion called *analytic* where the core problem is not asking if there is some stage where a computation halts, but whether there is a function $F : A \to B$ where for all n some computable relation for $F(n)$ holds. A classic example of this is isomorphism between structures. Using computability theory we can show that certain problems are as hard as the hardest analytic problems, and hence there *cannot* be a set of invariants to simplify the problem in the way that, for instance, dimension is an invariant for vector space isomorphism. Recently, Montalbán and I used this method to show that the isomorphism problem for torsion free abelian groups is analytic complete, and hence no reasonable set of invariants (in the sense of simplifying the problem) can exist for this problem.

For my own interests, behind all of this is a preference for studying fundamental logical questions about mathematics. This again is something I seem to have been drawn to from an early age. At high school I was able to study logic and philosophy as that was available in Queensland.

Actually, I believe that our working arena for all mathematics is computable mathematics in the sense that any function which occurs in "real life" likely will be computable. By the same token, one definition of a continuous function is one that is computable relative to some A. Thus computability is the study of continuity! Of course in the "really real" world, we "really" only deal with finite things, and these are abstracted to methods allowing us to analyze them. So number theory advanced when it was converted into analysis. We regard things as continuous to enable tools from continuous mathematics to attack them. Certainly being infinite is often a great approximation to the finite! Computation is at the core of a lot of mathematics.

Anyway, I have worked in and about these areas for the last three decades. A some stage I became very interested in computer science, and in that time developed *parameterized complexity* (a kind of complexity we feel is more attuned to practical computation) with Mike Fellows. At the same time I was working through Li and Vitányi's proofs of lower bounds for computability results using Kolmogorov complexity. For example the proof that two tapes have more power than one tape on a Turing machine. I really understood these proofs only very formally and put them on my large pile of "must look at again at some stage".

In January 2000, Denis Hirschfeldt and I organized a conference at Kaikoura, a small town on the east coast of the south island of New Zealand. This was under the auspices of the Marsden Fund for basic science of New Zealand. What we did was to bring in several overseas experts to speak about recent developments in mathematics in short graduate level courses. The resulting volume is called *Aspects of Complexity* and contains a number of fine short courses in and about complexity, and is inexpensive! This method being a great initiative of the founder of the NZIMA, Sir Vaughan Jones, whose overall goal was to raise the standard of New Zealand mathematics. At the Kaikoura meeting, there were some very illuminating lectures by Lance Fortnow on Kolmogorov complexity, and I came back with some fire in my belly to know some more.

At more or less the same time, one of my ex-postdocs, Richard Coles, was working at Auckland with Cris Calude. Coles visited me and asked me one of Cris's questions. The question was whether the Solovay degrees of halting probabilities were dense. What does this question mean? Well, in algorithmic randomness the halting problem is replaced by the halting probability. The halting problem is the collection of indices of programmes that halt. For a prefix-free machine (as discussed a little later), the domain of the machine is a prefix free subset of the collection of finite strings, and thus considered as a collection of cylinders will have measure under the uniform measure. Instead of the haling set, we have the halting probability which can be considered as the sum of the probabilities that the machine halts over all strings. We will discuss more on this topic in a little bit. Solovay reducibility is a continuous version of m-reducibility. The technical question was whether the resulting degree structure was a dense partial ordering. Hirschfeldt, Nies and I began to think about this question in depth, and eventually solve the question. At the same time, we were began a lot of background reading, as questions lead to other questions and we really knew little about the area. In doing so, we discovered a large, and for us hitherto unknown, body of work on algorithmic randomness. As Denis and I say in the introduction to our book: "We also found that, while there is a truly classic text about *general* Kolmogorov Complexity, namely Li and Vitányi (book), most of the questions we were interested in either were open, were exercises in Li and Vitányi with difficulty ratings of about 40-something (out of 50), or necessitated an archaeological dig into the depths of a literature with few standards in notation or terminology, littered with relentless re-discovery of theorems and a significant amount of unpublished material. Particularly noteworthy amongst the unpublished material was the aforementioned set of notes by Solovay, which contained absolutely fundamental results about Kolmogorov complexity in general, and about initial segment complexities of reals in particular. As our interests broadened, we also became aware of seminal results from Stuart Kurtz's PhD Dissertation, which, like Solovay's results, seemed unlikely to ever be published in a journal."

Anyway, about this time, we were able to prove a number of results in and around trying to understand what it meant to calibrate randomness, and begin the work on K-triviality. As our knowledge grew, we became aware of other work of Kučera and early work of Levin, Schnorr and Chaitin, with particularly attractive popular works by Chaitin. Perhaps foolishly, Hirschfeldt and I then decided to write a book organizing this material. We estimated that it would take 2–3 years.

So here we are nearly a decade down the track, and the book is just finished at the time of my writing this article. This is not simply laziness on our part, but also because there has been an explosion of results in this area, notably by lots of my postdocs such as Barmpalias, Bienvenu, Greenberg, Griffiths, Hirschfeldt, LaForte, Miller, Montalbán, Yu, and by other gifted authors including Kučera, Muchnik, Nies, Reimann, Shen, Slaman, Stephan, and Vereshchagin (and many others) and this necessitated endless re-writes.

Randomness is important in calculation and the like, but for me the fascination lies in the intuition that something fundamental is going on here. It is a striking fact that something, a string say, which is *algorithmically* random (meaning that it passes a bunch of computable tests) will act in some mathematical situation like the *expected statistical outcome*. Why should this be? Well, if the situation is normal then it will be computable, and hence if it did not act in the expected way, the very computational nature of the situation would allow us to compress the string, and hence it would not be random.

Personally, I am not driven by the thought that the material might be useful, but to try to understand what randomness means. I deal with question like: When is one real or string more random than another? How powerful as computational resources are randomness sources? If one string is more random than another how does that align to computational power? What does independence mean? What happens if we vary the machines? What are the correct notions of compressibility? These all seem fundamental and my intuition tells me they are important.

18.1. The Downey–Hirschfeldt Monograph

I have been asked to try to summarize what is in the Downey-Hirschfeldt book. The paragraph above is a good beginning.

The tools we will use will be based around computability theory. So we begin with a condensed course in "advanced computability theory" where we develop the tools we will need. The idea is that we will study *algorithmic* randomness. The theme of the book is that for such a notion of randomness, there are three natural approaches. In some sense they all derived from the intuition of von Mises. In a remarkable early paper von Mises argued that if we had a random real $\alpha = a_0 a_1 \ldots$ (which we will consider as an infinite sequence of 0's and 1's) then by the law of large numbers the first n bits of α, $\alpha \restriction n$ should contain as many 0's as 1's as $n \to \infty$. Moreover, if we *selected* n bits $a_{i_0}, \ldots, a_{i_{n-1}}$, with $i_0 < i_1 < \cdots < i_{n-1}$, then also we ought to have as many 0's as 1's. What selections should be possible? Well, if we are interested in *algorithmic* randomness then presumably the correct notion of selection should be more or less *computable* selection. It is important to realize that von Mises intuition was well before the development of computability theory, and hence the notion of using computable selections was due to Church many years later. Unfortunately, in its classic formulation, Ville showed that *no* collection of selection functions (computable or not) would suffice to characterize a reasonable notion of randomness as there would be natural effective statistical tests failed by some real, random relative to the selection functions.

Now we return to the three approaches. The first approach is the statistical one due to Martin-Löf. This views statistical tests as null sets (sets of measure 0), and asks that a real be random iff it avoids all *effective* null sets.

The second approach to defining randomness observes that if a string has patterns then that allows for compression of the string by a programme that exploits the patterns. This is the way commercial compression packages work. Kolmogorov defined a string σ to be random relative to a machine M if the shortest M-programme to generate σ has length $|\sigma|$ or longer. The Kolmogorov complexity C_M of a string σ relative to M is the *length* of the shortest M-programme for σ. It turns out that there is a universal M for the definition in the sense that for any other $\hat{M}$ there is a fixed constant c, and any σ, $C_{\hat{M}}(\sigma) \leq C_M(\sigma) + c$. For such M, we write C for C_M. Now the second approach asks that a real be random iff all its initial segments are incompressible. Unfortunately, using Kolmogorov's definition, no real is random since all will have C-compressible initial segments. The problem is that C does not really capture one of the key intuitions of the notion of *information content*. Namely, if $M(\tau) = \sigma$ our intuition is that the bits of τ encode the information of the bits of σ. Thus $|\tau|$ should be a reasonable notion of the information content of σ, and the shortest τ would give the Kolmogorov complexity of σ. *But*, plainly τ gives more information than just the bits of τ, it gives also *the length of τ* as well. That is we get the bits of τ and some kind of termination symbol saying that that is the end of the programme. First Levin and then a little later Chaitin suggested ways around this. Perhaps the most popular method is to use machines which act like telephone numbers[‡]: no programme is a prefix of another, and this gives a notion called *prefix-free* Kolmogorov complexity which we denote by K. It is also possible to use "continuous" complexity (of various types) where now if τ extends $\hat{\tau}$, then $M(\tau)$ should extend $M(\hat{\tau})$. This notion gives rise to several notions of complexity depending on the notion of allowable machine, and two main complexity are called Km and Km_P, monotone and process complexity.

Anyway our story has a happy ending, since for any of these notions, α is Martin-Löf random iff the all of the initial segments of α are random as strings. Even here we see the emergence another theme of the book: What are the appropriate measures of information content, and how do

---

[‡]Well, more or less. I know that in the US 0 is the operator, and 011 is the international prefix access code, but the spirit is correct.

they relate. For example, we include the difficult proofs that for strings σ,

$$K(\sigma) = C(\sigma) + C(C(\sigma)) + O(C(C(C(\sigma)))),$$

and that this $C^{(3)}(\sigma)$ is *sharp* in the sense that it *cannot* be replaced by $O(C^{(4)}(\sigma))$. This remarkable result is due to Solovay and no proof of it has appeared in print. We include many results about the relationships of the differing complexities, and see how it emerges that differing complexities can be appropriate for differing notions of randomness.

The final approach is the closest in spirit to von Mises. What we will do is bet on the "next bit" of α from $\alpha \restriction n$. Clearly, if α is random, if we use some kind of effective betting strategy, we should not be able to win infinite winnings. With the correct notion of effective betting, it turns out that we get the same notion of randomness. The notion of effective betting is one which corresponds to being effectively approximable from below.

Schnorr proved this coincidence of randomness notions, but then pointed out that if we were to intuitively think of effective betting strategies surely we would simply bet and then not later change our minds and maybe bet more, etc. This critique led to the development of other natural notions of effective randomness, such as computable, partial computable, weak and Schnorr randomness. I won't go into details (buy the book!), but these notions have complex interrelationships particularly when we look at them in relation to Turing degrees. For example, for a certain class of degrees with, in a quantifiable sense, little computational power, they all coincide; whereas in ones with high computational power they all separate.

Already we see the emergence of another theme. The computational complexity of a random set can affect its level of randomness and conversely. A classic illustration of this is Ω , the universal halting probability (the measure of the domain of a universal prefix-free machine). As shown by Chaitin, we know that this has very high computational power, like the halting problem and in a very compressed form. This fact does not accord with our intuition that random reals should have feeble computational power, in that whilst they might have lots of "information" none of it is usable.

What we now know in a very precise way is why this is all true. The following is a colourful way to think of this. Think of trying to pass a stupidity test. There are two ways to do this. One is to be the genuine article, but the other is to be so smart that you know the correct answers to appear stupid. We know that if we soup-up the randomness by asking that the real be (weakly) random given the halting problem as an oracle, then the computational power of the random real is very much weaker in many quantifiable ways. Moreover, Frank Stephan showed that random reals come in two varieties. There are those with very low computational power (in the sense that they cannot compute what is called a PA degree) which are the typical ones, and those that have some power in this sense, and all of these are the "false" randoms living above the halting problem. That is, almost all random reals have very little computational power and those that do are not very random, and all are basically the halting problem (or more) compressed. These PA degrees are those that are degrees of models of Peano arithmetic have remarkable interactions with random degrees, a fact first realized by Antonín Kučera. One recent gem is the result of Barmpalias, Lewis and Ng who showed that every PA degree is the join of two random degrees.

A great part of the middle of the book looks at interactions of notions of randomness and notions of computational power. It has become clear that what are called domination properties are key notions. For example consider the class of degrees **a** which have the property that every function f computed by **a** should be dominated by a computable function. This would seem a class of almost computable degrees with low computational power in some sense. But they can be random. And, indeed, for such degrees weak, computable, Martin-Löf and even weak randomness relative to the halting problem all coincide!

There are many such investigations. For example, the power to compute a PA degree is related to the initial segment complexity by looking at the growth rate of the initial segment complexity.

One striking phenomenon was the analysis of the K-trivial degrees. It is an old result of Chaitin, building on work of Loveland that α is computable iff $C(\alpha \restriction n) \leq C(1^n) + O(1)$, giving an information-theoretical definition of computability. Solovay showed that there are *noncomputable* reals β with $K(\beta \restriction n) \leq K(1^n) + O(1)$. Thus the characterization of computability fails for prefix-free complexity. Such reals β are called *K-trivial*. They have remarkable properties. I am sure that this class is explored by Nies in his contribution, which is highly apt as many of the basic properties of this class were discovered by Nies and his co-authors. They are very easy to construct, and give natural solutions to Post's problem. They lead us to explore notions of *lowness*. For example it can be shown that α is K-trivial iff α is low for K in the sense that it does not compress anything: $K^\alpha(x) = K(x) + O(1)$ for all x. This is the tip of the iceberg. There are also many other lowness notions we explore. These ideas are also explored in Nies' recent monograph called *Computability and Randomness*.

We also include analyses along the themes of calibrating randomness by various pre-orderings. Solovay reducibility is one, but you could ask that $\alpha \leq \beta$ if the initial segment complexities align in some way. For example, we can define $\alpha \leq_K \beta$ iff $K(\alpha \upharpoonright n) \leq K(\beta \upharpoonright n) + O(1)$. It makes sense that if a real is random iff $K(\alpha \upharpoonright n) \geq n - O(1)$ for all n, then the pre-ordering above should define some notion of relative randomness. Indeed we now know, that $\leq K$ can also be used to define higher levels of randomness such as randomness relative to the halting set and its iterates. You can also try to calibrate relative randomness according to derandomization power. Now $\alpha \leq_{LR} \beta$ if everything β makes non-random is also made nonrandom by α. How do such measures relate to each other and how do they relate to computational complexity? Myriad questions suggest themselves and there have been a lot of deep and unexpected results proven here.

The last part of our monograph is devoted to looking at other related ideas. For example, in the same way that classical Hausdorff and other dimensions refine measure 0, we can look at effective versions following the lead of Jack Lutz. We have a long section devoted to understanding algorithmic dimensions. For example, we address questions like: I have some source of partial randomness, say a real of dimension $\frac{1}{2}$, and ask can I extract a real of dimension 1 from this by some effective means? This question has a beautiful negative answer by Joe Miller, and we include this and other related results in and around partial randomness as defined via notions of effective dimension like Hausdorff, packing, box counting and the like. We even include Lutz's notion of dimension for *finite* strings. Additionally, we look at the notion of the halting probability as an operator, showing that almost every random real is a halting probability relative to some oracle. This is a contrast to Stephan's result that suggests that halting probabilities should be rare as they have high computational power. The reconciliation is that relativization here does not involve coding. Finally we look at fundamental sets like the collection of non-random strings and ask what kind of computational power these sets of strings have as oracles, according to differing access mechanisms. Such results have impact in complexity theory as they seem related to separation questions of complexity classes following the work of Allender and his co-authors. The biggest part of the story missing in our book is the part of the randomness story low down in classes like EXP or the like. There simply was not space.

WHAT WE LEFT OUT, AND WOULD HAVE LIKED TO INCLUDE

Actually one reviewer asked us to give an account of what we did *not* include in the 870 or so pages which constitute the book. I did prepare an account and sent it to Denis, but in the end we decided that we did not have enough time to polish the material[§]. However, I will list what I had proposed below, and it should give you a perhaps idiosyncratic view of areas of randomness I would like to explore, and see not covered by our monograph.

The concept of randomness has held a central place in mathematics and computer science in recent years. Whilst Hirschfeldt and I feel somewhat (completely?) inadequate in our knowledge of all of the recent developments we offer a few. In most cases, there is, as yet, no applications to or of algorithmic randomness. In most practical applications of randomness in algorithms, it seems that only weak random sources seem to suffice *in practice*. For example, the advent of the Metropolis-Hastings algorithm in Markov Chain Monte Carlo algorithms has revolutionized Bayesian statistics in recent years, and is now the mainstay of applications of statical applications in science. A comprehensive account of these applications to wildly diverse areas of mathematics can be found in Diaconis (Bulletin of the AMS article), and it would take another monograph to describe applications of this methodology to physics and biology. There is certainly no current work in algorithmic randomness trying to speak to this material and it would rather nice to see such a development.

In number theory, Terry Tao and others have made profound advances in the theory of distribution of the primes by the hypothesis that the primes are random. Now of course the primes are not random, but the intuition of generations of number theorists is that they are *random enough*, once you discount all the obvious facts about them. For instance, long ago van der Waerden proved that if you two colour the integers then one or the other colour will have arbitrarily long arithmetical progressions. In a very deep advance, Szemeredi proved that "random-ish" or "big" sets, sets of positive upper density[¶], have arbitrarily long arithmetical progressions. Now all known proofs of this fact filter through a lemma which says that either a big set resembles a random set, or it is highly structured and in either case there will be long arithmetical progressions.

[§]For a mixed metaphor, we needed to kill the albatross.

[¶]That is sets of integers A such that $\lim_{n\to\infty} \frac{\{z:z\in A \wedge z\leq n\}}{n} > 0$.

Using this idea, based in and around Szemeredi's techniques and their heirs, Green and Tao proved that the primes contain arbitrarily long arithmetical progressions. An abstract view of this methodology is provided by the *Dense Model Theorem* of Green, Tao and Ziegler; with an interesting view being presented by Trevisan, Tulsiani, and Vadhan in the Proceedings of the Annual Conference on Computational Complexity, 2009. It would be extremely interesting to try to make all of this precise, but the mathematics seems very deep. A very readable discussion of this can be found in Tao's talk at the Madrid International Conference of Mathematicians, and his "blog book".

In some sense we have seen similar ideas applied in combinatorial group theory where people have looked at average and *generic* case behaviour of decision problems for groups. That is for generic complexity, we can look at algorithmic which give correct answers on a positive upper density version (usually density 1) of finitely generated groups. In generic case complexity, we ask that the algorithm only be partial, always give the correct answer, and halt with upper density 1. The original work can be found in a paper in the Journal of Algebra by Myasnikov, Kapovich, Schupp and Shpilrain. A typical theorem says that in a given finitely presented group, the word problem is linear time generically computable. This accords strongly with computer experiments. Much of this work relies on the fact that almost all groups are hyperbolic as shown by Gromov. Such groups are "almost free" in the sense of the work of Martin Bridson, and the logic has been analysed by the wonderful work of Sela. A small amount of the computability here has been developed. The general theory of such structures remains to be developed.

It is clear that it all seems related to Ergodic Theory, certainly in the case of additive number theory and Ramsey Theory. This insight was first realized by Furstenberg. One can obtain things like van der Waerden's Theorem using methods from ergodic theory which clearly can be viewed as involving randomness. A good discussion of this can be found in in the work of Avigad. There has been some work quantifying the amount of randomness needed for the Ergodic Theorem to work by Reimann and by Hoyrup and Rojas. It seems that it is sensitive to the formulation, and either involves Schnorr randomness or 2-randomness.

Again the harmononious interactions within mathematics are revealed in that the Tao work and the Ergodic Theorem are related to results in analysis, such as the Lebesgue theorem saying that monotone functions are differentiable almost everywhere as articulated by Tao in his blog. The constructivist, Oswald Demuth, had a programme where he argued that random functions should behave well. Theorems such as the Lebesgue theorem beg for analysis in terms of algorithmic randomness, and recently Demuth's programme has been extended Brattka, Miller and Nies. They show that the level of randomness needed to obtain differentiability of a computable continuous function on the reals is precisely computable randomness. That is, every computable continuous function of bounded variation on the real interval is differentiable at each computably random point, and conversely there are computable monotone continuous functions on the interval differentiable only at the computably random points.

Notice that to even state such theorems we need the language of computable analysis. To treat such interesting results would have required Hirschfeldt and I to to develop quite a bit of computable analysis. Certainly to develop at least as much as long initial segments of the classic books of Albeth, Pour-El and Richards or Weihrauch. It is interesting that our original work in the area of algorithmic randomness appeared in computable analysis conferences, yet so far there have not been many applications to analysis. It seems that Demuth's programme has a number of significant implications for computable analysis, and believe there will be a lot more work in this area. The fact that it connects to ergodic theory is the icing on the cake. Perhaps also related is the recent work of Braverman and Yampolsky on computability and Julia sets. Again in that work left c.e. reals play a prominent role, and hence it is all related to algorithmic randomness. We can only say that we did not have enough space to treat such material properly. Similar comments apply to applying algorithmic randomness to fundamental concepts of physics. A very nice recent example is the work of Kjos-Hanssen and Nerode, and Kjos-Hanssen and Gács on Brownian motion and thermodynamics. Again not only would Hirschfeldt and I have needed to develop the computable analysis, additionally we would be forced to develop some more high powered probability theory. All of this would clearly have taken many pages to develop and Hirschfeldt and I leave this to Volume 2 (joke).

In our book we treat algorithmic randomness on Cantor space. This is measure-theoretically identical with the real interval, but not homeomorphic to it. A great omission is the treatment of other spaces, particularly non-compact spaces. In the case of the reals, we have a natural measure we can use, namely Weiner measure. But in more general cases, the situation is reasonably murky. Again a rigorous development requires a substantial amount of analysis and computable analysis. It would go significantly be-

yond the scope of the book. Very recent work of Hoyrup, Rojas and Gács
bases the whole thing on what are called Scott domains. This is another
area which is ready for development.

As a "big picture" view of algorithmic randomness in more general
spaces, it seems that analogs of the basic results seem to work. As a "big
picture" view, it seems that analogs of the basic results seem to work. These
results include existence of uniform tests, neutral measures, conservation of
randomness. The article of Gács 2005 Theoretical Computer Science article
has a good overview of the history. At Gács homepage there is a treatment
of this material in a set of online lecture notes.

Given the focus within our book on the interactions of randomness and
computability, Cantor space is a very natural domain. Even in the case
of Cantor space, there is more to be said in terms of non-uniform measures. We allude to some of this when we discuss the work of Levin, Kautz,
Reimann and Slaman. But the recent deep work of Reimann and Slaman
on *never continuously random* reals would need a substantial treatment of
set theory to do them justice and so in our book Denis and I only state
some results, and but do prove some of those that don't need overtly set-
theroretical methods. This is deep and important work I think.

In computational complexity theory there has been a number of major
developments regarding randomness[||]. To even state the results would need
quite a bit of development and we will give a breezy overview, referring the
reader to the (mildly) cited works for more details. referring the reader to
the cited works for more details. To treat these applications would need at
least another book. One strand began with the work of Valiant-Vazirani
who showed that SATISFIABILITY reduces to UNIQUE SATISFIABILITY
(i.e., at most one satisfying assignment) under randomized polynomial time
reductions. This attractive theorem led to Toda's Theorem showing that
$P^{\#P}$ contains the polynomial time hierarchy. Here $\#P$ is the counting analog of NP where we count the number accepting paths in a polynomial time
Turing machine. Toda's Theorem states that if we can have the advice of
such a counting oracle, then we can solve any problem in the polynomial
time hierarchy. All known proofs of Toda's Theorem work through the operator BP, which is the probability version of NP, namely that most paths
lead to the correct answer.

Another strand leads to two important generalizations of the notion of
"proof": interactive proofs and probabilistically checkable proofs. NP can

---

[||]I thank Eric Allender for some thoughts and corrections in this account be viewed as the class of sets A such that a prover can provide a short "proof of membership" for any element of A, which can be verified deterministically. A more general notion arises if one allows the verifier to be a probabilistic process; allowing this probabilistic verifier to hold a conversation with the prover (and allowing the verifier to be convinced by statistical evidence) leads to *interactive proofs* (and the class IP consisting of those languages where the prover can convince the verifier about membership). Initially, people suspected that IP would be a "small" augmentation to NP (for instance, there are oracles relative to which IP does not contain coNP by results of Fortnow and Sipser), and thus it came as a surprise when Babai, Fortnow, Lund, and Nisan showed that IP contains $P^{\#P}$ (and hence contains the polynomial hierarchy, by Toda's theorem). Shamir subsequently showed that IP is equal to PSPACE. The key step in the proof is to translate a PSPACE problem into a problem about polynomials over a finite field (via a process called *arithmetization*). The point of the large field is that two polynomials can intersect only rarely if they are not identical.

In an IP protocol, the prover is allowed to give different responses to identical queries from the verifier, along different computation paths. In the so-called multi-prover interactive proof model (MIP), this power is essentially taken away; this actually results in a *more* powerful system. Babai, Fortnow, and Lund showed that MIP = NEXP. This is extremely counterintuitive, because it means that an *exponential-size* proof can be verified by a probabilistic verifier who has only enough time to ask about a small part of the proof. The next step (in some sense, a "minimization" of the MIP=NEXP characterization) was an analogous characterization of NP in terms of Probabilistically Checkable Proofs; this is known as the PCP theorem of Arora, Safra, and Arora, Lund, Motwani, Sudan, Szegedy, and it is recognized as one of the crowning achievements of computational complexity theory (even thought it is only really a *reduction*!). The PCP Theorem states, roughly, that any problem in NP has a proof which is polynomial in length and can be verified in polynomial time with arbitrary high accuracy by checking a constant number of bits and running a logarithmic number of random tests on the proof. Stated another way, this says that *any* proof can be encoded efficiently in such a way, so that a referee need not read the entire proof, but can simply pick, say, 15 random bits in the paper and see if these bits satisfy some local consistency condition, and certify the proof as "correct" if no error is detected in these 15 bits; no correct proof will be rejected in this way, and the probability that an incorrect proof is accepted is less than the probability that an incorrect proof is accepted in the traditional method of journal refereeing.

The use of randomness has seen a lot of work to try to understand how it is possible to get sources of randomness. One idea is to take some weak source and try to extract nearly true randomness from it, or to extend the length of the source whilst still hoping to have the output "look random". This is an area of much deep work and Hirschfeldt and I briefly touch on this material when we look at packing and Hausdorff dimensions. The methods here are very sophisticated and again would need a lot of deep mathematics developed to treat them properly. Proper utilization of pseudo-random sources is especially important in cryptography.

Much work in probabilistic algorithms is centered on tasks for which we we have randomized algorithms, such as deciding polynomial identities, but for which we have absolutely no idea how to construct deterministic polynomial time algorithms. However, recently it has become apparent that perhaps this is merely because of our lack of brain power, rather than because deterministic algorithms are inherently weaker. That is, it is widely believed that BPP=P. Namely anything we have a randomized polynomial time algorithm for can actually be derandomized. This has happened in the case of primality testing, where the Solovay-Strassen algorithm gave a randomized algorithm for primality and subsequently Agrawal, Kayal, and Saxena gave a deterministic algorithm. Now it it is widely believed that not only is SATISFIABILITY hard, but *very hard*, in that it requires circuits of nearly exponential size (that is, for each input length, it is conjectured that the smallest circuit computing the function correctly for that input length has size $2^{\epsilon n}$ for some $\epsilon > 0$. Remarkably, Impagliazzo and Wigderson have shown that if there is any problem in Dtime(2^n) (such as SATISFIABILITY that requires circuits of size $2^{\epsilon n}$ then P=BPP; that is, *all* probabilistic algorithms can be derandomized.. Again it is not clear what the intuition really is connecting nonuniform complexity and randomness. Again this is way, way beyond the present book and we refer to Kozen's book for basic material and Wigderson's article in the proceedings of the Madrid International Congress of Mathematicians for an excellent overview.

Another nice application of the ideas of effective measure in computational complexity comes from the work of Lutz and his co-workers. The idea here is that if we believe, say, that P$\neq$NP then we should quantify how big P is within NP. Using these ideas, those authors have shown that various measure theoretical hypotheses on the structure of complexity classes in terms of small or large measure have significant consequences for separations. Here we refer to Lutz's or Hitchcock's web page for references and more details.

There are myriad other applications in computational complexity, not the least of which are those where combinatorial arguments are replaced by Kolmogorov complexity ones. For instance, constructing "hard" inputs on which any simple algorithm must make an error can be difficult, but frequently it is easier to show that all simple algorithms must make an error on *random* inputs. I think the intuition here is the following: We have some algorithm we wish to show has certain behaviour. We use that fact that a string of high Kolmogorov complexity should behave in the expected way as the process is algorithmic. If it did not then it could be compressed.

For example, we can use this idea to prove worst case running times for various sorting algorithms, or to prove the Hastad switching lemma. There are many, many applications of this form as can bee seen in in Li and Vitányi's book. Li and Vitányi's book is also an excellent source of other applications of Kolmogorov complexity to things like thermodynamics, computational learning theory, inductive inference, biology and the like. Again Hirschfeldt and I though this was beyond our ken and space or time available.

Another area we did not develop is the time and space bounded versions. This goes back to work of Levin on Kt complexity, and we refer to the excellent surveys of Allender here for more details and to Li and Vitányi's book for background and historical results.

We remark that Vitányi and other authors have used the notion of Kolmogorov complexity to explore the common information between strings in analysis of things in real life. For example in computational biology to try to compare two phylogenetic trees we would invent metrics such as what is called MAXIMUM PARSIMONY, and MAXIMUM LIKELIHOOD. The idea is to replace this with a more general measure of normalized relative Kolmogorov complexity. Of course this is great, except that computing the complexities is undecidable as we see in the present book. I applications the idea has been to use commercial text compression packages like GZIP and see what happens. Vitányi and his co-authors have has some success with music evolution. Again more works need to be done.

Finally while Hirschfeldt and I did develop some aspects of the Kolmogorov complexity of finite strings they were in support of our aims. It is certainly the case that there is a lot more, especially from the Moscow school flowing from the students of Kolmogorov. We can only point at the work of Shen, Muchnik, Vereshchagin, Uspenskii, Zimand and others.

THE MOST IMPORTANT OPEN PROBLEMS

What we have is a vast enterprise devoted to the themes above. The picture is still emerging with a number of very interesting technical questions still open. For example, if we allow "nonmonotonic" betting strategies, does computable randomness in this new sense coincide with Martin-Löf randomness? Is even a little bias allowable, etc? By this I mean suppose that I have a betting strategy which at some stage might favour some side, and if so is then committed to favouring that side. Is the resulting notion of randomness the same as Martin-Löf randomness**? As mentioned above, what about extensions to non-compact spaces such as the work of Peter Gács, and how should we understand independence as suggested by seminal works of Levin? (Actually, even reading some of those early works of Levin is challenging enough.) There are many questions in and around randomness, independence, computability and you should look at the survey of Miller and Nies or at either Nies' or our book here. Of course there are fundamental questions in complexity such as the conjecture that $BPP = P$. That is, there are a host of algorithms which efficiently solve problems using randomness as a resource. It seems that all of these can be derandomized, or at least that is the current thinking of people working in complexity. How should this be proven? What are the languages efficiently reducible to the collection of non-random strings. We know it contains PSPACE but is the intersection of such languages the same? Is the definition even machine independent? There are a host of questions, and I would hope they would be articulated by other contributors to this volume.

THE PROSPECTS FOR PROGRESS

In the future I would hope that this machinery would see applications in our understanding of things like physics and other sciences, as well as other parts of mathematics. We live in a world where most processes are finite and their approximations are computable. Thus the notion of expectation aligns itself to Kolmogorov complexity, as we mentioned in the beginning. Thus, one would think that algorithmic randomness should be an excel-

---

**The point here is that the usual proof that approximable effective betting strategies gives the same randomness as the Martin-Löf definition allows us first to favour $\sigma0$ and the perhaps later we might choose to favour $\sigma1$.

lent approximation to "true" randomness whatever that means, quantum or otherwise. Surely we could use algorithmic randomness to better understand physical processes. As mentioned above, Kjos-Hanssen and Nerode have some initial forays on this in Brownian motion. There is also work by people like Vitányi and his group on things like phylogeny and sequence matching using approximations to Kolmogorov complexity as a measure of common information rather than using things like minimum distance etc. The idea is that rather than figuring out the best metric to see how similar two things are, use the fact that something like a normalized version of the relative Kolmogorov complexity is a universal notion of information distance. The problem, of course, is that it is not computable. But we might be able to use compression algorithms like ZIP or the like to approximate. This method apparently works well for evolution of music. Even biological processes would seem to need randomness to understand them. But here I am into the world of wild speculation. Information is everywhere, and we are developing tools to try to understand it. I think the future of this area is fascinating.

Chapter 19

Is Randomness Native to Computer Science? Ten Years After[*]

Marie Ferbus-Zanda and Serge Grigorieff

LIAFA, CNRS and Université Paris 7

ferbus@liafa.jussieu.fr

seg@liafa.jussieu.fr

WHAT DREW US TO STUDY ALGORITHMIC RANDOMNESS

Authors: Hello! Long time no see. . .

Quisani: Nice to meet you again. Remember the discussion we had some years ago about randomness and computer science[8]? I would be pleased to go back to it. Before entering the many questions I have, let me ask you about your motivation to look at randomness.

19.1. Paradoxes Around Algorithmic Information Theory

A: Main motivation comes from paradoxes. They are fascinating, especially for logicians. Remember that the liar paradox is at the core of Gödel incompleteness theorem. Around the theory of algorithmic randomness there are (at least) two paradoxes.

Q: Let me guess. Long ago you told me about Berry's paradox: *"the least integer not definable by an English sentence with less that twenty words"* is just defined by this very short sentence. I remember that Kolmogorov built the mathematical theory of the so-called Kolmogorov complexity out of this paradox. The core idea being to turn from definability by sentences in a natural language (which is a non mathematical notion) to computability

---

[*]A sequel to the dialog published in Ref. 8 by the authors in Yuri Gurevich's "Logic in Computer Science Column", Bulletin of EATCS, 2001. The dialog involves Yuri's imaginary student Quisani.

and so replace English sentences by programs in any formal mathematical model of computability. What is the second paradox?

A: The distortion between common sense and a very simple mathematical result of probability theory. Namely, *if we toss an unbiased coin 100 times then 100 heads are just as probable as any other outcome!* Who really believes that the coin is fair if you get such an outcome? As Peter Gács pleasingly remarks (Ref. 14), *this convinces us only that the axioms of Probability theory, as developed in Kolmogorov, 1933,[18] do not solve all mysteries that they are sometimes supposed to.* This paradoxical result is really at the core of the question of randomness. Now, algorithmic randomness clarifies this paradox. There is no way to discriminate any special string using only probabilities: every length 100 binary string has the same probability, namely 2^{-100}. But this is no more the case with Kolmogorov complexity. Most strings have Kolmogorov complexity almost equal to their length whereas such strings as the 100 heads one have very low Kolmogorov complexity, much lower than their length. So getting a string with complexity much less than its length is a very special event. Thus, it is reasonable to be suspicious when you get an outcome of 100 heads out of 100 tosses.

Q: Of course, when you say "almost equal" and "low" it is up to a constant. Kolmogorov complexity is not uniquely defined but is really any function from a certain class $\mathcal{C}$ of so-called optimal functions *objects* $\mapsto \mathbb{N}$ such that any difference $|f - g|$, for $f, g \in \mathcal{C}$, is bounded.

A: Yes. For optimal functions obtained from enumerations of partial computable functions associated to particular models of computability, the so-called universal partial computable functions, the bound somewhat witnesses the particular chosen models. Let us quote what Kolmogorov said about the constant:[19] *Of course, one can avoid the indeterminacies associated with the [above] constants, by considering particular universal functions, but it is doubtful that this can be done without explicit arbitrariness. One must, however, suppose that the different reasonable [above universal functions] will lead to complexity estimates that will converge on hundreds of bits instead of tens of thousands. Hence, such quantities as the complexity of the text of War and Peace can be assumed to be defined with what amounts to uniqueness.*

19.2. Formalization of Discrete/Continuous Computability

Q: Great quote. Any other motivation to look at algorithmic randomness?

A: Fascination for the formalization of intuitive a priori non mathematical notions. Randomness is a topic in mathematics studied since the 17th century. Its origin is in gaming. Dice probability puzzles were raised by the Chevalier de Méré (1654). This lead to some development of probability theory. Observe that trying to formalize chance with mathematical laws is somewhat paradoxical since, a priori, chance is subject to no law. Understanding that, in fact, there were laws was a breakthrough. Nevertheless, what is a random object remained a long open standing problem.

Q: Finding the adequate mathematics to illuminate and make precise large parts of intuitive notions seems to be a long story in mathematics.

A: The most striking successes are the introduction of the logical language and of proof systems by Gottlob Frege, 1879,[12] bringing positive answers to Leibniz's quests of a "lingua caracteristica universalis" and of a "calculus ratiocinator".

Q: And how not to be excited by the convincing formalizations of the notions of computability which appeared in the twentieth century through the work of Turing, Church, Herbrand-Gödel? A real beacon of achievement.

A: You probably have in mind computability over discrete domains such as the integers or finite words. Computability over continuous domains such as the reals has also been convincingly formalized.

Q: Yes, using machines working with infinite discrete inputs and outputs representing reals. Big names being Turing, Grzegorczyk, Lacombe.

A: There is also a radically different approach which does not reduce computability in the continuous world to that in the discrete world. This is the analogic approach, due to Claude Shannon,[26] which is rooted in analysis and uses differentiation and integration of functions over the reals. This approach is less known, rather unrecognized, in fact, and this is a real pity. It is based on General Purpose Analog Computers (GPAC). These are circuits, with possible loops, built from units computing arithmetic and constant operations over the reals and the integration operator $(u, v) \mapsto \int u(x)v'(x)dx$. Shannon proved that the functions over the reals which are computed by GPACs are exactly the solutions of algebraic differential systems. Though such functions do not include all functions over reals which are computable in the Turing approach, a variation of Shannon's notion of GPAC computability allows to exactly get them, (cf. Bournez *et al.*[5] or Graça[16]).

Q: Oh, that reminds me of the bottom-up vs top-down phenomenon met when studying algorithms for trees and graphs. The Turing approach can be qualified as "bottom-up" since computability over reals is somewhat seen as a limit of computability of discrete approximations. On the contrary, what you tell me about GPACs is relevant of a top-down approach where computable functions are isolated among all ones via differential systems.

A: The paradigmatic example in computer science occurs in programming. In Ferbus-Zanda,[11] the bottom-up vs top-down phenomenon is viewed as a duality which can be pin-pointed in many places. For instance, in programming, an iteration (such as a loop) is relevant to the bottom-up approach whereas an induction can be qualified as top-down.

WHAT WE HAVE LEARNED? A PERSONAL PICK

19.3. From Randomness to Complexity

Q: Let us go back to our discussion some years ago about randomness and computer science.[8] You explained me how, in 1933, Kolmogorov founded probability theory on measure theory, letting aside the question "what is a random object?". And, up to now, probability theory completely ignores the notion of random object.

A: Somehow, probability theory deals about randomness as a global notion, finding laws for particular sets of events. It does not consider randomness as a local notion to be applied to particular events.

Q: Thirty years later, Kolmogorov went back to this question[19] and defined a notion of intrinsic complexity of a finite object, the so-called Kolmogorov complexity. Roughly speaking, to each map $\varphi : program \to object$ he associates a map $K_\varphi : object \to \mathbb{N}$ such that $K_\varphi(x)$ is the length of shortest programs p which output the object x (i.e., $\varphi(p) = x$). He proved that among the diverse K_φ with φ partial computable, there are minimum ones, up to an additive constant. To be precise, K_θ is minimum means that $K_\theta(x) \leq K_\varphi(x) + O(1)$ (that is, $\exists c \ \forall x \ K_\theta(x) \leq K_\varphi(x) + c$) is true for all partial computable φ. Any of these minimum maps is called Kolmogorov complexity. They are viewed as measuring the information contents of an object. I remember that the story about how to use Kolmogorov complexity to define random objects is that of a rocky road.

A: This is exactly that. Let us add that the question whether Kolmogorov complexity is a kind of intrinsic complexity of an object has been much discussed. An interesting connected notion emerged in 1988 with the work of Bennett. He introduced the so-called logical depth of an object which is the collection of maps D_s : *objects* $\to \mathbb{N}$ such that $D_s(x)$ is the shortest duration of the execution of a program p outputting x and having length in $[K(x)-s, K(x)+s]$, i.e., having length s-close to the Kolmogorov complexity of x. Bennett logical depth is much used, especially in biology.

19.4. Formalization of Randomness: Infinite Strings

Q: I also remember that you classified the diverse formalizations of the notion of random infinite binary sequence using the bottom-up and top-down paradigms. Let me recall what I remember. Per Martin-Löf top-down approach, 1965,[21] discriminates random infinite sequences in the space $\{0,1\}^\omega$ of all infinite sequences: random means avoiding all Π_2^0 subsets of $\{0,1\}^\omega$ constructively of measure zero (the so-called Martin-Löf tests). The bottom-up approach uses the prefix-free version H of Kolmogorov complexity. H is defined by considering the sole functions φ : *program* $\to$ *object* which have prefix-free domains: if $\varphi(u)$ and $\varphi(v)$ are both defined and $u \neq v$ then none of u, v is a prefix of the other. Using H, it has been proved that $\alpha \in 2^\omega$ is random if and only if $H(\alpha(0) \ldots \alpha(n)) \geq n + O(1)$.

A: For the bottom-up approach to randomness one can also use other variants of Kolmogorov complexity. For instance, with Schnorr process complexity S or Levin monotone complexity[20] Km, the inequality $\geq n + O(1)$ can be replaced by an equality $= n + O(1)$. This is false for prefix-free Kolmogorov complexity: one can show that (with random sequences) the difference $H(\alpha(0) \ldots \alpha(n)) - n$ grows arbitrarily large. In fact, one can also use plain Kolmogorov complexity C: $\alpha \in 2^\omega$ is random if and only if $C(\alpha(0) \ldots \alpha(n)) \geq n - g(n) + O(1)$ for all computable $g : \mathbb{N} \to \mathbb{N}$ such that the series $\sum_i 2^{-g(i)}$ is convergent. The proof of this result, due to Miller and Yu, 2004,[21] has since be reduced to a rather simple argument, cf. Bienvenu *et al.*[4]

Q: So, Kolmogorov's original idea relating randomness and compression does work. Even in the naive way with Schnorr complexity and with Levin monotone complexity.

A: There is another interesting way to look at randomness with the idea
of compression. Recently, Bienvenu & Merkle[4] obtained quite remarkable
characterizations of random sequences in the vein of the ones obtained using
Kolmogorov complexity. Their basic idea is to consider Kolmogorov complexity in a reverse way. Instead of looking at maps F : *programs* → *strings*
they look at their right inverses Γ : *strings* → *programs*. Those are exactly
the injective maps. If F has prefix-free domain then Γ has prefix-free range.
The intuition is that we associate to an object a program which computes
it. Hence the denomination "compressors", and "prefix-free-compressors"
when the range is prefix-free. It seems clear that the theory of Kolmogorov
complexity and the invariance theorem can be rewritten with compressors.
In particular, if Γ is partial computable then the map $x \mapsto |\Gamma|$ is greater
than Kolmogorov complexity up to a constant. The same holds with prefix-
free compressors and the prefix-free version of Kolmogorov complexity. Surprise: it is sufficient to consider *computable compressors* rather than partial
computable ones to characterize randomness. An infinite binary string α
is random if and only if $|\Gamma(\alpha(0) \cdots \alpha(n))| \geq n + O(1)$ for all computable
compressors having prefix-free ranges. A version à la Miller & Yu with no
prefix-freeness also holds.

Q: Any other characterization of randomness?

A: There is a very important one which deals with martingales and constitutes another top-down approach to randomness. It was introduced by
Peter Schnorr. A martingale is just a map $d : \{0,1\}^* \to [0, +\infty[$ such that
$d(u) = d(u0) + d(u1)$ for all $u \in \{0,1\}^*$. Suppose you are given an infinite
binary sequence α and you successively disclose its digits. A martingale
can be seen as a betting strategy of the successive digits of α. Your initial
capital is the value of d on the empty string. After digits $\alpha(0), \ldots, \alpha(n)$
have been disclosed, your capital becomes $d(\alpha(0) \ldots \alpha(n))$. So, the addi-
tivity condition on d is a fairness assumption: the expectation of your new
capital after a new digit is disclosed is equal to your previous capital. Let
us say that d is winning against α if $d(\alpha(0) \ldots \alpha(n))$ takes arbitrarily large
values. Which means that its sup limit is $+\infty$. Schnorr[22] proved that α is
random if and only if is no c.e. martingale wins against α.

Q: What is a c.e. martingale?

A: A martingale is computably enumerable, in short c.e., if its values are
computably approximable from below. Technically, this means computable

enumerability of the set of pairs (u, q) such that u is a finite string and q is a rational less than $d(u)$.

19.5. Random versus Lawless

Q: This relates randomness to unpredictability. So random sequences are somewhat chaotic and do not obey any law.

A: No, no. Random sequences do obey probability laws. For instance the law of large numbers and that of the iterated logarithm. Though they are unpredictable, random sequences are not lawless. An interesting notion of lawless sequence has been introduced by Joan Moschovakis, 1987–94.[23,24] She developed it in the framework of constructive mathematics, so her work has not received in the randomness community the attention it deserves. She deals with infinite sequences of non negative integers but her ideas apply mutatis mutandis to binary sequences. The notion she introduces is relative to a given family of so-called "lawlike" sequences and sets of integers which has to be closed under relative computability. Let us describe her ideas in the simplest framework where lawlike means computable. A binary sequence α is *lawless* if for any computable injective map $\gamma : \mathbb{N} \to \mathbb{N}$ and any computable map $\beta : \{0, 1\}^* \to \{0, 1\}^*$ there exists a prefix u of α such that the string $u\beta(u)$ (obtained by concatenation) is also a prefix of α. Think of γ as selecting and permuting an infinite subsequence of α and think of β as a predictor function: its role is to guess a string that comes next to a prefix. Thus, α is lawless if any computable predictor is correct for at least one prefix of any permuted infinite subsequence of α (obtained via a computable process). Of course, "correct for at least one prefix" implies "correct for arbitrarily large prefixes" and also "incorrect for arbitrarily large prefixes".

Q: This selection process has some common flavor with von Mises' notion of "kollectiv". What is known about lawless sequences?

A: You are right. It is easy to see that the family of lawless sequences is a Π_2^0 subset of $\{0, 1\}^\omega$. A simple construction shows that this family is non empty. In fact, it is dense in $\{0, 1\}^\omega$. Finally, it is constructively of measure zero hence is disjoint from the family of random sequences. To prove this last assertion, let $A_k(n)$ be the set of infinite strings such that all digits of ranks n to $n + k$ are zeroes. Observe that the union A_k of $A_k(n)$'s, $n \in \mathbb{N}$, has measure $\leq 2^{-k}$ hence the intersection A of all A_k's (which is a Π_2^0 set) is a Martin-Löf test. Letting γ be the identity and β_k map a string x to a string of $k + |x|$ zeroes, observe that the lawless condition insures that any lawless sequence is in some $A_k(n)$ hence in A_k for all k, hence in A.

19.6. Randomness and Finite Strings: Incompressibility

Q: Going back to the 100 heads outcome, the intuition of randomness is related to Kolmogorov complexity being close to the length of the string. In other words, for finite strings randomness means incompressibility. That incompressibility is a necessary condition for randomness seems clear: a compressible string has redundant information and this contradicts the idea of randomness. But why is it a sufficient condition? Is this taken for granted or is it possible to get strong arguments in favor of such an identification?

A: Martin-Löf gave a convincing argument showing that failure of randomness implies compressibility. Let us illustrate it on an example. Fix some real r such that $0 < r < 1/2$ and consider the set A_n of all strings of length n with $< rn$ zeros. Some calculation shows that the cardinal $N(n)$ of this set is asymptotically dominated by 2^n, which means that $\rho(n) = N(n)/2^n$ tends to 0 when n increases to $+\infty$.

Q: Well, this is essentially the proof of the law of large numbers, is not it?

A: Sure. Now, let us represent a string in A_n by the binary representation of its rank for the lexicographic ordering on A_n. This gives a way to describe any string in A_n by a binary "program" with length $\log(N_n) = \log(2^n) + \log \rho(n) = n + \log \rho(n)$. Since $\rho(n)$ tends to 0, the logarithm tends to $-\infty$.

Q: Wait, to describe a string of A_n in this way, you also need to know the set A_n. Which reduces to know the length n of the produced string. Thus, such a description amounts to a program which produces a string using its length as an input. So, this involves conditional Kolmogorov complexity and only proves that the length conditional Kolmogorov complexity of any string in A_n gets arbitrarily less than its length n when n grows. In other words, for any $c \in \mathbb{N}$, when n is large enough, all strings in A_n are c-length conditional compressible. This relates failure of equidistribution of zeroes and ones to length conditional compressibility. Not to compressibility itself!

A: It turns out that compressibility and length conditional compressibility are tightly related: Martin-Löf proved that c-length conditional compressibility implies $(c/2 - O(1))$-compressibility.

Q: Is this particular example an instance of a general result?

A: Yes, Martin-Löf developed a notion of statistical test for binary strings quite similar to that for infinite strings. This is a family $(V_i)_{i \in \mathbb{N}}$ of sets of binary strings (also called "critical sets") which satisfies three properties: 1) it is decreasing with respect to set inclusion, 2) the relation $\{(i, u) \mid u \in V_i\}$ is recursively computable and 3) for all n, the proportion of strings of length n in V_i is at most 2^{-i}. Intuitively, V_i is the set of strings which fail randomness with significance level 2^{-i}.

Q: So, failure of equidistribution can be turned into being in some V_i for some statistical test. Now, this should be turned into compressibility.

A: Yes, Martin-Löf proved is that there is a largest statistical test $(U_i)_{i \in \mathbb{N}}$. Largest up to a shift: there is some d such that $V_i \subseteq U_{i+d}$ for all i. And this largest test can be chosen so that being in U_i implies being i-compressible.

19.7. Representation and Kolmogorov Complexity

A: Kolmogorov wanted a universal notion of complexity of finitary objects which would therefore be robust. Nevertheless, it turns out that Kolmogorov complexity does depend on the representation of objects.

Q: Can there be different interesting representations of integers which would not be essentially equivalent as concerns Kolmogorov complexity? Wait, if f is computable then the complexity of $f(x)$ is bounded by that of x up to a constant.Thus, for an injective f, we have equality up to a constant.

A: That is right. Let us consider non negative integers. Suppose you represent them as words so that you can computably go from that representation to the usual unary representation and vice versa. Then your argument proves that the associated Kolmogorov complexity is equal (up to a constant) to the usual one. But there are many ways to represent integers for which there is no computable translation with the unary representation.

Q: You mean mathematical representations like Russell representation of an integer n as the family of all sets with exactly n elements?

A: Exactly. Such a mathematical representation deserves to be called a semantics. There are other ones. A variation of Russell semantics is to consider n as the family of all equivalence relations with exactly n equivalence classes. A very interesting semantics, due to Alonzo Church, views n as the functional which iterates a function n times.

Q: Such semantics are set theoretical and deal with classes of sets. You need some effectivization.

A: Sure. Instead of all sets, consider the sole computably enumerable subsets of $\mathbb{N}$. Similarly, consider the sole computably enumerable equivalence relations on $\mathbb{N}$. Finally, for Church, consider the sole functionals associated to terms in lambda-calculus.

Q: So, a representation is now a partial computable function which maps a program to a code for a computably enumerable set or relation or a lambda term which is considered as a functional! OK, one can surely prove a version of the invariance theorem and define the Kolmogorov complexity for Russell as the length of a shortest program mapped by a universal map onto a code for a c.e. set with exactly n elements. For the index semantics, we just replace c.e. sets by c.e. relations and for Church we consider lambda terms.

A: You got it. Now, what do you think? More complex the semantics, higher the associated Kolmogorov complexity of the induced representation of integers?

Q: I would say that Russell semantics is less complex than the index one and that Church is the most complex one.

A: Surprise! Ferbus-Zanda & Grigorieff proved in Ref. 9 that Church semantics leads to the usual Kolmogorov complexity C. The index semantics leads to that with the first jump $\emptyset'$ oracular Kolmogorov complexity $C^{\emptyset'}$. As for Russell, it leads to something strictly in between.

Q: How do you interpret such results?

A: A semantics for integers can be viewed as an abstraction of the set of integers. In some sense, such results allow to measure the abstraction carried by the diverse semantics of integers.

Q: What about negative integers?

A: If you represent them just as positive integers augmented with a sign, you get the same Kolmogorov complexities. Now, you can use the usual representation of an integer as the difference of two non negative ones. For Church, we again get the usual Kolmogorov complexity. But for Russell and index we get the oracular Kolmogorov complexities with the first and second jumps respectively.

19.8. Prefix-freeness

Q: I am still puzzled about the prefix-free condition. Plain Kolmogorov complexity is so natural. Restriction to partial computable functions with prefix-free domains seems quite strange. What does it mean?

A: Chaitin argued about it as self-delimitation: the program stops with no external stimulus. This is a common feature in biology. For instance, your body grows continuously while you are a child and then stops with no external signal to do so. Why is it so? Biological experiments have shown that the genetic program which rules the body growth contains a halting command: *programmed cell-death* or *apoptosis* not governed by the outside, a kind of self-delimitation.

Q: You mentioned Miller & Yu's result which characterizes randomness with plain Kolmogorov complexity instead of the prefix-free version. The price being a correcting term involving convergent series. What about other results in the theory. For instance, what about Chaitin Omega number?

A: There is a version of Omega which works with plain Kolmogorov complexity. Let us first recall what is known with Kolmogorov prefix-free complexity. Recall that $U : \{0,1\}^* \to \{0,1\}^*$ is prefix-free optimal if U is partial computable with prefix-free domain and, up to a constant, the Kolmogorov prefix-free complexity of a string x is equal to the length of shortest programs p such that $U(p) = x$. The original result by Chaitin (cf. the footnote on page 41 of his 1987 book) states that if U is optimal and A is any computably enumerable non empty subset of $\{0,1\}^*$ then the real

$$(*) \quad \Omega[A] = \mu(\{\alpha \in \{0,1\}^\omega \mid \exists i \; U(\alpha(0)\ldots\alpha(i-1)) \in A\})$$

is random. This real is the probability that a finite initial segment of α is mapped in A. This has be extended in Becher & Figueira & Grigorieff & Miller, 2006,[2] and in Becher & Grigorieff,[1] (cf. Theorem 2.7 and the addendum on Grigorieff home page) to randomness with iterated jumps as oracles. Namely, if A is Σ_n^0 many-one complete then $\Omega[A]$ is n-random (which means random with the $(n-1)$-th jump as oracle. The same results hold with the probability

$$(*)_k \quad \Omega[k, A] = \mu(\{\alpha \in \{0,1\}^\omega \mid \exists i \geq k \; U(\alpha(0)\ldots\alpha(i-1)) \in A\})$$

that an initial segment of length $\geq k$ of an infinite string is mapped in A.

A: Wait. If U has prefix-free domain then an infinite string has at most one prefix in the domain of U. So the definition you consider for $\Omega[A]$ coincides with the usual one which is the sum of all $2^{-|p|}$ such that p is a finite string and $U(p) \in A$.

Q: Sure. But with plain Kolmogorov complexity, when U is partial computable with a non prefix-free domain, we have to stick to definitions $(*)$

and $(*)_k$. Recall that usual optimal maps for plain Kolmogorov complexity are obtained from enumerations of partial computable maps and are universal "by prefix-adjunction". That means that $U(0^e1p) = \varphi_e(p)$ if φ_e is the e-th partial computable map. All this being said, the same randomness and n-randomness results are proved in Ref. 1 for the real $\Omega[k, A]$ with the condition that k is large enough and that U is universal by prefix-adjunction. None of these conditions can be removed.

19.9. Approximating Randomness and Kolmogorov Complexity

Q: What about applications of randomness?

A: Most obvious topic to use random sequences is cryptography . But there is a problem: no random real is computable! Von Neumann, 1951[28] pleasingly stated the problem : *"Anyone who considers arithmetical methods of producing random reals is, of course, in a state of sin. For, as has been pointed out several times, there is no such thing as a random number there are only methods to produce random numbers, and a strict arithmetical procedure is of course not such a method.* Clearly, von Neumann implicitly refers to computable things when he says "there is no such thing as". To get simultaneously computability and randomness, one has to lower the randomness requirement with time or space bounds. But cryptography has a new requirement: encryption should be easy whereas decryption should be hard. This involves problems which are quite different of those of algorithmic information theory.

Q: So, is there any concrete application of Kolmogorov complexity?

A: Yes. Cilibrasi & Vitányi, 2005,[6] developed a very original use of approximations of Kolmogorov complexity to classification via compression (see also Ferbus-Zanda[11]). Such approximations are those given by usual compressors like gzip,. . . It gives spectacular results. We cannot enter the details but let us say that this approximation keeps the basic conceptual ideas of Kolmogorov complexity.

19.10. Randomness, Kolmogorov Complexity and Computer Science

Q: Is randomness native to computer science?

A: Going back to your question about our motivation, the last exciting thing about randomness is that its formalization is rooted in computer science. As Leonid Levin claims on his home page, *while fundamental in many areas of science, randomness is really "native" to computer science.*

Q: This is quite a definite assertion! Maybe too much definite?

A: Look, all known formalizations of the intuitive notion of random object go through computability and/or information theory. Computability for the approach using Martin-Löf tests, that is Π_2^0 subsets of $\{0,1\}^\omega$ which are constructively of measure zero. Computability plus information theory for the approach using Kolmogorov complexity in the prefix-free version by Chaitin and Levin. The denomination *algorithmic information theory* fully witnesses this double dependence. These two subjects, computability and information theory, are largely relevant to computer science and are, indeed, central in computer science.

Q: But computability and information theory are also mathematical topics using ideas having no relation with computer science. Would you consider that Turing degrees or developments in the theory of finite fields are fully relevant to computer science?

A: Of course, computability and information theory have their own life and take ideas, intuitions and methods outside computer science. But, take a historical point of view. Algorithms were developed more than 2000 years ago in Mesopotamy and Ancient Egypt. Euclid algorithm to compute the greatest common divisor of two integers is still commonly used today. For centuries, computability remains a collection of algorithms. The sole theoretical results relevant to a kind of computability theory are about the impossibility of geometrical constructions with ruler and compasses (Gauss and Wantzel works). But computability, as a mathematical theory, emerges (let us say that it wins its spurs) only with the development of machines. And machines are at the core of computer science and become a real discipline with computer science.

In a similar way, randomness is three centuries old but the central notion of random outcome has been clarified only with computability theory and machines. This has shed some light on randomness completely different from that given by probability theory.

Q: OK, randomness is pervaded by computer science. As for being native to computer science, hum.... Is randomness native to anything?

A: One last point. You mentioned the theory of Turing degrees as a subject having few to do with computer science. We disagree with that opinion. Turing degrees involve methods like the priority methods which bring deep knowledge of asymmetry. Indeed, they deal with the construction of computably enumerable sets which are highly asymmetrical objects. If an object is in the set, one eventually knows about that using a simple loop program. But there is no general algorithmic method to insure that an object is not in the set. Now, asymmetry is also one of the big problems in computer science which comes from non determinism. Think of the P=NP problem. Up to now, no use of priority method has entered computer science problems around non determinism. But who knows....

Also, there are deep results in randomness theory involving Turing degrees. Let us mention a result due to Kučera and Gács[15] which insures that every non computable sequence is equicomputable with some random one (i.e., each sequence can be computed with the other one as an oracle).

19.11. Kolmogorov Complexities and Programming Styles

Q: Again, about the question whether randomness is native to computer science. The most important topic in computer science is programming. But Kolmogorov complexity does not care about the diverse programming paradigms: the invariance theorem collapses everything.

A: One can see things differently. Kolmogorov complexity is really about compiling, interpreting and executing programs. There are more than the plain and prefix-free Kolmogorov complexities, cf. Ref. 10. And the diversity of Kolmogorov complexities corresponds to different situations which are met with programming. Let us look at four Kolmogorov complexities: the original plain one, the prefix-free one, Schnorr process complexity[26] and Levin monotone complexity.[20]

In the assembly languages, the programmer has to explicitly manage the memory and the indirect access to it. Fortran and Algol and other imperative languages are more abstract : memory management is a task devoted to the system. Still, there is a notion of main program and input/output instructions are part of the programming language. Interaction between the user and the machine is done through physical device such as screen, keyboard, mouse, printer, and the programmer has to manage everything explicitly. Such languages are not interactive and lead to a family of prefix-free programs. In fact, instructions are executed sequentially with possible loops or jumps due to goto instructions. But the last instruction, if it does not lead to some jump (if part of a loop or a goto) halts the program. Thus, the status of last instruction is really a marker which does not occur anywhere else in the program and is the source of a prefix-free character of the language. Observe that this marker is explicit in some languages like Pascal where it is the "end" instruction followed by a dot. With such programming languages, viewed as universal maps, the associated Kolmogorov complexity is the prefix-free one.

Q: What about plain Kolmogorov complexity and programming style?

A: Well, it has to do with more abstract programming languages. Languages like LISP (John McCarthy, 1958), ML (Robin Milner, 1973), or PROLOG (Alain Colmerauer, 1972). Such languages are executed through a *Read-Eval-Print* loop which is not an explicit instruction but a meta loop during the execution. This avoids explicit instructions in programs for input/output management. These languages are interactive. A program is just a family of definitions of functions (in functional programming) or relations (in logic programming). One can always add some more definitions to a program, there is no explicit nor implicit end marker. The family of programs of such languages are intrinsically *not prefix-free*. With such programming languages, viewed as universal maps, the associated Kolmogorov complexity is the plain one.

Q: Fashionable languages like Java, C♯ are not of that form. Is that related to object orientation?

A: Absolutely not! OCAML is of that form, it is interactive and compiled. Fashion is sometimes very disapointing. A pity that languages with such solid mathematical foundation are unrecognized by most programmers. This is all the more pitiful that these languages all come from UNIX and that UNIX does contains a Read-Eval-Print loop!

Q: Well. If I understand correctly, plain Kolmogorov complexity is related to the best programming styles since they are the most abstract ones. So, in this perspective, it should deserve more consideration than the prefix-free version which is related to less elegant programming! Now, what about Schnorr process complexity?

A: For Schnorr process complexity we consider partial computable functions $F : \{0,1\}^* \to \{0,1\}^*$ which are monotone increasing with respect to the prefix ordering. So, if p is a prefix of q and $F(p)$ and $F(q)$ are both defined then $F(p)$ is a prefix of $F(q)$. This is called *on-line* computation. The obvious version of the invariance theorem holds and the definition of Schnorr complexity is similar to that of plain Kolmogorov complexity. Schnorr complexity comes in when looking at the system level. If h and h'

are user histories and h is h' up to a certain time t, hence is a prefix of h', then the respective reactions r and r' of the system are such that r is r' up to time t, hence r is a prefix of r'. What is ordered by the scheduler up to time t obviously cannot and does not depend on what happens afterwards!

Q: It seems that there is always a current output: F is total!

A: No, the system can be blocked, waiting for some event. In that case the current output is not to be considered.

Q: OK, now with monotone complexity.

A: Recall that for Levin monotone complexity we consider partial computable functions $F : \{0,1\}^* \rightarrow \{0,1\}^{\leq \omega}$ which are monotone increasing with respect to the prefix ordering. Such an F maps finite strings into finite or infinite strings. The obvious version of the invariance theorem holds and monotone complexity of a string x is defined with an optimal F as the length of shortest programs p such that x is a prefix of $F(x)$ (equality is not required, only to be a prefix). This can be interpreted in many ways. In some sense, a program for x is really a program for both x and a possible future of x. This has to do with Kripke semantics for intuitionism or Everett theory for quantum mechanics: an event x has a lot of possible continuations. We can also consider the output x of a program p as an incomplete information about the true output of p. Programming with incomplete information is essential in artificial intelligence and expert systems. With the monotone complexity, we take into account that the information x is an incomplete one, consider all possible futures of x and minimize length among programs for such futures. This is related to denotational semantics for incomplete information such as lazy integers.

Q: What are lazy integers?

A: Consider the family $\mathbb{N} \cup \{S^n(\perp) \mid n \in \mathbb{N}\}$. The intuition of $S^n(\perp)$ is that of an integer $\geq n$. And you order lazy integers according to the information they carry (not according to their size): $x < n$ if and only if $x < S^n(\perp)$ if and only if $x = S^m(\perp)$ for some $m \leq n$. In particular the true integers are pairwise incomparable since they carry incompatible informations.

19.12. Computability versus Information

Q: Mathematical algorithms exist since Ancient Times. What about the management of information? There were census in the Roman Empire. But no mathematics was involved in it, except fastidious counting.

A: Yes. Do you know that IBM (International Business Machines) has to do with census? Hermann Hollerith created the so-called Hollerith machines around 1889 in order to efficiently tabulate statistics coming from the US census and allow to deliver the results of a census in reasonable time. In particular before the next census is started! Hollerith created a company developing punch-card machines, which eventually became IBM. It is with Hollerith like machines that the key notion of memory first appeared in information processing and computing (it was already present in Jacquard machines). Observe that, from the origin in the 1940's, up to the 1970's, all programs were written on punched cards, in the vein of what was done with Hollerith machines. Of course, such machines require sophisticated technology. This did not exist in Ancient Times: papyrus and clay plates were clearly not suitable to manage huge quantities of information. This is in sharp contrast with algorithms which could be run without machines.

Q: Oh! So about fifty years of manipulation of information in machines with memory preceded the treatment of algorithms in machines.

A: Yes. This had some consequences: since information was managed through machines, it was an engineering world. In fact, information management was not a theory, it was the world of technical tricks and engineers. The picture is quite different with algorithms which were mostly a mathematical subject and were developed long before there were machines. As soon as there was a mathematical language (Frege), mathematicians also looked at computability. This may be related to Leibniz famous dream of a calculus ratiocinator which should allow to ease any quarrel: "calculemus (let us compute)...". On the contrary, information did not fascinated mathematicians.

Q: So you tell me that mathematicians were fascinated by Turing machines which were theoretical machines at a time when there were real machines with memory (à la Hollerith)... Rather ironic!

19.13. Kolmogorov Complexity and Information Theories

Q: Information theory is not exclusively in computer science, it has many facets. Shannon and Kolmogorov are not looking at it in the same way.

A: Right. We can see at least five approaches which have been developed. Different approaches which focus on different aspects of information.
First approach. Shannon, in his famous 1948 work, looks at it from an engineering point of view. Remember, he was working at Bell Labs. So

information is a message, that means a word coded letter by letter, and the problems are related to the physical device transmitting it. He introduces a quantitative notion of information content in transmitted messages. To measure variation of this quantity, he borrows to thermodynamics the concept of *entropy* and bases his theory on it. Cf. Ref. 10.

Q: Yes, I heard about that. The main problem are how to optimize the quantity of information transmitted through a channel and how to deal with lossy channels. His approach is a purely syntactic analysis of words which makes no use of semantics.

A: *Second approach.* Wiener cybernetics and the Macy interdisciplinary conferences (1946–1953) looked at communication and interaction, feedback and noise, how information is learned. This prefigured much of Shannon's work and lead to what is now known as cognitive sciences.

Q: *Third approach.* I know another approachindexsemioticssemiotics. It takes into account the context in which an information is known. In other words, it differentiates the semantics of a message and its information content: information depends on the source which sends the message and the information content of a message, its pertinence, depends on the context in which the message is considered. Umberto Eco gives a simple and illuminating example to make clear this distinction: the message "tomorrow it will snow in Paris" does not have the same meaning in December than in August! Was it one of the approaches you were considering?

A: *Fourth approach.* Yes. Now, the fourth approach. Solomonoff and Kolmogorov brought the biggest abstraction to the concept of information, mixing it with general computability and introducing a measure via Kolmogorov complexity. The information content of an object is independent of any consideration on how this information is used (as a message for instance). This is a static vision of information. Introducing a conditional version of Kolmogorov complexity, he refines this notion of intrinsic complexity of an object by relativizing it to a context (which can be seen as an input or an oracle, etc. for the program) carrying some extra information. This exactly matches the problem pointed by Eco about the necessity to distinguish signification and information contents.

Q: Let me guess. The fifth approach is about databases.

Q: *Fifth approach.* You are right. The last approach to information is that brought by Codd (1970) with relational databases. For Codd the fundamental feature of information is its structuralization. In the relational model, information is organized in tables. Each line in a table gives the values of some fixed attributes. Tables are related when they share some attributes. Codd's theory relies on mathematical logic and the mathematical theory of relations. The choice to create tables with such and such attributes is done with consideration to the semantics of the modeled system. Thus, the distinction raised by Eco between semantics and information content is taken into account in the construction of the relational schema of a database following Codd's theory. As Kolmogorov did, Codd also makes complete abstraction of the physical device carrying the information. Codd was working at IBM and his theory was such a revolution in information management that it took many years to be accepted. Surprisingly, it is not IBM but another company, namely Oracle, that built the first relational DBMS (DataBase Management System, that means the system behind a software to manage databases). Nowadays, all DBMS are relational... Let us quote the dedication of his book, 1990:[7] *"To fellow pilots and aircrew in the Royal Air Force during World War II and the dons at Oxford. These people were the source of my determination to fight for what I believed was right during the tens or more years in which government, industry, and commerce were strongly opposed to the relational approach to database management"*.

Q: Strange that information theory, especially the theory of relational databases, though involving non trivial mathematics, is rather unrecognized among mathematicians.

A: Worse than that. Even in theoretical computer science, the mathematical theory of relational databases is rather a marginal topic. This may be related to the new challenge coming from the huge amount of information of the web that relational databases are not appropriate to manage.

WHAT ARE THE PROSPECTS FOR PROGRESS?

Q: AIT (Algorithmic information theory) is now quite fashionable. What you told me shows that AIT should enter more deeply into computer science. Let it be through programming or through information management. The more than, with the web, information is overwhelming.

A: Yes. One can expect some formidable impetus to AIT. One last thing. Let us view an algorithm as a black box. A conceptual point of view introduced by the Macy group (Norbert Wiener and al.) What Kolmogorov did was to take from the black box the sole length of the program. A very abstract notion and a rudimentary look at operational semantics.

Q: This distinction denotational vs operational goes back to Church?

A: No, no. This goes back to Frege with the distinction between Sense and Reference. It gave birth to two main traditions: First, Tarski semantics and model theory. Second, Heyting-Brouwer-Kolmogorov semantics which is really proof theory.

Q: Does Kolmogorov also enters this subject?

A: Yes, in 1953 Kolmogorov looked at the notion of algorithm, its operational aspects. This eventually led to the notion of Kolmogorov-Uspensky machines. Which were extended by Schönhage. What Kolmogorov looked for has been successfully done by Yuri Gurevich with the notion of Abstract State Machine (initially called Evolving Algebra, cf. Ref. 17). Gurevich succeeded to formalize the notion of algorithm. Yet another intuitive notion getting a mathematical status. But this you know first-hand being Yuri's student.

Now, knowing what is an algorithm, other features than the mere length of a program can be considered. This could lead to other forms of AIT.

References

1. Becher, V. and Grigorieff, S. "Random reals à la Chaitin with or without prefix-freeness". *Theoretical Computer Science*, 385:193–201, 2007.
2. Becher, V. and Figueira, S. and Grigorieff, S. and Miller, J. "Randomness and halting probabilities". *Journal of Symbolic Logic*, 71(4):1394–1410, 2006.
3. Bienvenu, L. and Merkle, W. "Reconciling data compression and Kolmogorov complexity". *ICALP 2007, LNCS 4596*, 643–654, 2007.
4. Bienvenu, L., Merkle, W. and Shen, A. "A simple proof of Miller-Yu theorem". *Fundamenta Informaticae*, 83(1-2):21–24, 2008.
5. Bournez, O. and Campagnolo, M.L. and Graça, Daniel S. and Hainry, Emmanuel. "Polynomial differential equations compute all real computable functions on computable compact intervals". *Journal of Complexity*, 23(3):157–166, 2007.
6. Cilibrasi, R. and Vitányi, Paul M.B. "Clustering by compression". *IEEE Trans. Information Theory*, 51:4, 1523–1545, 2005.
7. Codd, Edgar F. "The Relational Model for Database Management (Version 2)". Addison Wesley Publishing Company, 1990.
8. Ferbus-Zanda, M. and Grigorieff, S. "Is Randomness native to Computer Science". *Bulletin of EATCS*, 74:78–118, June 2001. Revised version reprinted in "Current Trends in Theoretical Computer Science", vol.2, 141–179, World Scientific Publishing Co., 2004.
9. Ferbus-Zanda, M. and Grigorieff, S. "Kolmogorov complexity and set theoretical representations of integers". *Math. Logic Quarterly*, 52(4):375–403, 2006.

10. Ferbus-Zanda, M. and Grigorieff, S. "Kolmogorov complexity in perspective. Part I: Information Theory and Randomness". *Synthese*, to appear.
11. Ferbus-Zanda, M. "Kolmogorov complexity in perspective. Part II: Classification, Information Processing and Duality". *Synthese*, to appear.
12. Frege, G. "Begriffsschrift: eine der arithmetischen nachgebildete Formelsprache des reinen Denkens". Halle, 1879. English translation: "Concept Script". In Jean Van Heijenoort, ed., "From Frege to Gödel: A Source Book in Mathematical Logic, 1879-1931". Harvard Uni. Press, 1967.
13. Frege, G. "Über Sinn und Bedeutung" (On Sense and Reference). *Zeitschrift fr Philosophie und philosophische Kritik C:2550, 1892*,
14. Gács, P. "Lectures notes on descriptional complexity and randomness". *Boston University (Peter Gàcs' home page)*, pages 1–67, 1993.
15. Gács, P. "Every sequence is reducible to a random one". Information and Control, 70:186–192, 1986.
16. Graça, Daniel S. "Computability with Polynomial Differential Equations. PhD thesis, Technical University of Lisbon, 2007.
17. Gurevich, Y. "Evolving algebras: an attempt to discover semantics". *Bulletin of EATCS*, 43:264–284, June 1991.
18. Kolmogorov, A. "Grundbegriffe der Wahscheinlichkeitsrechnung". Springer-Verlag, 1933. English translation: "Foundations of the Theory of Probability". Chelsea, 1956.
19. Kolmogorov, A. "Three approaches to the quantitative definition of information". *Problems Inform. Transmission*, 1(1):1–7, 1965.
20. Levin, L. "On the notion of a random sequence". *Soviet Mathematics Doklady*, 14:1413–1416, 1973.
21. Martin-Löf, P. "On the definition of random sequences". *Information and Control*, MIT, 9:602–61, 1966.
22. Miller, J.and Yu, L. "On initial segment complexity and degrees of randomness". *Trans. Amer. Math. Soc.*, MIT, 360:3193–3210, 2008.
23. Moschovakis, J.R. "Relative Lawlessness in Intuitionistic Analysis". *Journal of Symbolic Logic*, 52(1):68–88, 1987.
24. Moschovakis, J.R. "More about Relatively Lawless Sequences". *Journal of Symbolic Logic*, 59(3):813–829, 1994.
25. Schnorr, C-P. "A unified approach to the definition of random sequences". *Math. Systems Theory*, 5:246–258, 1971
26. Schnorr, C-P. "A Process complexity and effective random tests". *J. of Computer and System Sc.*, 7:376–388, 1973.
27. Shannon, C. "Mathematical theory of the differential analyser". *Journal of Mathematics and Physics*, MIT, 20:337–354, 1941.
28. Von Neumann, J. "Various techniques used in connection with random digits." In Monte Carlo Method, Householder, A.S., Forsythe, G.E. & Germond, H.H., eds. National Bureau of Standards Applied Mathematics Series (Washington, D.C.: U.S. Government Printing Office), 12:36-38, 1951.

PART V

Computational Complexity, Randomized Algorithms and Applications

Chapter 20

Randomness as Circuit Complexity
(and the Connection to Pseudorandomness)

Eric Allender*

*Department of Computer Science
Rutgers University*
allender@cs.rutgers.edu

WHAT DREW ME TO THE STUDY OF COMPUTATION AND RANDOMNESS

I started out my undergraduate studies majoring in theater. However, I also knew that I enjoyed math and computers, thanks to a terrific high school math teacher (Ed Rolenc) who installed a computer in one of the classrooms in my high school in Mount Pleasant, Iowa in the 1970's. Thus, when I was an undergrad at the University of Iowa and I decided to pick a second major that might make me more employable in case theater didn't work out, I picked Computer Science. My initial computing courses didn't really inspire me very much, and I continued to think of computing as a "reserve" career, until two things happened at more-or-less the same time: (1) I worked in summer theaters for a couple of summers, and I noticed that there were *incredibly talented* people who were working at the same undistinguished summer theaters where I was working, leading me to wonder how much impact I was likely to have in the field of theater. (2) I took my first courses in theoretical computer science (with Ted Baker and Don Alton), which opened my eyes to some of the fascinating open questions in the field. They gave me some encouragement to consider grad school, and thus (after taking a year off to "see the world" by working as a bellhop in Germany) I entered the doctoral program at Georgia Tech.

At that time, in the early-to-mid 1980's, Kolmogorov complexity was finding application in complexity theory. (I'm referring to the material that's now covered in the chapter called "The Incompressibility Method"

*Supported in part by NSF Grants DMS-0652582, CCF-0830133, and CCF-0832787.

in the standard text by Li and Vitányi.[11]) Also, there were influential papers by Mike Sipser[13] and Juris Hartmanis[8] that discussed resource-bounded Kolmogorov complexity. Also, the theory of pseudorandom generators was just getting underway, with the work of Yao[16] and of Blum and Micali.[5] Thus Kolmogorov complexity and randomness were very much in the air.

Please recall that, at the time, most people believed that BPP and RP were likely to contain problems that could *not* be solved in *deterministic* polynomial time. (BPP is the class of problems that can be solved efficiently if we have access to randomness, and RP is the natural way to define a probabilistic subclass of NP.) I was struck by the observation that there was some tension between the popular conjectures relating deterministic, probabilistic, and nondeterministic computation. For example, if nondeterministic exponential time has no efficient deterministic simulation, it means that there are sets in P that (for many input lengths) contain no strings of low resource-bounded Kolmogorov complexity. However, if efficient pseudorandom generators exist, then every set in P *must* contain strings of "low" resource-bounded Kolmogorov complexity, *if* it contains many strings of a given length. However, if one quantifies "low complexity" as meaning "complexity $O(\log n)$," then it would follow that RP = P (and now we know that this same hypothesis implies BPP = P). So just what is going on, when one considers the resource-bounded Kolmogorov complexity of sets in P? Can an efficient computation say "yes" to a large number of strings, without accepting strings with low resource-bounded Kolmogorov complexity? My first STOC paper grew out of precisely these considerations.[1]

WHAT WE HAVE LEARNED

At a basic level, the most important lesson this field gives us is conceptual. Until the mathematical framework that we now call Kolmogorov Complexity was established, there was no meaningful way to talk about a given object being "random". Now there is. And it is crucial that the key ingredient is the notion of *computability*. Until we grappled with computation, we could not understand randomness.

Starting from that flash of insight and a few simple definitions, the field has grown remarkably. You are asking me "What have we learned?" My initial reaction is to simply point you to a standard textbook such as the one by Li and Vitányi[11] or the soon-to-be-released volume by Downey and

Hirschfeldt.[6] But instead, my answer will concentrate on some material that is not emphasized in these volumes.

For me, one of the most surprising and simple insights that has come out of the study of randomness recently, is the fact that there is a very natural and close connection between Kolmogorov complexity and circuit complexity.

At first blush, these topics seem to have nothing to do with each other, for several reasons:

1. *Kolmogorov complexity measures the complexity of strings, while circuit complexity measures the complexity of functions.* This, of course, is no significant difference at all. Any string x of length m can be viewed as the initial part of the truth table of a function f_x having size $2^n < 2m$. Thus we can define Size(x) to be the size of the smallest circuit computing f_x. Viewed in this way, Kolmogorov complexity and circuit complexity each measure the complexity of strings.

2. *Kolmogorov complexity is not computable, in stark contrast to circuit complexity.* This, on the other hand, would seem to to present an insurmountable barrier to making any meaningful connection between Kolmogorov complexity and circuit complexity. The way around this barrier is to employ one of the oldest tricks in the toolkit of complexity and computability: oracles.

The function Size(x) gives the size of the smallest circuit computing f_x, where the circuit is made up of the usual circuit components: AND and OR gates. But there is a natural and well-studied notion of providing oracle access to a circuit. An oracle gate for a set A has some number of wires (say, m) as input, and produces as output the answer to the question "*Is z in A?*" where z is the m-bit string that is fed into the input wires of the gate. We define Size$^A(x)$ to be the size of the smallest circuit having oracle gates for A that computes f_x.

It's not too hard to show that both $K(x)$ and $C(x)$ (the usual Kolmogorov complexity measures) are polynomially-related to Size$^H(x)$, where H is the halting problem (or any other set that is complete for the computably-enumerable sets under polynomial-time reductions).[4]

Similarly, Kt(x) is polynomially-related to Size$^A(x)$, where A is any set complete for exponential-time. (Here, Kt is a time-bounded notion of Kolmogorov complexity defined by Leonid Levin.[10]) In a paper that I wrote with Harry Buhrman, Michal Koucký, Dieter van Melkebeek, and Detlef Ronneburger,[4] we defined another notion of time-bounded Kolmogorov complexity in the style of Levin's definition, that is polynomially-related to Size(x) (with no oracles). Thus we can see that the measures that are central to the studies of complexity and to the study of randomness are, in fact, reflections of each other.

For me, the most exciting thing about this connection is that it enabled amazing techniques from the field of derandomization to be applied to questions about Kolmogorov complexity. Derandomization, and the study of pseudorandom generators, has been an incredibly active and productive field in the last few decades. Advances in that field have completely changed the way that many of us think about probabilistic computation.

As I've already mentioned, back when I was in grad school, most people in the field used to think that BPP and RP were probably stricter larger than P. Now, the situation is completely reversed. What happened to turn things around? Two things. There had previously been some interesting results, showing that the existence of hard-on-average one-way functions would imply somewhat fast deterministic simulation of probabilistic algorithms. But then Nisan and Wigderson came up with a new class of pseudorandom generators, based on the seemingly weaker assumption that there is a problem in exponential time that is hard on average in the sense of circuit complexity.[12] The real tide change came in 1997, when Impagliazzo and Wigderson weakened the assumption even further, by showing that BPP = P if there is any problem computable in exponential time that requires circuits of exponential size.[9] A number of additional important papers on derandomization followed soon thereafter.

But all of this exciting work showing how to eliminate the need for random bits in simulating BPP seemed divorced from the study of randomness in the sense of Kolmogorov complexity. By making use of the connection between Kolmogorov complexity and circuit complexity, we were able to bridge this divide, and present some interesting reductions. For instance, consider the set of random strings (using K or C complexity). Of course, the set of random strings is undecidable, and is Turing-equivalent to the Halting problem — but it is far from obvious how to make use of this set using an *efficient* reduction. We were able to show that is poly-time Turing reducible to the set of random strings,[4] and is in NP relative to this set.[3] Even the Halting problem itself is efficiently reducible to the set of random strings — if one allows reductions computable by poly-size *circuits*, instead of poly-time machines.[4]

Similar techniques show that the set of random strings in the sense of Levin's Kt complexity is complete for EXP (under reductions computed by poly-size circuits, as well as under NP-Turing reductions).

WHAT WE DON'T YET KNOW

Continuing in the same thread from the previous question, here are some annoying open questions:

- *Is the set of Kt-random strings in P?* Of course, the answer has to be "no", or else EXP has poly-size circuits and the polynomial hierarchy collapses. But I know of no reason why it should be hard to *prove* unconditionally that this set is not in P.
- *Is EXP poly-time reducible to the set of random strings?* I think that it's reasonably likely that there is some complexity class larger than that is poly-time reducible to the set of random strings, but it is far from clear to me that *every* decidable set should be poly-time reducible to the set of random strings. Right now, we cannot even rule out that the Halting problem is poly-time reducible to the set of random strings. The fact that there *is* such a reduction computable by poly-size *circuits* indicates some of the subtleties involved.

Perhaps the most glaring hole in our edifice of knowledge relating to randomness, is the fact that we have no real proof that randomness even *exists* in our universe. There is no proof that, say, quantum mechanical phenomena can be exploited in order to provide a satisfactory source of randomness at the macro level, and certainly there is no really satisfactory source of randomness currently available for industrial applications that would dearly love to have a perfect source of randomness. For instance, in order for public-key cryptography to work at all, it is essential that secret keys (such as pairs of large primes for the RSA cryptosystem) be selected more-or-less uniformly and independently from a large space. If there is only a small set of likely keys, then the system is vulnerable to cryptanalysis. Thus there is a lot of money riding on the hope that keys are being generated at random — but it is not clear that there will *ever* be, in principle, a way to *prove* that a process is truly random. Even worse, it is not clear that one can ever fully disprove the theory of Laplacian Determinism — which means that we'll never really know if randomness exists at all in our universe. So we'll just have to keep muddling along, with regard to that question, relying on "faith" that enough randomness is available when it is needed.

Even here, the theory of derandomization has provided a number of useful tools. Trevisan[15] showed that the Impagliazzo-Wigderson generator can be used as a "randomness extractor" that can take input from a "bad" source of randomness and produce a small list of samples (on a somewhat smaller space) that approximates what one would obtain if one had access to the uniform distribution on this space.

THE MOST IMPORTANT OPEN PROBLEMS

We can't understand randomness without grappling with computation — and we can't really claim to understand computation until the P vs NP question is settled. Thus I'll start with the completely non-controversial choice of naming the P vs NP question as the most important open problem in the field.

And yet, for practical matters — as well as for the questions that drive the study of randomness — the P vs NP question is probably not as relevant as the question of the *circuit complexity* of various problems. To see what I mean by this, consider the problem faced by people who recommend what key size to use for various cryptosystems. Even if someone were able to prove P $\neq$ NP, and were furthermore able to prove a superpolynomial run-time lower bound for the problem the cryptosystem is based on, it would still be *impossible* to choose a secure key size, if no circuit size lower bound were in hand. For instance, consider any problem that is complete for EXP; we know that every program solving such a problem must have a huge run-time for *large* inputs, but there is *absolutely* no way to say how large the inputs have to be, before the run-time must be large. There's no guarantee that there can't be a C++ program that can run on your laptop and solve the problem for inputs of 10,000 bits in a few seconds — precisely because we don't know how to prove that EXP requires large circuits. In contrast, building on circuit lower bounds, Stockmeyer and Meyer showed that there are some natural and interesting problems that can't be solved for inputs of size 400 by any circuit that will fit in the galaxy.[14]

So *circuit size* is even more important to understand than program run-time. For the study of randomness, I'd have to say that one of the most important open questions is the question of whether there is any problem computable in time 2^n (e.g., SAT) that requires circuits of size $2^{\epsilon n}$ for some $\epsilon > 0$. This turns out to be *equivalent* to the question of whether pseudorandom generators exist (with certain parameters); this is surveyed nicely by Fortnow.[7]

THE PROSPECTS FOR PROGRESS

I try to be optimistic. I wrote a survey recently,[2] outlining some of the approaches that have been proposed, to try to overcome the barriers that seem to block progress toward proving circuit lower bounds (and let me repeat that I think that circuit lower bounds are really the most important goal to strive for, if we want to understand randomness). In the survey, I tried to make the case that, although there is certainly a great deal of pessimism about the prospects for quick resolution of any of these problems (such as the P vs. NP problem), there is nonetheless some reason for hope.

More generally, the historical record shows that our understanding of this topic has evolved significantly with each passing decade. Viewed in this light, I'm not only optimistic about the prospects for progress — I think that progress is all but inevitable.

References

1. E. Allender. Some consequences of the existence of pseudorandom generators. *Journal of Computer and System Sciences*, 39:101–124, 1989.
2. E. Allender. Cracks in the defenses: Scouting out approaaches on circuit lower bounds. In *Computer Science – Theory and Applications (CSR 2008)*, volume 5010 of *Lecture Notes in Computer Science*, pages 3–10, 2008.
3. E. Allender, H. Buhrman, and M. Koucký. What can be efficiently reduced to the K-random strings? *Annals of Pure and Applied Logic*, 138:2–19, 2006.
4. E. Allender, H. Buhrman, M. Koucký, D. van Melkebeek, and D. Ronneburger. Power from random strings. *SIAM Journal on Computing*, 35:1467–1493, 2006.
5. M. Blum and S. Micali. How to generate cryptographically strong sequences of pseudo-random bits. *SIAM Journal on Computing*, 13:850–864, 1984.
6. R. Downey and D. Hirschfeldt. Algorithmic randomness and complexity. To be published, 2010.
7. L. Fortnow. Comparing notions of full derandomization. In *IEEE Conference on Computational Complexity '01*, pages 28–34, 2001.
8. J. Hartmanis. Generalized Kolmogorov complexity and the structure of feasible computations (preliminary report). In *IEEE Symposium on Foundations of Computer Science (FOCS)*, pages 439–445, 1983.
9. R. Impagliazzo and A. Wigderson. $P = BPP$ if E requires exponential circuits: Derandomizing the XOR lemma. In *ACM Symposium on Theory of Computing (STOC) '97*, pages 220–229, 1997.
10. L. Levin. Randomness conservation inequalities; information and independence in mathematical theories. *Information and Control*, 61:15–37, 1984.
11. M. Li and P. Vitányi. *Introduction to Kolmogorov Complexity and its Applications*. Springer, third edition, 2008.

12. N. Nisan and A. Wigderson. Hardness vs. randomness. *Journal of Computer and System Sciences*, 49:149–167, 1994.

13. M. Sipser. A complexity theoretic approach to randomness. In *ACM Symposium on Theory of Computing (STOC)*, pages 330–335, 1983.

14. Larry J. Stockmeyer and Albert R. Meyer. Cosmological lower bound on the circuit complexity of a small problem in logic. *J. ACM*, 49(6):753–784, 2002.

15. L. Trevisan. Construction of extractors using pseudo-random generators. *Journal of the ACM*, 48(4):860–879, 2001.

16. A. Yao. Theory and application of trapdoor functions. In *IEEE Symposium on Foundations of Computer Science (FOCS)*, pages 80–91, 1982.

Chapter 21

Randomness: A Tool for Constructing and Analyzing Computer Programs

Antonín Kučera

Institute for Theoretical Computer Science
Faculty of Informatics, Masaryk University
Brno, Czech Republic
tony@fi.muni.cz

WHAT DREW ME TO THE STUDY OF COMPUTATION AND RANDOMNESS

I think that the main reason is completely non-pragmatic — probability theory is a fascinating and beautiful branch of mathematics which is full of surprises. One is easily tempted to formulate "must hold" hypothesis based on intuitively clear reasons, which turn out to be false after a careful analysis. This naturally leads to more rigour and humility, which I find very healthy. Another "personal" reason is that many clever and nice people I knew for a long time started to study automata-theoretic models enhanced by randomized choice. As I wanted to keep talking to them, I joined the club.

It is also remarkable how broad and diverse the practical applicability of abstract concepts and results of probability theory is, and how simple explanations for seemingly complicated phenomena can be found using this theory. For example, some aspects of animals' behaviour may appear strange until a faithful stochastic model of the situation is constructed, where the impact of possible decisions available to the animals is explicitly quantified. Then, it often turns out that the "strange" behaviour is actually a combination of rational decisions that maximize the probability of survival.

Considering the elegance, power, and wide applicability of probability theory, it is not very surprising that this theory plays an important role also in computer science. Information technologies are developing very fast, and sometimes substantially faster than the formal foundations they are based on. This fact surely has its negative aspects (personally, I am not very happy to see the progressive electronic "virtualization" of my money in the situation when the security of widely used cryptographic technologies is based on uncertain and unproven hypotheses). However, this also motivates and stimulates the development of new models, analytical techniques, and the corresponding algorithms. There are new challenges reflecting the practical problems of state-of-the-art technologies (related to security, safety, reliability, performance, etc.), which are attacked from various sides using different approaches. New exciting lines of research are initiated, and randomization plays an important role in many of them. It is nice to take part in all that.

WHAT WE HAVE LEARNED

My personal (and hence necessarily incomplete) view of the role of randomness in computer science is reflected in essentially three sentences.

(1) Randomization is a very powerful tool for designing efficient and simple algorithms.

It can be documented by many examples that if a computer is equipped with a source of randomly generated bits, some problems suddenly admit an amazingly simple and efficient solution. In some cases, one can even prove that a given problem cannot be solved without randomization, or at least not with the same level of efficiency. An ordinary deterministic program must always produce the same output for a given input, and some problems (such as leader election in distributed algorithms) are therefore provably unsolvable by deterministic algorithms. However, if an algorithm "flips a coin" at an appropriate moment, it can behave somewhat differently when it is run repeatedly, and this can be used as an advantage. Another demonstrative example is the problem of comparing two files stored at different computers connected by a network. If the efficiency of a given solution is measured by the number of bits transferred by the network, one can show that without the use of randomness, the best solution is to transfer the whole file from one computer to the other and compare them locally. However, with the help of randomization, one can design a simple solution where only a few bits are transferred and still one can tell with a very high confidence whether the two files are the same or not.

Randomized algorithms are typically simple, easy to implement, and efficient. They are used in distributed systems, communication protocols, cryptography, data management, and other fields. In some cases, their role is indispensable. For example, randomized algorithms for primality testing that are widely used in public-key cryptography are simple and completely satisfactory from the practical point of view. It is not likely that they will be replaced by the recently discovered deterministic polynomial-time algorithm in a near future.

The price to pay for the efficiency and simplicity of randomized algorithms is uncertainty. Randomized algorithms can make errors and produce wrong answers. However, the probability of producing an error can usually be made arbitrarily small, much smaller than the probability of hardware error. Hence, this issue is not very limiting from the practical point of view.

(2) Randomness is very useful for modelling some behavioural aspects of computational systems

Obviously, the behaviour of "inherently" stochastic systems such as randomized algorithms or quantum computers cannot be faithfully modelled and analyzed without taking randomness into account. However, stochastic models can also be used to analyze systems with deterministic design whose behaviour is to some extent unpredictable because of errors, exceptions, user actions, or unpredictable scheduling of computational threads. The associated probabilities can be approximated by performing tests and measurements, or simply "guessed" by employing some empirical knowledge. After creating a stochastic model of such a system, one can evaluate the probability that the system behaves "badly". For example, one can compute the probability of all runs of a given parallel program that do not satisfy some liveness property. The program may have uncountably many bad runs, but still be considered as correct if the total probability of all such runs is zero. Generally, qualitative results (telling that the probability of some event is positive or equal to one) tend to be more stable than quantitative ones. Qualitative properties are often not influenced by small changes in the underlying stochastic model, it is only important which of the alternatives are assigned positive/zero probability.

Randomization can also help to build a simplified formal model in situations when the behaviour of some system component is inherently deterministic, but its construction details are unknown or too complex. The behaviour of such a component can be approximated by assigning probabilities to the actions taken by the component under specific conditions. The resulting model can be substantially smaller but still relevant.

(3) Probabilistic analysis provides unique information which cannot be discovered by other methods.

Probabilistic analysis can be applied also to systems which are not inherently randomized. For example, one can try to evaluate the expected running time of a given algorithm for a randomly generated input (i.e., the expected time complexity), which is usually more relevant than the worst-case time complexity. Another example is performance evaluation, when we try to analyze the system's behaviour under a given load. For example, we may wonder what is the average length of an internal queue when the average frequency of incoming user requests reaches a given threshold. Such a system can be modelled by a continuous-time Markov chain, and hence the question can be answered by applying the abstract results of Markov chain theory. Without employing probabilistic analysis, the only known method for answering the question (and other questions related to long-run average behaviour) is "wait and see".

In this context, one should also mention the Internet search engines. The original Pagerank algorithm used by Google computes a stationary distribution in a certain Markov chain obtained by modifying the web graph (the nodes of the web graph are the individual www pages and the edges are the links). Roughly speaking, this formalizes the idea of measuring the relevance of www pages by the expected frequency of visits by a randomly clicking surfer.

WHAT WE DON'T YET KNOW AND THE MOST IMPORTANT OPEN PROBLEMS

One of the fundamental open problems in the area of randomized computations is the exact classification of the computational power of randomness. Some of these questions can be formalized as hypothesis about the relationship between various complexity classes which consist of problems that can be solved on randomized, deterministic, and non-deterministic computational devices, such as random access machine (RAM) or Turing machine. In the case of randomized computations, the classical RAM or Turing machine is equipped with a source of random bits, which can influence the individual computational steps. This formalizes the "coin flip". Now we can ask what class of computational problems can be solved by these devices if their resources (time and space) are bounded by some functions, and what is the relationship between these complexity classes. One concrete example is the question whether P = BPP. Here P is the class of problems that can be solved in polynomial time by a deterministic algorithm (i.e., there is a deterministic algorithm which for every concrete instance of the problem of size N computes the correct answer after at most POLY(N) computational steps, where POLY(N) is some fixed polynomial in N). The class BPP consists of problems solvable by a randomized algorithm which always halts after a polynomial number of steps, but can produce a wrong answer with probability at most $1/3$ (the constant $1/3$ is no "magic number", it can be safely replaced by an arbitrary constant strictly less than $1/2$). Such an algorithm must be run repeatedly, and the final answer is produced by taking a "majority vote" of the produced answers. In this way, the probability of producing an error can be "quickly" pushed below an arbitrarily small (but still positive) given threshold.

A closely related open question is the characterization of the actual computational power of quantum computers. There are problems like integer factorization, discrete logarithms, or solving Pell's equation, which are solvable on a quantum computer in polynomial time, but are believed to be computationally infeasible with an ordinary computer. The best known deterministic algorithms for these problems require exponential time. However, we do not have any proof showing that a polynomial-time deterministic algorithm cannot be discovered. So, it is not clear whether quantum computers are indeed "substantially" (i.e., exponentially) more powerful than ordinary computers. Nevertheless, it has been shown that quantum computers provably offer polynomial speedup for some problems. For example, searching an unsorted database with N entries provably requires c.N computational steps on a classical computer, where c is some constant. On a quantum computer, only d.SQRT(N) steps are needed, where SQRT(N) is the square root of N and d some constant. So, a quantum computer can solve this problem "quadratically faster" than a classical computer.

A big (though perhaps less spectacular) challenge is the development and advancement of the existing techniques for formal modelling and analysis of computational systems, which could reflect different types of behavioural aspects within a single formalism. For example, the current system load can be modelled by continuous-time Markov chains, which can faithfully approximate history-independent random events such as requests for action from incoming users. Real-time systems can be modelled by timed automata, and thus one can prove that certain actions are completed within a given time limit. The interaction between a system and its environment can be understood as a game, and hence one can borrow the analytical techniques of game theory. Now, if we want to study the performance of a real-time system which communicates with its environment, none of the above formalisms is sufficient on its own. The question is to what extent one can combine and integrate the existing formal models and the corresponding analytical techniques, which are often quite different. Ideally, this study should result in an efficient framework for formal modelling and analysis of complex systems, and thus establish a solid basis for a new generation of software tools for formal verification.

For example, a very interesting class of models is obtained by combining the paradigms of randomized and non-deterministic choice. Non-deterministic choice is performed in situations where multiple different alternatives are possible, but there is no fixed probability of which one will be taken. Thus, one can model stochastic systems with decision-making. The decisions are either controllable or adversarial, depending on whether the choice is determined by the system or its environment. This can be viewed as a stochastic game played by two players with antagonistic objectives (the controller wants to achieve a "good" behaviour, the environment aims at the opposite). Almost every class of stochastic process with discrete state-space can be extended into a corresponding game model. If this is applied to discrete-time Markov chains, one obtains competitive Markov decision processes, which are relatively well understood (particularly in the special case where all non-deterministic states are controllable; this corresponds to "ordinary" Markov decision processes). However, the same extension is applicable also to other classes of stochastic processes, particularly to continuous-time Markov chains. Although the games determined by continuous-time Markov chains provide a convenient modelling language, the corresponding theory is much less developed and many fundamental questions have not yet been answered.

Another important question related to formal verification of stochastic systems is whether the scope of algorithmic analysis can be extended to certain classes of infinite-state systems. Most of the existing algorithms are applicable only to systems with finitely many states. However, many systems use unbounded data structures such as stacks, counters, or queues, which makes them infinite-state. There are some encouraging results, particularly about discrete-time stochastic systems with unbounded recursion (i.e., systems with a finite control unit and an unbounded stack), but in general this question is open.

THE PROSPECTS FOR PROGRESS

One successful method for formal verification of randomized systems is stochastic model-checking combined with specific methods for symbolic state-space representation, which is implemented in some experimental verification tools. This technique surely has a big potential to attack real-world problems.

Efficient quantitative analysis of stochastic systems is impossible without efficient numerical methods. When analyzing a given finite-state stochastic system, one typically needs to solve a system of recursive linear equations where the number of variables is proportional to the number of states of a given system. Such large equational systems can rarely be solved explicitly, and one has to rely on the available approximation methods. Interestingly, certain classes of infinite-state systems (e.g., the discrete-time stochastic systems with unbounded recursion mentioned earlier) can also be analyzed by solving a finite system of recursive equations, but these equations are no longer linear. It has been recently shown that Newton's approximation method is applicable also to this type of equational systems, which provides a potential basis for efficient quantitative analysis of infinite-state stochastic systems.

There are many other exciting ideas which deserve to be listed and discussed in here. I do not dare to judge which of them turns out to be most influencing and useful in the future. Let's see.

Chapter 22

Connecting Randomness to Computation

Ming Li

School of Computer Science
University of Waterloo
mli@uwaterloo.ca
http://www.cs.uwaterloo.ca/~mli

WHAT DREW ME TO THE STUDY OF COMPUTATION AND RANDOMNESS

A disciple once told the great Chinese philosopher Zhuangzi [369–286 BCE]: I have a giant tree whose trunk is so rugged and random and whose branches are so gnarled and irregular, that no carpenter can measure it and it is useless for anything. Zhuangzi replied: Its uselessness is exactly its great usefulness.[7]

In the early 1980's, Kolmogorov complexity was like such a giant tree. It was investigated by few great minds, but disconnected from the applications and people generally did not know what to do with such a beautiful and fundamental notion of complexity. These Kolmogorov random strings are so rugged and gnarled that they are not computable and not even approximable. No one will ever be sure if a string is random. How can we use such phantom strings meaningfully? Other contemporary "trees", such as the computational complexity invented during the same period in the 1960s, were blooming and the researchers were harvesting great fruits from the new notion of time and space complexities of Turing machines.

At the time, I was doing my Ph.D. under the supervision of Juris Hartmanis, who together with Richard Stearns initiated the formal study of computational complexity. Juris told me about Kolmogorov complexity during one of our little meetings in his office, 4th floor Upson Hall, often accompanied by his dog Kiki who knows when not to bark. I failed to appreciate the new concept and did not know what to do with it. A year later, I was working on the problem of simulating Turing machines with two

283

work tapes by one work tape, both with an additional one-way read-only input tape. The well-known upper bound is the quadratic simulation of Hartmanis and Stearns 20 years before. In 1963, M.O. Rabin proved that two tapes cannot simulate one tape in real time. In 1974, this argument was slightly improved by S.O. Aanderaa to k work tapes. Then in 1981, P. Dúris and Z. Galil, and later jointed in by W.J. Paul and R. Reischuk, obtained a $\Omega(n \log n)$ lower bound. A big gap had remained open. While I was working on this problem, I suddenly remembered what Juris told me about Kolmogorov complexity. A Kolmogorov random string was a perfect input for my Turing machines. It possesses all properties I wished an input to have, and it was just one fixed single string to work with. I quickly obtained the tight $\Omega(n^2)$ lower bound. Independently, Wolfgang Maass at Berkeley (earlier than I) and Paul Vitányi at CWI also obtained similar results. We all used essentially the same argument, with Kolmogorov complexity. The argument was previously introduced and used by Wolfgang Paul, Joel Seiferas, Janos Simon, S. Reisch, and G. Schnitger to simplify or improve known results. Our results further demonstrated the power of this argument by settling a long standing open question.

In 1986, Berkeley STOC conference, Paul Vitányi and I met, for the first time. We drove to the beautiful Napa Valley and decided that this subject was so beautiful that we should write a book about it. The book[5] was published 11 years later.

In this article, I will write about two research topics which stem from the seemingly uselessness of Kolmogorov complexity and which have contributed to the current popularity of Kolmogorov complexity.

WHAT WE HAVE LEARNED

The Kolmogorov argument, later we called it the "incompressibility method" in Ref. 5, that was used to prove Turing machine lower bounds, turns out to be a general method that was successfully used in a wide range of problems in combinatorics, graph theory, algorithm analysis (for example for sorting algorithms), string matching, routing in computer networks, circuit lower bounds, formal language and automata theory, to name just a few. The incompressibility method consists of the following steps:

(1) Fix one Kolmogorov random input.
(2) Analyze the algorithm with respect to this single input.
(3) Show that if the time/space complexity (on this single input) does not hold, then the input can be compressed.

To help the reader to appreciate the method, we describe below two examples of the incompressibility method.

22.1. Average-case Analysis of Algorithms

A primary goal of Computer Science is about designing and analyzing computer algorithms in applications. While the "worst case" analysis is more of a classroom teaching tool, it is the average-case complexity of an algorithm that is usually more relevant to practical questions. Yet, analyzing average case complexity of an algorithm is often difficult and messy. By the definition of it, it involves averaging the performance of all inputs, of length n, according to some distribution. Some brute force counting argument and some more advanced probabilistic arguments sometimes work, but there are very few general tools for analyzing average case complexity of algorithms, especially, those that are easy-to-use, easy-to-teach, and easy-to-understand.

The incompressibility method is a natural tool for average-case complexity of algorithms. It has some good properties:

(1) The method is input independent. It always uses one fixed Kolmogorov random input that has all the desired properties. Since most strings are Kolmogorov random, the complexity on the single Kolmogorov random input is the average case of the algorithm (under the uniform distribution, or under a computable distribution where one random string of each length has its fair share of probability).
(2) The method is algorithm independent for lower bounds.
(3) The method is problem independent — it is always the above 3 steps.

Additionally, the method is powerful, simple, easy to use and teachable to undergraduate classes. Unlike other methods, such as the probabilistic method, it does not require advanced mathematical tools.

Let us consider an example. In 1959, D.L. Shell published what we now call the Shellsort algorithm. It uses p increments $h_1, \ldots, h_p$, with $h_p = 1$. At the k-th pass, the array is divided in h_k separate sublists of length n/h_k by taking every h_k-th element. Each sublist is then sorted by Insertion Sort.

Shell's original algorithm runs in $\Theta(n^2)$ time, using $p_k = n/2^k$ for step k. Papernow and Stasevitch in 1965 gave an $O(n^{3/2})$ time version. Pratt improved this to $O(n \log^2 n)$ in 1972 by using $\log^2 n$ increments of form $2^i 3^j$. Incerpi-Sedgewick, Chazelle, Plaxton, Poonen, Suel proved in the 1980s the worst case bound $\Omega(n \log^2 n/(\log \log n)^2)$. There are relatively fewer work for average case analysis. In the 1970's Knuth gave a $\Theta(n^{5/3})$ bound for 2-pass Shellsort. Yao in the 1980's analyzed 3-pass Shellsort. Janson and Knuth in 1997 give $\Omega(n^{23/15})$ lower bound for 3-pass Shellsort.

Using the incompressibility method, together with Tao and Paul Vitányi, we proved a general average-case lower bound for Shellsort for all p:[2] p-pass Shellsort on average takes at least $pn^{1+1/p}$ steps.

Since the proof takes only a few lines, let's describe the proof here. Fix a random permutation Π such that

$$C(\Pi) \geq n \log n. \tag{22.1}$$

For pass i, let $m_{i,k}$ be the number of steps the k-th element moves. Thus the run time

$$T(n) = \sum_{i,k} m_{i,k}. \tag{22.2}$$

From these $m_{i,k}$'s, one can reconstruct the input Π, so

$$\sum_{i,k} \log m_{i,k} \geq C(\Pi) \geq n \log n. \tag{22.3}$$

Maximizing the left, subject to (22.2), all $m_{i,k}$ must be equal, say m. Then

$$\sum_{i,k} \log m = pn \log m \geq \sum_{i,k} \log m_{i,k} \geq n \log n. \tag{22.4}$$

Thus

$$m^p \geq n,$$

hence $T(n) = pnm > pn^{1+1/p}$. Since most permutations are random, this lower bound *is* the average-case lower bound.

22.2. Low Probability Events

The incompressibility method may appear to work only for high probability events, giving an average-case analysis naturally. It is a pleasant surprise that this argument also works wonderfully for low probability events, due to the work of Robin Moser in STOC'2009.[6] The following was the essence of Moser's (best paper award) talk at STOC'2009. Lance Fortnow formulated Moser's talk in the usual incompressibility language, and I follow Fortnow's writing here. This is not the full Lovász Local Lemma but it captures the main principle. Robin Moser and Gábor Tados later proved the original Lovász Local Lemma ($r < 2^k/e$).

(The Lovász Local Lemma) Consider a k-CNF formula ϕ with n variable and m clauses. Assume each clause shares a variable with at most r other clauses. Then there is a constant d such that if $r < 2^{k-d}$, then ϕ is satisfiable. Moreover, we can find that assignment in time polynomial in m and n.

Again, since the proof is just a few lines, let me give the complete proof here. Fix a Kolmogorov random string x of length $n + sk$, where the value of s will be determined later, such that the conditional Kolmogorov complexity of x

$$C(x|\phi, k, s, r, m, n) \geq |x| = n + sk. \tag{22.5}$$

Algorithm Solve(ϕ)

1.   Using the first n bits as assignment for ϕ.
2.   While there is an unsatisfied clause C
3.   Fix(C);

Fix(C)

4.   Replace the variables of C with next k bits in x;
5.   While there is a unsatisfied clause D that shares a variable with C
6.   Fix(D);

Suppose the algorithm makes s Fix calls. If we know which clause is being fixed, we know the clause is not satisfied, hence we know all k bits corresponding to this clause. Hence we can describe x by its first n bits, plus the description of the clauses being fixed.

For each of the m Fix calls from **Solve**, it takes $\log m$ bits to indicate the clause. For other (recursive) Fix calls from Fix, it takes only $\log r + O(1)$ bits since there are only r other clauses that share a variable with the current clause. Thus, we must have

$$m \log m + s(\log r + O(1)) + n \geq n + sk.$$

We must have $r \leq 2^{k-d}$, and letting $s \geq m \log m$ finishes the proof.

It is important to emphasize two issues: (1) The Lovász Local Lemma is about low probability events. This proof presents a way of using the incompressibility method for low probability events. (2) The original Lovász Local Lemma is non-constructive. This proof actually introduces a method of giving constructive proofs (randomized algorithms) by the incompressibility method.

22.3. What Have We Learned: Information Distance

While Kolmogorov complexity measures the information contents within one string, information distance[1]

$$d(x, y) = \max\{K(x|y), K(y|x)\} \qquad (22.6)$$

measures the information contents shared by two strings. While this metric is universal, "provably better" than any other computable measures of similarities between two objects, it is not computable. However, unlike the incompressibility method, this time, we really have to compute it, if we wish to use Formula (22.6).

In 1997, when I returned to Waterloo from my sabbatical in Hong Kong, I was thinking about two problems:

• Many prokaryote genomes had been completely sequenced, can we find a general method to build a whole genome phylogeny that tells a more faithful story than the classical 1-gene phylogeny? The classical methods depend on certain models of protein evolution via multiple alignment. There were no clear and accepted evolutionary models for whole genomes.

• The second problem is totally irrelevant from the first. On a hiking trip to the Lion Rock, in Hong Kong, Charles Bennett told me that he had a collection of chain letters (real letters, not emails) he had collected during the past 20 years. We wanted to figure out the evolutionary history of these letters, by an automatic and simple method. These chain letters, although they are at the similar length of a gene sequence, their mutation models are somewhat different from that of a gene. For example, there are paragraph swaps that defy the multiple alignment method.

Can we find a robust method that would give the objective "distance" between any pair of two objects, whether they are genomes, chain letters, emails, music scores, programs, pictures, languages, or even an internet query and an answer? Especially, we are interested in a method that is not dependent on our (possibly biased) knowledge of the data, to avoid overfitting. The information distance is application-independent, universal, and theoretically optimal. It is the natural candidate. However, it is not computable. Can we simply use a good compression algorithm to "approximate" it? Will this be a problem in practice? I decided to test this idea by experimenting with real life data in different domains. In order to

compensate the sequence length differences, we used normalized versions of information distance: $d_{\text{sum}}(x,y) = K(x|y) + K(y|x)/K(xy)$,[3] and later,[4] $d_{\text{max}}(x,y) = \max\{K(x|y), K(y|x)\}/\max\{K(x), K(y)\}$.
The normalized information distances worked very well for both whole genome phylogeny[3] and chain letter phylogeny (C.H. Bennett, M. Li and B. Ma, Chain letters and evolutionary histories, *Scientific American*, 288:6 (June 2003), feature article, 76-81). We further tested this idea to plagiarism detection when I was at UCSB, during 2000–2002. The whole (mitochondrial) genome phylogeny[3] of 20 mammals is given in Figure 22.1.

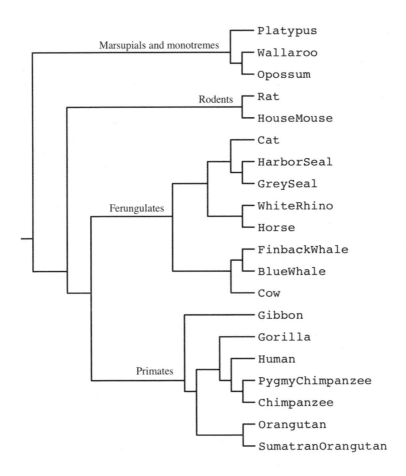

Fig. 22.1. The evolutionary tree built from mtDNA sequences.

Many further applications and justifications have been carried out by many other research teams (for references, see Ref. 5): clustering analysis, software design, software metrics and obfuscation, web page authorship, web security, topic and domain identification, protein sequence/structure classification, phylogenetic reconstruction, hurricane risk assessment, SVM kernel for string classification, ortholog detection, clustering fetal heart rate tracings, analyzing worms and network traffic, picture similarity, language classification, music classification, authorship attribution, stemmatology, parameter-free data mining paradigm, internet knowledge discovery, network structure and dynamic behavior, gene expression dynamics in macrophase, and question and answering. Stimulated by this work, a competitive approach based on compression has been developed to Pearson-Neyman hypothesis testing (null-hypothesis versus alternative hypothesis), tests for randomness of strings generated by random number generators, and lossy compression and denoising via compression.

In a SIGKDD04 paper (pp. 206–215), Keogh, Lonardi and Ratanamahatana demonstrated that this method was better than all 51 other methods for time series clustering proposed in the field from the top conferences SIGKDD, SIGMOD, ICDM, ICDE, SSDB, VLDB, PKDD, and PAKDD. The power of method also rests in the fact that the method is almost blind to the applications (although this is not completely true), avoiding overfitting and over tuning parameters. This also makes the application of the method easy: often one does not even need domain knowledges (such as mutation models) to apply the method.

THE MOST IMPORTANT OPEN PROBLEMS AND THE PROSPECTS FOR PROGRESS

We have discussed two applications of randomness to computation. The incompressibility method takes the advantage of one typical random input that cannot be found. The information distance applications, on the other hand, need to compute what's not computable, but the universality of the theory, although approximated by non-optimal practical compressors, has helped us to avoid the pitfalls of overfitting.

There are several interesting open questions or research directions. I describe them below in random order.

The simple and elegant constructive proof of Lovász Local Lemma shows the high potential of incompressibility method. We expect more surprises. Bill Gasarch recently sent me an article describing several generalizations from this work. The incompressibility method is elementary, widely applicable (for example, average-case analysis of most sorting algorithms), powerful, and easy to use. It is suitable for undergraduate courses for computer science students, and can be used by non-theoreticians to analyze their systems. I experimented in teaching the method in my CS341 (Algorithms) third year undergraduate class (Spring, 2009) at the University of Waterloo (lecture notes in ppt files available at request). As a result, one undergraduate student improved an $\frac{1}{2} \log n$ lower bound for (sequential) stack sorting by Tarjan and Knuth (Knuth *Art of Computer Programming*, Vol. 1) over 40 years ago to $0.513 \log n$, using the incompressibility method.

The incompressibility method is actually not always better and easier. Quicksort is one example. Currently, the probabilistic method gives better and (perhaps) simpler average case bound for Quicksort. Can we improve and simplify [B. Lucier, T. Jiang, M. Li, Average-case analysis of Quicksort and binary insertion tree height using incompressibility. *Inform. Process. Lett.* 103(2007) 45–51]?

Can we improve the Shellsort average case bound we have presented here? At least, for small p, the bound is not tight.

Applications of information distance have remained to be somewhat superficial. Although some systems, such as the QUANTA Question and Answering system (X. Zhang, Y. Hao, X. Zhu, M. Li, Information distance from a query to an answer. In *Proc. 13th ACM SIGKDD*. pp. 874–883, Aug. 12–15, 2007), have been built based on such a theory, much more and deeper analyses on the practical data are needed. The obvious question is: what if the other metrics are used instead? Keogh, Lonardi and Ratanama-hatana's KDD04 work has provided one such thorough study for clustering time series. More such studies are needed, for each of the applications we have mentioned.

References

1. C.H. Bennett, P. Gács, M. Li, P. Vitányi, and W. Zurek, Information Distance, *IEEE Trans. Inform. Theory*, 44:4(July 1998), 1407-1423. (STOC, 1993)
2. T. Jiang, M. Li, P. Vitányi, A lower bound on the average-case complexity of Shellsort. *J. Assoc. Comp. Mach.*, 47:5(2000), 905-911.
3. M. Li, J. Badger, X. Chen, S. Kwong, P. Kearney, H. Zhang, An information-based sequence distance and its application to whole mitochondrial genome phylogeny, *Bioinformatics*, 17:2(2001), 149-154.
4. M. Li, X. Chen, X. Li, B. Ma, P.M.B. Vitányi, The similarity metric, *IEEE Trans. Information Theory*, 50:12(2004), 3250-3264.
5. M. Li and P. Vitányi, *An introduction to Kolmogorov complexity and its applications*, Springer-Verlag, 3rd Edition 2008.

6. R.A. Moser, A constructive proof of the Lovász Local Lemma. *Proc. STOC'2009.*

7. Zhuang Zhou (Zhuangzi), Xiao1 Yao2 You2. Chapter 1. 369-286 BCE.

Chapter 23

From Error-correcting Codes to Algorithmic Information Theory

Ludwig Staiger

Institut für Informatik, Martin-Luther-Universität
Halle-Wittenberg, Germany
staiger@informatik.uni-halle.de

WHAT DREW ME TO THE STUDY OF COMPUTATION AND RANDOMNESS

As a postgraduate I stayed with R. R. Varshamov at the Computing Centre of the Armenian Academy of Sciences (Yerevan) and became acquainted with the theory of error-correcting codes. I tried to combine this with my knowledge on regular ω-languages (sets of infinite strings accepted by finite automata in the sense of J. R. Büchi and R. McNaughton). I observed that convolutional codes can be described as such sets of infinite words (Subspaces of $\mathrm{GF}(q)^\omega$ and convolutional codes, *Inform. and Control* 59, pp. 148–183.). In this respect the rate and the error-correcting capability of convolutional codes were a major issue. They can be described with the help of the concept of entropy of languages, a combinatorial concept specifying the amount of information which must be provided on the average in order to specify a particular symbol of words in a given language.

In 1979 I spent half a year in Moscow at Academy of Sciences and visited regularly the information and coding theory seminars of the Institute of Problems of Information Transmission where I met the most influential Soviet researchers in classical information and communication theory, like Blokh, Dobrushin, Pinsker. Thus I became interested in the question 'How can one measure information?' Shortly after I remarked that the entropy of (ω-)languages is connected to description complexity, a result which I presented at the 1981 conference Mathematical Foundations of Computer Science (published in *Lecture Notes in Comput. Sci. No. 118*, J. Gruska and M. Chytil (Eds.), Springer-Verlag, Berlin, pp. 508–514.). Here I derived upper and lower bounds on the description complexity of infinite words in constructively specified sets of infinite words.

After a talk given in late 1981 at the Institute of Information Theory and Automation of the Academy of Sciences in Prague Igor Vajda drew my attention to Hausdorff dimension as a measure for information finer than entropy (or entropy dimension, box counting dimension etc.*).

These measures of information were all known as dimensions from Fractal Geometry but until the 1990s there were only a few papers which put them into relation to the description complexity. The investigations in these papers had in mind to study infinite words or sets of infinite words which are random only to some degree — they were later on referred to as partially random or ε-random. In 1993 I published in a first larger collection of results relating bounds on the description complexity of infinite sequences to the dimensions of sets containing these sequences (Kolmogorov complexity and Hausdorff dimension, _Inform. and Comput._ 103, pp. 159–194).

1993 was also the year when I met Cris Calude first. He was the one who draw my attention from the mere interest in the amount of information to questions related with randomness, though I regard both directions as integral parts of AIT. From that time we met more and more often which resulted in a fruitful cooperation.

WHAT WE HAVE LEARNED

As I explained above I am mainly interested in partial randomness. In this respect some progress was made using the combination of martingales and order functions — this combination was already used by Schnorr in his 1971 book (Zufälligkeit und Wahrscheinlichkeit, _Lecture Notes in Mathematics No. 218_, Springer-Verlag Berlin) to classify random sequences. Surprisingly, the first papers using this concept did not mention it. In 1998 I contributed a talk named 'How much can you win when your adversary is handicapped?' to the Festschrift in honour of the 60th birthday of Rudolf Ahlswede (_Numbers, Information and Complexity_, I. Althöfer, Ning Cai, G. Dueck, L. Khachatrian, M. S. Pinsker, A. Sarközy, I. Wegener and Zhen Zhang (Eds.), Kluwer, Dordrecht 2000, pp. 403–412.) where I found a correspondence between the exponent of the growth of the capital of a gambler playing against an adversary who has to choose his outcomes,

---

*Later I became aware that this very dimension figures under at least twelve different names. See C. Tricot, Douze définitions de la densité logarithmique, Comptes Rendus Acad. Sci. Paris, Série I 293 (1981), 549–552.

heads or tails, in a restricted way. The restriction was measured by Hausdorff dimension and turned out to coincide with the exponent of growth.

Two years later Jack. H. Lutz coming from computational complexity presented results relating gambling systems and Hausdorff dimension (Gales and the constructive dimension of individual sequences, In *Proc. 27th International Colloquium on Automata, Languages, and Programming 2000*, Springer-Verlag, Heidelberg, pp. 902–913). As a description of the gambling systems he used a combination of martingales and exponential order functions for which he coined the term s-gale. For semi-computable martingales his concept yields the constructive dimension which coincides with the (asymptotic) description complexity. In the sequel Jack Lutz and his collaborators, most notably his student John M. Hitchcock developed this approach to great extent involving also as a more recent development of Fractal Geometry the packing dimension.

A quite different approach which resulted in two definitions of ε-random sequences ($0 < \varepsilon < 1$) was pursued by Kohtaro Tadaki in 2002 (A generalisation of Chaitin's halting probability Ω and halting self-similar sets, *Hokkaido Math. J.* 31, pp. 219–253). He generalised Chaitin's random Omega numbers to ε-random Omega numbers in different ways. These ε-random Omega numbers, in some sense, exhibit the properties of random sequences in a scaled down by a factor of ε ($0 < \varepsilon < 1$) manner.

Together with Cris Calude and Sebastiaan Terwijn we finally arrived at essentially three definitions of ε-random sequences (On partial randomness. *Ann. Pure and Appl. Logic*, 138 (2006), pp. 20–30). Later Frank Stephan and Jan Reimann (On hierarchies of randomness tests, In *Proceedings of the 9th Asian Logic Conference*, Novosibirsk, World Scientific, 2006) showed that these three notions of ε-random sequences define in fact different classes of partial random sequences.

WHAT WE DON'T (YET) KNOW

Each of the three different notions of ε-randomness reflects one feature of the various but equivalent definitions of (non-partial) randomness like Martin-Löf tests, Solovay tests, prefix complexity or a priority complexity. Which one among these is the 'true' ε-randomness, that is, reflects most naturally the behaviour of sequences which satisfy a scaled down randomness?

In the theory of s-gales the behaviour of the gambling strategy, that is, the growth of the gambler's capital is compared against the family of exponential functions. Which results are valid if we leave this in some respects convenient family and compare the growth of the gambler's capital against more general families of functions. Here, to my knowledge, since the appearance of Schnorr's book no substantial progress has been made.

THE MOST IMPORTANT OPEN PROBLEMS

As each researcher probably will favour problems related to his own work, I can present only a very subjective point of view. Some years ago I found it very interesting that the very reputable journal for problems of classical Information Theory *IEEE Transactions on Information Theory* published a paper named 'Information Distance' by Charles H. Bennett, Péter Gács, Ming Li, Paul M. B. Vitányi and Wojciech H. Zurek dealing with a subject concerning AIT. This seemed to be a start in closing the gap between classical Information Theory and AIT. Up to now there seems to be not much progress. To my opinion there should be a greater exchange between classical Information and Communication Theory and AIT, in particular, AIT should try to give ideas appearing in the classical part a constructive shape.

THE PROSPECTS FOR PROGRESS

Once David Hilbert said: "Wir müssen wissen, wir werden wissen.[†]" His wish was contrary to Kurt Gödel's incompleteness results. Nevertheless, the intellectual curiosity, the scientists' thirst for knowledge will provide good prospects for an ongoing progress in all parts of AIT.

[†]We must know, we will know.

Chapter 24

Randomness in Algorithms

Osamu Watanabe

Dept. of Mathematical and Computing Sciences
Graduate School of Information Science and Engineering
Tokyo Institute of Technology
watanabe@titech.ac.jp

WHAT DREW ME TO THE STUDY OF COMPUTATION AND RANDOMNESS

I love programming; I have been writing programs since I first read a Fortran programming text when I was in junior high school. In particular, I
was very amused when I found that it is possible to make a program that
understands and executes a certain set of instructions designed by myself;
yes, I discovered the notion of "interpreter" and wrote my own interpreter
in Fortran at junior high, which I am a bit proud of ;-). From this experience, I started thinking, though very vaguely, that there may be some way
to discuss the notion of "computation" formally.

Then I came across the book of Davis[1] (Japanese translation) in the
univ. library when I was searching for a book to study during my first
univ.'s summer vacation. I knew nothing about Turing machines, etc.,
and the things written in the book looked very strange to me; but what
was explained there was *very* attractive and fun to read. In fact, I was
surprised to see the notion of "universal Turing machine", which is similar
to the interpreter I made. What was more shocking was to see that the
universal Turing machine is a key to proving the limits of computation,
i.e., the existence of problems nonsolvable by computers. Since then I have
been captivated by "computation."

Much later, when studying computation more seriously for my master
thesis project, I learned Rabin's randomized primality testing algorithm,
one of the first examples of remarkably efficient randomized algorithms,

and got interested in the fact that randomness can be used for designing efficient algorithms. It was an interesting surprise to me. In general "randomness" is regarded as an obstacle. As a source of uncertainty, randomness is something that makes our life difficult; though, of course, it makes our life interesting from time to time. But "randomness" indeed helps us in some serious situations like solving the primality testing problem. I felt that several interesting and important aspects of "randomness" could be revealed by investigating it through "computation", my favorite notion. This was my starting point for investigating randomness and computation. While studying randomness and computation, I became more and more confident about this idea, and I now strongly believe that there are some important features of randomness that can be seen only through a computational perspective. Below I would like to share some of these features with the reader. In this article I focus on randomness related to algorithms. There are some other aspects of randomness that can be understood through computation; but I would like to leave them to other articles.

WHAT WE HAVE LEARNED

One of the important findings in computer science is that randomness is useful in computation, or more precisely, randomness is an important computational resource. Almost all of us (not only computer scientists) would agree that randomness is useful in computation. But I think that not so many know that randomness has been used in several different ways in algorithms or in computation in general. I think that such a variety of usages is due to, at least superficially, different features of randomness. Let me explain, in the following, several typical examples of using randomness in computation.

We will see, in the following, four usages of randomness; to be concrete, we will use some problems and explain how randomness is used for solving them. Here by "problems" we mean computational problems or tasks to compute a certain result/output for a given input instance.

We begin with a standard sampling problem.

Example #1: **Approximate Estimation Problem**
given: A set B of black and white balls in a bag, in total n balls.
task: Estimate the ratio of white balls to total balls in B
approximately.

Of course, one can get the exact answer by checking all balls in B, and this is the only way if we are asked for the exact answer. But if a reasonable approximation is sufficient, then it is not necessary to see all the balls. Sampling some number of balls and estimating the ratio from the observed balls is enough.

Let us consider this more precisely. Let n_0 be the number of white balls in B; then the correct ratio is $p_0 = n_0/n$. Suppose that we want to approximate p_0 within $1 \pm \varepsilon$ relative error; that is, our task is to compute some number between $(1 - \varepsilon)p_0$ and $(1 + \varepsilon)p_0$. Then it can be shown that sampling $c/(\varepsilon^2 p_0)$ balls randomly from B is sufficient. One should be careful here; there is some chance that the sampling is biased and sampled balls are all black although B contains a good number of white balls. That is, we have a possibility that some undesirable result is obtained. But taking $c = 30$, for example, we can show that this probability is less than 0.1%.

This is a standard way of using randomness and this type of sampling has been used very widely in all sorts of research fields. Maybe there are some who think that this is essentially the only way of using randomness. But we have more ways that look very different!

Example #2: **Sorting Problem**
given: A sequence of n numbers $a_1, ..., a_n$.
task: Rearrange them so that $a_{i_1} \leq \cdots \leq a_{i_n}$ holds.

This sorting problem is well-known, and `quick sort` is known as an efficient algorithm for solving this problem. It is also well-known that the standard "deterministic" `quick sort` has bad input instances. To see this, let us recall `quick sort`. For a given n numbers, the algorithm first picks one number, e.g., a_1, as a *pivot*. Next it classifies numbers into three groups: those less than a_1, those equal to a_1, and those larger than a_1. Then it sorts the first and the third groups, and the result is obtained by putting those sorted numbers in order of the first, the second, and the third groups. This process runs fast if the first and the third groups are of almost the same size recursively; in contrast, the running time gets increased if these two groups are not evenly divided many times during the computation. We can select pivots appropriately by looking through all numbers to be sorted and selecting one close to their median; but this is time-consuming, and unfortunately, there is no simple/quick way to select pivots to prevent uneven grouping. For any such simple pivot selection, one can always define worst-case input instances for which the algorithm needs an unusually long computation time.

Randomness gives a smart solution to this problem. We simply choose pivots randomly. Then it can be proven that, no matter which input instance is given, the probability of selecting a bad pivot is small; from this we can then show that the number of bad pivot selections during the whole execution is small with high probability. As illustrated in this example, randomness can be used to avoid worst-case instances and to convert an algorithm that is good on average into one good even in the worst-case.

In computer science, we often consider "an adversary" when analyzing a given algorithm. An adversary is someone who gives the most problematic situation to the algorithm. We may imagine that worst-case instances are given by such adversary; that is, this adversary examines the execution of the algorithm carefully to create worst-case input instances. Here the role of randomness is to make the computation unpredictable to adversaries, thereby preventing them from creating bad instances. Similar usage of randomness can be found in the design of algorithms for computer games, crypto systems, and so on, where we may assume some sort of adversaries or opponents.

The next example is not a problem. It is a randomized process that reveals an interesting feature of randomness.

Example #3: **Monopolist Game**
given: n players, each having w one dollar coins.
game: Play the following in rounds until all but one player becomes bankrupt.

- In each round, every active player puts $1 on a table, and we spin a roulette wheel to determine a winner. The winner takes all the money on the table.
- A player who loses his money must declare bankruptcy and becomes inactivate.

A simple program can simulate this game and it is easy to see that with high probability the game terminates with a *monopolist*, one player taking all nw dollars. In fact, the *length* of the game, that is, the number of rounds executed until a monopolist emerges, is almost always $(nw)^2 \pm \Delta$.

This shows one feature of randomness. That is, randomness is not even. Though every player wins with equal probability, still a monopolist emerges. Suppose, for example, that each player wins in turn instead of randomly. Then no one gets bankrupt and the number of active players is the same forever. The fact that a monopolist emerges with high probability is due to the bias that true randomness possesses. This property of randomness causes some problematic situations from time to time, but it can also be used for designing some algorithms. For example, a similar idea can be used to select a leader processor among a collection of processors that look exactly the same.

There is an interesting story to this example. I considered this game in order to discuss a biological phenomenon called "orientation selectivity" from an algorithmic viewpoint. When I formalized the process in this way, I thought that I could easily find some analysis in the literature. In fact, we can easily find the analysis for the case $n = 2$ in standard textbooks on probability theory; see, e.g., the famous one by Feller.[2] It seemed, however, that a process like this had not been analyzed, and we worked for some time to get some partial bounds. The problem then was completely solved by Eric Bach, essentially in one night! I explained this game and our open problem over a dinner I had with him and Jin-Yi Cai, one of my good friends and strong research collaborators, at the end of a visit to Jin-Yi. Then when I flew back to Tokyo next day, I found Eric's email with an idea for the solution waiting for me!

Finally, let me give an example not belonging to the above three types. It is quite unlikely that this is the last type; there should be more. On the other hand, I think that this example is one of the most interesting ways of using randomness.

Example #4: **Problem of Finding Smallest Enclosing Disk**
given: A set S of points in the Euclidean plane.
task: Find the smallest disk (or circle) $D(S)$ containing all the points of S.

Note that a disk in the Euclidean plane is (usually) determined by any three points on its edge; hence, S has at least three points, say, a, b, c, on the edge of $D(S)$ (see (1) of the following figure), which we call *extreme points*. Thus, our task is to find these three extreme points.

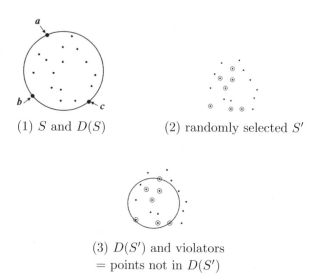

(1) S and $D(S)$ (2) randomly selected S'

(3) $D(S')$ and violators
= points not in $D(S')$

A naive way is to consider all sets of three points of S and examine whether the disk determined by these three points contains S. By this approach, the desired disk $D(S)$ can certainly be found. But this computation needs $O(n^3)$ time, i.e., time proportional to n^3. On the other hand, E. Welzl proposed a randomized algorithm for this problem running in $O(n)$ time on average. The algorithm executes as follows. First select c_0 points randomly from S, where c_0 is a constant determined in a certain way; for our problem we may use $c_0 = 20$. Let S' denote the set of these selected points. Precisely speaking, we assume some weight for each point, and points are selected with probability proportional to their weights. Since the weights are all equal initially, our first selection of S' is uniformly at random. Then we compute the smallest disk $D(S')$ containing all points of S'. We may use the above naive method for computing $D(S')$; since S' consists of only 20 points, we do not have to worry about the computation time. Clearly, if $S' \supseteq \{a, b, c\}$, then we are done. Otherwise (and if $D(S') \neq D(S)$), there must be some points not contained in $D(S')$. We call these points *violators* and double their weights. Now we proceed to the second round and select again 20 points; note that the selection is no longer uniform. This process is iterated until $D(S)$ is obtained as $D(S')$. Note that unless $D(S) = D(S')$, at least one of the three extreme points becomes a violator and gets its weight doubled. Thus, soon or later, all three extreme points have large weights and it is quite likely that they are all selected in S' and $D(S')$ becomes our desired solution $D(S)$. It can be shown that with high probability $D(S)$ is obtained as $D(S')$ in $O(\log n)$ rounds.

In the field of computational/algorithmic learning theory, researchers have been working hard to design efficient algorithms for finding some rules from observed data. In this field, Y. Freund and R.E. Schapire introduced a very powerful method — boosting — to design good learning algorithms based on weak learning algorithms. Though the context is very different, boosting has a similar flavor to the above randomized algorithm.

In this example, randomness is used to get around some complicated analysis for designing algorithms. For designing an efficient algorithm for our problem, we might have to develop deep understanding of enclosing disks and extreme points. But using randomness, we designed an efficient algorithm without such understanding. It should be noted that, for our problem, some important properties have been found more recently that enable us to design nonrandomized algorithms. But the randomized algorithms still perform better than those nonrandomized ones. Anyway, we cannot always obtain such knowledge, and randomness remains important.

WHAT WE DON'T (YET) KNOW

As we have seen, researchers have found that randomness is useful in several different ways for designing algorithms. We do not know, however, which aspects of randomness are essential and when they become essential for designing efficient algorithms. Note that the answer differs depending on the way one defines "efficient algorithm." For example, if we think that even constant factors are important, then there are many cases that randomness will be essential.

On the other hand, if we consider more relaxed notions of "efficient algorithm", we may have different situations. In the extreme case where "polynomial-time computability" is used for the efficiency criterion, we may be able to claim that randomness is not essential for designing standard algorithms. We have some strong evidences supporting this claim based on a sequence of great results* by M. Blum, O. Goldreich, J. Hastad, R. Impagliazzo, L. Levin, A.C. Yao, A. Wigderson, and others. Roughly speaking, in these results researchers have developed techniques to use the

---

*This topic is one of my favorite research topics, and I love these results. They are really deep and interesting. But I would like to skip explaining them in detail here. For this topic, I recommend the reader a textbook such as Ref. 3.

computational hardness of certain problems for generating a pseudorandom sequence that looks random to all polynomial-time algorithms. By using these techniques, we can prove, for example, that if some exponential-time solvable problems are indeed very hard in the circuit computation model, then we can simulate any polynomial-time randomized algorithm without using randomness.

This is about the extreme case. While these results are important and interesting, a stronger notion of "efficiency" than polynomial-time computability may be more reasonable for many practical situations. For example, we may regard $O(n^2)$ time or even $O(n^3)$ time as computationally efficient, but maybe $O(n^4)$ time computation is too expensive. Then randomness may still be essential for some sorts of problems. Also we often need to consider "nonstandard algorithms", algorithms that make use of some subroutines as a black box or that need to deal with input instances containing some randomness. For such "nonstandard algorithms", we may not be able to use the above strong techniques and it seems quite likely that randomness is necessary. But then we do not know when and how is randomness important.

THE MOST IMPORTANT OPEN PROBLEMS

As explained above, in many practical situations, randomness may be necessary for efficient algorithms, but we do not have any universal technique like the one using pseudorandom sequence generators. Maybe what we need to develop instead is a collection of techniques for understanding random executions of algorithms.

For example, there are several local search algorithms solving satisfiability problems such as the 3SAT problem. For analyzing these local search algorithms, it may be likely that both randomness in execution and randomness in input instances are essential. Then since randomness is unavoidable, what we need are techniques for analyzing the (random) execution of algorithms on (random) input instances; that is, techniques for tracing the value of variables used in the execution. Note that once a target algorithm and its work variables are fixed, its execution can be regarded as a finite Markov chain. But the number of states becomes exponential in the number of variables; one challenging technical project is to develop a formal method for the approximate analysis of huge Markov chains. Here understanding of the role of randomness relevant to the current computation may be important; it may help us to reduce the complexity of the target Markov chain.

Target algorithms could be anything so long as they are important in computer science in general. It does even make sense to consider algorithms using no randomness if our target is the analysis of their executions on random input instances. For some concrete open problems, I would like to propose the analysis of various local search algorithms for NP-hard problems, such as 3SAT or MAX3SAT. For example, for the 3SAT problem, it is important to compare various randomized local search algorithms theoretically and give some reasoning why, e.g., `Walksat` outperforms the other similar local search algorithms. The relationship between various message-passing algorithms (including, Pearl's belief propagation) and local search algorithms for solving, e.g., MAX3XORSAT (in other words, MAX3LIN) is also an important target of our analysis.

THE PROSPECTS FOR PROGRESS

I claimed that it is important to develop techniques for analyzing (random) executions of a given algorithm on (random) input instances. Such techniques should be quite useful for analyzing random phenomena in general. Movements/changes of status of huge number of proteins in some solution can be regarded as the variable value changes in a random execution of a certain local-search algorithm. Changes of people's preference through network interactions may be formulated as variable value changes in a random execution of a certain message-passing algorithm.

Yet a more challenging and important project is to extend these analysis techniques for analyzing massively heterogenous computations. Since the improvement of computational devices is getting close to the limit, we may not be able to expect the speed-up that we experienced in the past by simply improving CPU chips; other methods for improving the power of supercomputers are getting more and more important. One possible solution is to combine huge numbers of computational devices into cluster machines. Furthermore, it is getting popular to use mixtures of various computational devices to compose a cluster machine. This approach seems reasonable to meet various resource constraints and to maximize the machine's performance under given constraints. On the other hand, it would be very difficult to analyze and control systems consisting of huge numbers of computational devices with various different characteristics–much more difficult than analyzing the change of a huge number of variables in randomized algorithms. But I hope that current and future techniques for analyzing randomized computations can be used as a basis for attacking this challenging problem.

Acknowledgments

I would like to thank Prof. Lane Hemaspaandra and Mr. Mikael Onsjö for their comments for improving the presentation of this article.

References

1. M. Davis, *Computability and Unsolvability*, McGraw-Hill Book Co., Inc., New York, 1958.
2. W. Feller, *An Introduction to Probability Theory and Its Application*, John Wiley & Sons Inc., 1968.
3. O. Goldreich, *The Foundations of Cryptography - Volume 1*, Cambridge Univ. Press, 2001.

PART VI

Panel Discussions
(Transcriptions)

Chapter 25

Is the Universe Random?

Cristian S. Calude, John L. Casti, Gregory J. Chaitin,
Paul C. W. Davies, Karl Svozil and Stephen Wolfram

PANEL DISCUSSION, NKS SCIENCE CONFERENCE 2007. UNIVERSITY OF
VERMONT, BURLINGTON, USA. JULY 15, 2007.[a]

John Casti (as moderator): Some weeks ago when I was recruited for this
job by Hector (Zenil) and I first agreed to look after this session I had
some interaction with Stephen (Wolfram), and others – about what this
panel discussion might address and how. And I think that the consensus
that emerged was that it should be something as vague as possible but
no vaguer [laughs]. I believe that everybody in here has some pretty well-
established ideas on most of the issues we want to discuss, certainly on
randomness, complexity, physics, mathematics, etc., Stephen (Wolfram) at
one point put forth the idea, which he formulated as question when he
asked: "Is the Universe random?" I found that to be a pretty interesting
question, although phrased not quite in the way that I would have phrased
it and for this reason I want to try, a bit later, to sharpen that a little bit
if possible.

COMPLEXITY VS. RANDOMNESS

John Casti: Now I wanted to actually focus not on randomness but on
complexity, which is in fact my manifesto, because I think that complexity
is a deeper notion than the concept of randomness. And I certainly regard
randomness in some sense as the degenerate case of complexity, because I
see it as a notion of extreme complexity. And just to give a little bit of a
starting point on the idea of complexity I'll start by rephrasing Stephen's

---

[a]Transcript and footnote comments by Adrian German (Indiana University Bloomington). Recording: Jeff Grote. Audio available at:
http://forum.wolframscience.com/archive/topic/1429-1.html.

question as follows: "Is the Universe (or is the real world) too complex for us?" And in order to even make sense of that question I think that one has to de-construct it and ponder almost every word in the question and allow it to be sufficiently vague to the point of having almost any kind of interpretation. And to further illustrate that point I am going to give you a little list that I got from Seth Lloyd. Years ago, when I first went to the Santa Fe Institute, before Seth Lloyd became "Seth Lloyd" he was a post-doc there (at the Santa Fe Institute) and for fun and just to illustrate the point that complexity is as much in the eye of the beholder as it is inherent in the object itself he wrote an article (I doubt it was ever published) mimicking the Baskin-Robbins story of 31 flavors of ice-cream. It was called: "Thirty-one flavors of complexity." And he identified 31 different notions and definitions that people had used in literature to refer to complexity. Let me just quickly run through the list, I'm not going to say anything about them individually, I just want to enumerate them to you. So here's one set, having to do with the difficulty of description, and there you see algorithmic complexity right near the top of that list! Here's another category, that had to do with the difficulty of creation – and here are, listed, several other notions of complexity.

And I don't think there's anybody in this room, certainly not me, that can actually tell you what each one of these things *actually is*–without going to the literature and looking it up. And that's not very important, what matters is that there is somebody out there who's thought seriously about complexity and put forth this list to prove that there exist so many different concepts that can be taken as a meaningful notion of complexity. Here's some more: degree of organization, things like excess entropy, sophistication, metric entropy and so on. And this list even goes on a little bit longer: ideal complexity, hierarchical complexity, correlation, stored information, algorithmic mutual information, etc. Now if you kept track of the slides as I ran through them you would have found not just 31 flavors of complexity, but now more than 40 flavors of complexity, not counting that one [he points to the slides] and this just goes to prove the point that what we mean by complexity is really something that has to be (in my opinion) kind of content-dependent.

And if you ask the question: "Is the real world too complex for us?" then you have to either add something to this list, as partly to answer your question or make something of this list, that seems to capture the essence of what you mean by complexity in that particular situation. Now let me just show you a couple of artistic examples: this first one shows you a picture that I got from Karl Sims and below it you can see the Lisp code for that picture: if you had a on your computer and would run that code on it, you would get what you see on this screen. Now this is very much in the spirit of Greg(ory) Chaitin's notion of "shortest possible program that can be used to describe a particular object." You can count the number of characters in that Lisp code and that's not necessarily to be construed that that is the shortest such program – as in general that is unprovable, but it's probably a pretty good approximation to that–but certainly serves a purpose of being an algorithmic description of that particular object. Below it you see what looks like a random mishmash of dots, but in fact there is a pattern in that picture. This is an example of a very popular thing, it's called stereography (or stereoscopy) and if you sort of de-focus your eye and stare at this for a sufficiently long period of time a three dimensional object will emerge out of it. You won't see it now while I am projecting it to you because this picture is probably too small but the point is that it illustrates a different kind of complexity and it has to do with a notion that I personally believe to be the most essential ingredient in a complex adaptive system, which is the idea of *emergent phenomena*: that the objects that make up the system interact in order to create something that is not there in the objects themselves at first.

Now, here's another example, and I'll just put it up as a question: which of these two do you think is the most complex? Is it this human on the left[b] (those of you who are involved in any way in the arts world will recognize this face, it's the face of a famous British artist named Damien Hurst) and on the right there is a skull[c] which is now the world's most expensive skull in the history, probably. You will have the opportunity to possess this object for a cool one hundred million dollars[d] so – what do you think? Why is this thing so expensive? What is it even? It's just a skull! But the artist Damien Hurst, who might even argue that that aspect is in fact secondary, platinum plated it and encrusted it with diamonds and put it into this gallery in London, and said that for a hundred million dollars it can be yours. Now, from a standpoint all I ask here is a real world example of a system, two systems in fact – which one of the two do you think is more complex: the artist or the creation? If you can call this a creation – although it is a creation, but a pretty expensive one actually. But does it

---

[b]http://www.hemmy.net/images/arts/diamondskull01.jpg
[c]http://www.hemmy.net/images/arts/diamondskull02.jpg
[d]http://www.hemmy.net/2007/06/04/50m-diamond-studded-skull/

really have any actual complexity? Now that's an open question, I think, but an interesting one.

So, just to close this little part, my manifesto, a notion of complexity that I like a lot and is one that is relatively recent is the notion that complexity is not something inherent in the object itself – whether it's the Universe or anything else – instead it has very much to do with the observer. That complexity is a concept that has to be interacted with – by the observer. And just to give you a trivial example, think about a stone in the street. You go up and pick up that stone and wonder, while you look at it, to see how complex it is. Well, if you are a typical person like me, with no special training [in geology,] you probably say that the stone is not very complex, because there are only a limited number of ways you or I as non-specialists can have interaction with the stone. On the other hand if you were a geologist you would have a whole bunch of additional ways in which you can interact with the stone that are inapplicable to me and that stone becomes vastly more complex for a geologist than for somebody who is untrained and doesn't have these observing capability powers.

So complexity might be in this definition the number of inequivalent ways in which you can interact with an object (and the emphasis is on *can*). And I don't mean to imply or claim that this is a universally accepted or vague-enough definition of complexity but it's one that *at least acknowledges* that complexity is not just inherent in the object itself[e]. This brings us, finally, to my version of Steve's question which I formulate as follows:

"Is the real world (also known as the Universe) too complex for us?"

And my argument is that this question is at least as much – if not more – a question of philosophy than it is a question of physics or mathematics so, with that as my manifesto I would like to ask for your opinions on the matter.

CAN WE CONTROL THE UNIVERSE?

Paul Davies: Well, I think it's clear that this introduction has been provocative enough to everyone present in this room and that there are at least 2–3 speakers ready to go, by my count.

---

[e]Unlike information, apparently. If information is intrinsic to an object (e.g., algorithmic information, in fact, is the very definition of the object,) complexity seems to be the ability to extract that information from the object. This ability depends on the observer. The more the observer can extract the more complex both the observer and the object are.

Greg(ory) Chaitin: I agree. From my point of view the question is: if the Universe is too complicated for us then obviously we have to re-engineer the human being until the Universe is no longer complex for us!

Stephen Wolfram: I too, have several things to say. One is that it is perhaps fortunate that the field of biology does not necessarily have, or does not a priori start from a definition of life– otherwise biologists wouldn't be able to even get started doing biology!

There's indeed much to say about these issues of complexity but I think that this is a good question to start with: "Is it inevitable that the Universe is too complex for us to the extent that we would have to re-engineer ourselves?" Because one could just as easily ask: "Could there be situations where it's inevitable that we have the same complexity as the Universe?"

Greg Chaitin: So – is it conceivable that if we invented it and if it is conceivable that it has the same complexity as us we just forgot that we invented it? That's one question that follows from both your questions.

Stephen Wolfram: Well – but we didn't invent the Universe, did we?

John Casti: This question is a little bit similar to something you see in system theory where the complexity of the controller has to be at least as great as the complexity of the object being controlled and so. Thus, one might argue by saying: "look, however complex we are as human beings or whatever notion of complexity you choose to use, you can only access that part of the Universe which is as complex as (or less complex than) you are." This would be the biological correspondent for the definition of complexity based on the number of inequivalent ways of interaction. But what if the Universe has indeed lots of degrees of complexity and we only have one?

Paul Davies: So then does that mean, if we ask a question much vaguer, that we will never be able to control the whole Universe? That there's a limit to technology? If we imagine there's no limit to money – or time ahead of us – is there nevertheless a bound on what is technologically feasible, and we could never conquer the Cosmos. Is that what you're asking?

Stephen Wolfram: I have a thing to say about that. Some day, perhaps soon, we may have a fundamental theory of physics. We may know everything about how the Universe has been built. So the only question is: even getting that, what can we do with it? Where does it take us? So an obvious question is: if you know the fundamental laws of physics can tell you whether such and such difficult problem can be solved? It's not obvious, but depending on the problem we might be able to tell. In general however, it may be undecidable whether a solution might be possible for any given hard problem, even knowing all the fundamental laws. So this question is an important one because we know that it's not sufficient to just know the laws of physics in order to be able to predict nature better.

Paul Davies: But within this realm of asking what complexity is, and given that as John is saying there are certain several degrees of complexity, and if we may be able to manipulate the complexity of the unknown then my question would be: is there a sort of a bound on the Universe? I'm talking about a complexity bound that acts on our Universe and that says that you can only achieve so much in controlling the Universe – after that there will be some insurmountable boundary on capability that we'll hit.

Stephen Wolfram: Maybe something like the second law of thermodynamics?

Paul Davies: Yes, like the second law of thermodynamics but not quite the same, because, you see, by doing some things very cleverly – and this is what truly interests me – it's not just any sort of information, it's very specific sorts of information that enable us to manipulate our environment. So if we don't figure out the motions of the asteroids in the solar system we can still produce some changes, e.g., with a little push from one asteroid we can cause it to hit a planet, which is now a planet with a different trajectory, and by a sufficiently clever accumulation of changes I think we could configure our galaxies eventually. But there must be a limit to what we can do! We couldn't be able to work on the Universe and, still not be able to actually explain it, don't you think?

Stephen Wolfram: But, you see, there might be two issues: one of them is that there might be undecidability and/or irreducibility, to figuring out what would happen...

Paul Davies: None of them would be undesirable!

Stephen Wolfram: Right. So then the question is if there exists a way so that we can configure the asteroids to post a big message in the Universe.

Greg Chaitin: Let's say we're in a Universe which is a piece of software running on a computer ... It might be possible for the software to have a routine somewhere where it rewrites itself. Well, then, it might be that if you understood the Universe sufficiently well you might be able to change the laws of physics, or modify them in the future like through a piece of self-modifying code!

Stephen Wolfram: If I can just interrupt – it seems to be within the general acceptance that we experience the Universe as just an emulation of reality. So presumably between the fundamental laws of physics and what we actually experience there are lots of layers between us and them because we're not experiencing things down at their level, we're experiencing some effective theory of what's going on. So maybe what you're saying is that one could rewrite the lower-level laws of the operating system so that what we experience is, kind of like in the movie "The Matrix", something that is very different from the current laws of physics, and then the question is – given the fundamental laws of physics, is there a way to assemble the matter in the universe so that the effective behavior of the Universe is very different from the behavior that we currently experience?

Greg Chaitin: But then how would you define a bug in the design? Because one might crash the Universe by getting the code to go the wrong way. In general you want something constructive, but this can turn destructive, so it's going to get harder to say what's going on.

Stephen Wolfram (answering to Chaitin): Well, for example, it could be a new state of the Universe. To take a very simple possibility it used to be said when I was younger, about the nuclei, that pions make the nucleons subtract, raw mesons make them repel, maybe [half-]mesons make them protract more...

This whole idea turned out to be nonsense but the point is that if you could ever squash a nucleus hard enough, suddenly it would become a different kind of nucleus. Maybe there is a different state of the Universe which has never been triggered.

MINDS, OBSERVERS AND THE UNIVERSE

Paul Davies: It seems though that there could be a mind in the outer space some other kind of mind than ours, some other kind of beings than us, and they would inhabit the same Universe and for them the complexity might be different than what we perceive. Because when we think of complexity we always have this prejudice we always only accept and only assume our position and think that we're the only ones around. But we can imagine that the complexity of the outer space as seen through the eyes of some other beings might be different, although in a sense it's the same world, but seen through a different perspective – perhaps computational or some other type – it might look vastly different.

Stephen Wolfram: But then would you expect that those "beings"... I mean – why do you even call them "beings"? Maybe only because you'd have to identify them as *some* kind of intelligence.

Paul Davies: Yes, yes... they are observers ...

Stephen Wolfram: What counts as an observer?

Paul Davies: Well, "What is it like to be?" as famously asked by Thomas Nagel in his paper. You can try to imagine what is it like to be *them*. Some kind of existence... What is it like to be *different* than what we are now?

Greg Chaitin: So that's the basic need for a foundational model.

Stephen Wolfram: But there are many times when people say "imagine what's like to be a computer" or "imagine what it is to be like [I don't know what else]" and so on. And just trying to imagine is not going to do the job. My favorite of these is: "The weather has a mind of its own" But maybe that doesn't pass your test. Most people can't really imagine with accuracy what it would be like to be something else...

Paul Davies: That's exactly my point.

Stephen Wolfram(continues): ... but still, we feel that this other existence has to have something like a *mind*.

Paul Davies: Yes. But I am not sure about the weather. (Laughter)

John Casti: So back to your original question about perhaps existing inherent limitations on our ability to, in some sense, manipulate the Universe...

Paul Davies: Well, I think that could be made into a good science question and a Ph.D. student would be very happy to work on it.

John Casti: What is your belief on the outcome of that research?

Paul Davies: I think there is indeed a limit.

Greg Chaitin: Well, who cares if there's a limit?! Just imagine that there is...

John Casti (interrupting): No, I think it's interesting because the question isn't as much as to whether there is indeed a limit, or who cares if there's a limit. The question is: "Where is the limit?" And "What do you have to do to push that boundary out?"

Karl Svozil: Well, I think that my position is that – I truly believe that our ability to manipulate the Universe is only limited by our fantasy and our fantasy alone. So I think that anything which we can fantasize about we can do something about, and we can produce it in some way or the other.

Stephen Wolfram: So as a simple minded analog to this we can think about the second law of thermodynamics and we can ask the question: given some energy in the form of molecules and random dots in a vacuum can we successfully extract something that we care about from that random configuration of bits? And I suppose one of the lessons of technology is that it used to be the case that people would say: once this thing is thermalized all is lost we can't extract anything useful from it, but as we get to be able to compute more we get to be able to set up some very close approximations of it and you can really extract something that way – well, actually (now speaking to Davies) you must have thought the same thing, because your first book was along these lines wasn't it?

Paul Davies: Right, right, I think – I am assuming that there is some thermodynamic price to be paid and you can approach some idealized goals in there but I don't think it's negative entropy per se, because by being clever, you can achieve a lot, by having a strategy, you can minimize the thermodynamic damage in order to achieve some of your goals.

Stephen Wolfram: But we're sort of dealing with the goal of decoding the Universe, and not that of being able to know where to push it.

Paul Davies: Right, right, right – but that is why I think that this is directly related to the goal of a mind (as opposed to just crude intentions) and that's why I think that observers are significant and not an optional exercise or to be viewed simply as constraints. Because, in a sense, I think that an observer, or a mind if you like, could, in principle decrease the relative complexity of the Universe and I wonder actually what the limits are.

Stephen Wolfram: But some of us think that much of what the Universe does is computationally just like what the mind is doing.

Paul Davies: I know that.

Stephen Wolfram: So it is not the case that the mind is above the Universe and because we have minds, you know – (he decides to reformulate:) Let's say that we are the first "intelligence," in our sense, or the first intelligent things to arise in our Universe, then it could be that the whole future of the Universe is going to depend a great deal on that and that one day, you know, our Universe will look vastly different precisely because of the minds in it. It could be – but I think that's extremely unlikely.

Paul Davies: Do you think that it would achieve that without minds, or that can't be achieved at all?

Stephen Wolfram: There's a great deal of stuff already existing, all these things in the Universe, and you look around and you ask yourself: Did they require a mind to make? What special thing would a mind decide to make? If there were an arbitrarily powerful future technology – one that could move stars and galaxies around with – what would that future mind make with it?

Paul Davies: Right, well, so let me ask not your question but another question, which is related, because as a hobby of mine I'm chairman of the SETI detection. So the question that is arising is if we pick up a signal from E.T. on my watch – is it going to be anything like a sentient being, or is it possible that we're getting something like the appearance of intelligence, as an emergent phenomenon?

Greg Chaitin: Let me see if I can synthesize the NKS position on this. The NKS position is that any universal computer is a mind and universal computation is everywhere in the Universe. So the entire Universe has a mind!

Stephen Wolfram: Or, more operationally, think about pulsars. When pulsars were first detected the first thought was that this very special signal must be of intelligent origin and that later gave way to other interpretations – and now we know that pulsars are simply not an intelligent being in your sense of the word. But now if you look at the details of pulsar you'll see that they indeed present a very interesting, fascinating type of complexity and my claim is that that sequence that you see, from the blips, in the magnetosphere and so on, is every bit as mind-like as anything else you will ever see anywhere else you might decide to look.

COMMUNICATING MINDS

Paul Davies: That might suggest that we could try to look for a message from the pulsars.

Stephen Wolfram: Well – what's a message?

Paul Davies: It would be something that we could recognize as conveying information even if on an infinitesimal scale.

Stephen Wolfram: But that's a very inefficient message!

Paul Davies: May very well be, but that's what we've been practicing.

John Casti: The point is that your perspective basically dramatically anthropomorphizes ...

Paul Davies: I totally agree.

John Casti: You are saying: "a message is something that we would be able to recognize as meaningful, etc." And what I'm asking is indeed similar: if you make sense of it you're at least as complex as it is, and in that very sense. However in my definition I am not excluding other possible uses or meanings that the observer, lacking the enabling knowledge, or the necessary complexity to relate to that, can't see yet. And it all revolves around the idea of a mind that would be the support for that emergence. If complexity is in the eye of the beholder, the mind is the beholder and we need at least one.

On the other hand the idea that everything is a mind – that's a little bit too generous, perhaps: it's as if every statement is a theorem which is not very helpful, because you can't distinguish between minds and non-minds if everything is a mind.

Stephen Wolfram: But there's a sharp distinction between... (pauses for a moment then starts again:) Listen, most of what's been studied in physics in the past is not mind-like, it's in fact too simple to be mind-like. The idea that people have had since ancient times (and that was the predominant thought about things and I don't think that it's wrong, as modern scientists would have people believe) was to say that there are many things that you can predict on the basis of thinking that the weather has a mind of its own – you can determine the extent to which something is conceivable or predictable and so on.

But, with respect to this thing about SETI, I'd be really curious to push on and find out: *Is there* a message?

Paul Davies: Right!

Greg Chaitin: Can I say what I think that the right message to send would be? It wouldn't be a TV program or anything like that. Fred Hoyle had another idea, in another book – which was a pretty bad book, I preferred the "A for Andromeda[f]" myself – had the idea that the kind of message that you would send would be a piece of software that wanted to propagate. So if you send a computer program and obviously an instruction manual for the computer/machine you would run it on, and then the people in this story of Fred Hoyle's – they construct a computer and start to run the program on it, and it's an artificial intelligence. So this would be one way for organisms which would be software to send themselves across the Universe.

Stephen Wolfram: That's the ultimate computer worm.

Greg Chaitin: Yes, this would be a life form that would transmit itself as a message. And it seems to me that that would be the only interesting message: a message that wants to come here and start running here because that's the only interesting way to send yourself across the Universe: as a message, if you're software. You're an AI program that comes to Earth and starts running. And now you're here!

Paul Davies: This raises an interesting question regarding what the most efficient way of decoding the message would be.

Greg Chaitin: But the message is – but the way you tell that it's a message, the viewpoint is that you start running the software and if it starts having a conversation with you and it wants to take over the government, then that was a meaningful message ... otherwise it would look random ...

Stephen Wolfram: I see what you're saying. But there's an interesting fact: in the past when you would scan radio bands you would hear all kinds of obviously meaningful stuff. Nowadays with all the cell phones and all the digital stuff and so on increasingly you can't hear anything any more

---

[f]From http://www-users.cs.york.ac.uk/susan/sf/dani/028.htm: "A for Andromeda" was coauthored with John Elliot, a television writer. It has a premise which other writers have borrowed since: A new radio-telescope picks up what has to be a message coming (for a long time, obviously) from the direction of Andromeda. When the message is finally decoded (it was designed to be decodable) it turns out to include a very complex computer program, the design information needed to run the program, and data. Naturally the computer is built and the program is run. It then provides the researchers with instructions for creating a human being to serve as its interface.

because it's all been compressed. And also the whole idea of broadcasting signals seems to be old news.

Paul Davies: Perhaps I happen to be a very skeptic about the more classical SETI approaches. I think we need radical new approaches to address the question.

Stephen Wolfram: But do you think it even makes sense, I mean does the question even make sense? Because if it is the case that minds are everywhere what does it mean ...

Paul Davies: That's what shocked me in this discussion because if you think that that is the case then there's one or more of these things out there and they are trying to communicate ...

Stephen Wolfram: But I think the issue is just like AI. What does it mean to have AI?

Paul Davies: Well, we know what it means to have a conversation with another human being.

Stephen Wolfram: We know what it means to have human-like AI.

Paul Davies: Yes.

Stephen Wolfram: It's something that is being very much like us.

Greg Chaitin: More like a historical accident.

Stephen Wolfram: So the question is how much you can abstract away from the details of human-ness and still have something that you know is intelligent.

Paul Davies: Let me answer this question. We can imagine life on another planet, classical SETI, but can we imagine getting a message from a physical system that did not go through this process of emergence of organic complexity and evolution of culture? Are we in any way prepared to receive and understand that?

Stephen Wolfram: So here's a case that might be relevant: you know whales' songs, right? Whales songs presumably are communicating something of relevance. Presumably – because it's not clear, I don't think anyone is completely sure either way. I mean nobody knows if there is a correlation between what one whale says, or sings, and what that whale does. So let's assume there is a message. When we listen to whales' songs it's really hard to say what they're about. Here's an amusing historical anecdote: in the distant past, when the radio was new, Marconi who was one of the developers of radio had a yacht and he went to cross the Atlantic in his yacht, and in the middle of the Atlantic he was listening on his radio mast and he heard this kind of ... woo-eeeo-ooaah. I can't really imitate the thing but that's what they sound when you listen to them – and he got them from his radio mast. OK, so what did he conclude that these things were? His best hypothesis was that they were the martians ... That's what he concluded that they were: radio messages from the martians.

They turned out to be ionospheric whistler bars from magnetohydrodynamics. So this story about Marconi's reception is a case where we hear something which is at least seemingly undistinguishable from what is a mind-like physical system and presumably a mind-like biological system.

John Casti: Actually I think that this whale business raises a very interesting and important message for SETI and that is: we have, through the years, seen a lot of different experiments – people trying to talk to dolphins and chimpanzees and octopuses and whatever – and by and large my impression has been that in most of these experiments communication was so limited that one would be very justified to simply call them failures, all of them. Failures in the sense that the level of actual communication was pretty minimal. So here we are trying to communicate with organisms that we have shared a very long evolutionary history with, for so many centuries, and we do a very poor job in establishing any meaningful communication – even if we believe that we are indeed communicating. Why do we ever imagine that if we ever get a signal from a beam coming from a distant civilization – that we have nothing in common with – that we'd in fact be able to understand the message?

Paul Davies: Well, you are raising a valid point, but ...

Greg Chaitin: Can I invert this question, if I can interrupt, I think that you should go about this the other way, Paul: we should be the SETI, we should be the message, so what we should do would be to send a message and hope that somebody that receives it would run it on the computer, and the program's output should be us – or a copy of us. It would be an AI exploratory program which will look around and see the planet it's on and then send us the information about it in a message that we could understand.

Paul Davies: It's been done.

Greg Chaitin: Oh, it has?

Paul Davies: Yes, but not entirely like that...

Stephen Wolfram: Well, not like that, nothing like that! What was sent out was an *absurd* thing.

Greg Chaitin: OK, to be more precise let me explain what I would send out: I would be sending out examples of very simple lists, so that it becomes clear what the semantics of your programming language is. Then I would start sending out pictures of the list in the box interacting with the outside world, because I would like the list to have a camera and ... I know it's difficult, this is not a complete program or I would have done it already. But then you send out this long program that is in fact you, that would start going around and interact with the natives trying to find information and send it back to us. And there we go: zoom! We're the SETI and the idea is to attract attention to yourself so that you will get to be run. You want to send a piece of code that people would be curious to try running on the computers and see what happens! That's how we get there: we put ourselves into this other environment ...

Stephen Wolfram: By the way I think that this thing about communicating with/to animals and so on, my own guess with respect to that (I guess I should be warning you I will say something slightly outrageous) but my own guess is that in the next few years if anybody is indeed truly working on this with modern techniques I think that it would be possible to do a lot more than has been done so far, and here's how: right now, (when we try to talk to a chimpanzee or any other animal we're trying to talk to) we don't record everything in the chimpanzee's life and every kind and/or way that they interacted so far and so on. I bet that if that was actually recorded and correlated with its experiences that one could actually start to figure out a reasonable communication mechanism. I mean even now there are some promising devices, and my favorite example is (and I saw this as a product marketed under the name) Bow-Lingual[g], which some of you may have seen, which is this thing where you type what you want to say in English and it comes out in dogs' barks. And then there's also a thing where it listens to the barks of the dog – and you have to adjust it, adjust it to the particular type of dog you have – and it analyzes based

---

[g]http://www.crunchgear.com/2009/07/13/bowlingual-portable-dog-language-translator/

on you know the format of a selection out of a small collection of mostly emotional states kind of communication things and it tells you what the dog had to say when it barked.

John Casti: Does this sound vaguely like the famous New Yorker cartoon[h]: two dogs sitting in a room, one of them sitting at a computer terminal typing away and the other one the other one down the floor, when the one doing the typing says: "You know, on the Internet nobody knows that you're a dog!"

THE ISSUE OF AI

Stephen Wolfram: Let's say you succeed to establish at some level a communication with whatever type of being. The issue then is: what are the topics you talk about and, what are the purposes that seem to make sense in the context of this conversation? You talked about meaning, and meaning gets closely associated with purpose: how do you know if the thing has a purpose, that it has a meaning and a purpose?

What are the premises and/or purposes of the E.T.'s we're looking for?

Paul Davies: I understand your question. Well, the thinking behind classical SETI, if we can go over that a bit, is incredibly parochial. It goes along the lines that they are like us but they have been around a lot longer and have got more money (laughs) and so they will send us messages and they will take it as a matter of fact, that is, they will assume that we have just sort of stumbled on the radio technology, so they will make it very obvious that this is a message and – in other words, they will take the lead!

John Casti: Maybe acting a little bit like missionaries when they were going to recruit the natives for the church, and we know that this has not been an especially successful experiment, for the natives especially.

Stephen Wolfram: Well, as successful as Greg said, simply. But I think that the best argument that I've heard from the SETI community is: we might as well do what we *can* do.

Paul Davies: Yes, it's true that essentially we're just listening out there ... But what we saw and maybe I'll go off track, but my point of view on this is that there is no reason that E.T. would beam messages out until they can be sure that we're here and they don't know we're here beyond about a hundred light years away and so it's most unlikely that there will be any

---

[h]`http://eitransparency.org/UserFiles/SocialMedia/images/Dog.jpg`

advanced civilization within that distance. Thus the strategy may appear be sound but it's maybe a few hundred years too early to start with.

Stephen Wolfram: I think honestly, and ultimately ... I have to say, I think that all these things – I mean this question about what counts as a mind and what purposes ...

Paul Davies: Well, that's what interests me!

Stephen Wolfram: ... well, yes – but I think that this issue is deeply muddled. I mean, I used to be a big enthusiast of SETI and I figured out all sorts of things about how you could use commercial communication satellites to do wonderful radio telescope data processing by using spares that got used – there's a lot of great things one can do. But after I started working on NKS, I kind of became much more pessimistic about it ...

Paul Davies: Well, but then you'd have to – because if your view is that the Universe is full of mind and maybe communicating minds, and maybe this would occur in systems as diverse as pulsars as you mentioned, and so on – then you'd have to ...

Stephen Wolfram: OK, but what is a communicating mind? Whenever a physical system has an effect on another ... it is a communicating mind!

Paul Davies: That's right. But then we would have to be the rather impoverished and idiosyncratic example as the one mind of the Universe otherwise you can show that if we represent a whole class, you know, of biologically instantiated intelligence, or something, whatever you call it, then there may be superminds, or bizarre minds – just as are our minds – swirling around too, so if there are some guys like us out there then all of it is worth doing ...

Stephen Wolfram: So then the question is what is the probability that there exist minds – well, minds may be common – the question would be "How human like are those minds in the distribution of all minds?"

Paul Davies: Well this is where having an ensemble of one is a terrible thing because it is very easy to think of all sorts of reasons as to why it had to be this way: so, if you play out evolution somewhere else you say "Well, yes, of course it makes sense to have information processing and sense organs, and then obviously you want that one off the ground and then in case you get hit, you need something that will prevent damage, etc." and you can easily talk the argument into a humanoid type of entity – if we're not careful!

Stephen Wolfram: That's because we can't figure out something else – although that, in and of itself, is very embarrassing!

Paul Davies: But I think that, having followed SETI myself over 20–30 years, I've been amazed how even minor changes in technology have reflected on the change of the basic SETI strategy. Do you remember how in the early days the main question was what frequency they should use? The main issue was to try to guess a range to cover! Nobody talks about that any more because you can cover billions of channels simultaneously. So the technology just in 10–20 years has evolved so much that it changed the whole search strategy. So in another thousand years we may say "Obviously we have been using such and such..."

Stephen Wolfram: I think the invariant of SETI is that their search effort is centered/focused just around the most significant technological breakthrough on the Earth at the time; so, for example, the canals on Mars where just around the canal building period on the Earth and in general there's a big influence from from the technology of the moment on Earth.

Greg Chaitin: There's a wonderful SF story about this. The idea is that some people are doing a SETI experiment on Earth, and it's their big project, and so on. Meanwhile there are people on other planets, these are E.T.'s also trying to communicate – so here we are, having discovered absolutely nothing using radio waves! Meanwhile, neutrino beams which this other planet considers to be the only logical possible mode of communication are flashing out with these urgent message we can't see, and the sunset colors are modulating with some other rays, in ways that another group of beings think that would be the most obvious of all possible ways of communicating with another rational human being (which, as they happen to think, would be by colors). So the sky is flashing with neutrino beam broadcasts from one civilization and this meaningful message at sunset from somewhere else from a second other civilization and while we keep listening in the radio frequency spectrum *nobody* is communicating with anybody else!

Stephen Wolfram: Very interesting but let's come back to the reality of our discussion. Is the Universe random or not?

Is Mathematics Inevitable?

Cris(tian) Calude: May I? ... I think that the Universe is random. And even if it is not true it is still productive for the human mind to believe in the randomness of the Universe and I will give you my very short explanation as to why that may be so. All the laws of nature are mathematical idealizations – essentially ways to approximate the finite by the infinite. The power of the mathematics as a modeling tool is based on the fact that it introduces the infinite as the approximation of the very large finite. All the equations that we use in physics introduce essentially the infinity which is very nice and perfect from a mathematical point of view but cannot be checked in reality except for a very small sample. So I think that it is more appropriate to not even speak about the Universe in general, but to speak about only local rules. Probably the simplest example is to give you one million of digits of Ω and pretend that we have a law of the Universe – but this sample is just a local regularity in the infinite randomness.

Stephen Wolfram: There's something (one part of) what you said that I don't quite understand, or maybe I just don't agree with it. Which is the claim that physics somehow talks about the infinite. If this may appear to be so, it is only because the traditional physics as it has been formulated in the last three hundred years happens to use the ideas of calculus – and has been developed around the ideas from calculus. Had CAs[i] been developed before calculus I think history would have been entirely different. That is to say that physics, as we currently see it, happens to be using differential equations but nothing in physics says that's the reality. There's nothing in physics that is precisely a differential equation, absolutely nothing. You're looking at fluid dynamics or something, you see a bunch of molecules bouncing around and they happen to be well approximated by Navier-Stokes. The motion of planets is not differential equations but they happen to be well described by them. There's a lot of detail that is not that. I think that the laws of physics as they happen to have been formulated in the last 300 years in terms of this infinite mathematics happen to be just a coincidence of history. I don't think it reflects something (or anything) core about physics.

John Casti: I want to comment on that because it would be meaningful to do the following thought experiment: imagine that Newton had had a super computer. What would mathematics have looked today? I think that

---

[i]Cellular automata.

all sorts of things that were part of the traditional mathematics, like all the stuff about infinite series and limits and calculus and all the traditional mathematics as we know it today, and calculus itself, would have never developed at all, or to the level where they are today and instead we would have had developed finite theories and finite mathematics, combinatorics various kinds of things having to do with computing so I completely subscribe to that view that it was an accident of history. Even the notion of the continuum is simply a matter of computational convenience ...

Stephen Wolfram: I think this is a controversial view! But I'd be fascinated in what other people have to say about this, or think about this question: "Is the continuum indeed an accident of history?"

John Casti: Or a matter of computational convenience?

Paul Davies: Maybe it is a matter of platonic idealization.

Greg Chaitin: Well, don't some people think that geometry is built into the human brain, and in the eye and its intrinsic in the way images are processed by humans? That the human brain has tremendous intuition for geometrical processing, for images and a lot of the brain is processing this through dedicated hardware...?

Stephen Wolfram: So you're saying that the reason we think about the continuum is because our brains are built in such a way that makes it a useful way for us to think. Which would make the continuum necessary, in some sense, so that we can think about the surrounding world.

Greg Chaitin: Perhaps – I'm sorry, I am giving you a parody of Kant.

Cris Calude: I agree with what you said, both about you know calculus being a historical accident for physics, and the other things you pointed out – but I don't see how it changes my point, because CAs are still useful to the degree that they model the properties of the infinite by the finite and then if you think that most non-trivial properties of CAs in general are undecidable then this reduces you immediately to your program where you have small interesting CAs that display high complexity. But this in a sense says "Well I accept to downgrade my grandiose interest and I am happy to look at the local instead of global."

Stephen Wolfram: But I think that it is the case that if we finally find a fundamental theory of physics which ...

Greg Chaitin: You will, Stephen, you will...

Stephen Wolfram: Well, thanks... Greg thinks if we just keep trying harder (people laugh). Or, if I am not making *Mathematica* v. 8 or, whatever. No, let's just go back to what we were saying, so the question was: if we find a simple, sort of, ultimate rule for physics and maybe it is a discrete program and we just run this thing and if we believe that when we find this thing and we run it we will see every single detail of the Universe in this program's run. And maybe it's a practical matter not relevant but I don't see how the infinite comes into it ...

Cris Calude: But if you run it for a very short time this will only give you a very primitive description and definitely you would leave aside many interesting properties of the Universe. How would you know, for example, that there is a finite limit for running this program?

Stephen Wolfram: But this program is not a model in the same sense that Navier-Stokes is, in the numerical analysis sense, a model. This program would be a precise representation of the Universe of every second and every detail and feature in the Universe. So it's not that as you run it you're getting closer and closer to what the Universe is (in the sense that you try to obtain a better approximation of it). No! If you run it for 10^{-200} you will see exactly the first 10^{-200} second of the Universe. If you run it 10^{200} times longer then you get the first second of the Universe, the whole second but I mean absolutely exactly, just as ...

John Casti: ... in every little detail ...

Stephen Wolfram: ... that's right, in every single little detail, so I don't see how the infinite gets in there at all! For example, if you ask the question: "Can there ever arise in the Universe (to intelligent beings, for example, like us) a super intelligence" then that question, if the Universe is unbounded, could in principle be an undecidable question, and then the infinite is in there. But when it comes to the basic description of the Universe, I don't see how the infinite enters in that idea of how physics will play out.

Greg Chaitin: Yes, Stephen, but Cris is only saying that the infinite was for a while a convenient approximation to the finite, because it was just that continuous mathematics was more developed than combinatorial mathematics which is a little harder to work with, and so it's a historical accident. He's just saying it's a convenient approximation. He's not saying that the Universe is continuous.

Cris Calude: Even worse: I say it is random. Or that it is more productive to be thought as random for the human mind so that we try to extract as much information about it as possible. It makes you a little bit modest in what you try to achieve ... and then perhaps ...

IS THE UNIVERSE COMPUTABLE?

Stephen Wolfram: I think one of the key questions that we would certainly like to know the answer of, and one that I mentioned earlier: "Is the Universe like π or is it like Ω?" This is clearly a question that is of great importance if you're interested in the foundations of physics.

Paul Davies: Could I sharpen that by giving a very clear metaphor? Imagine that we – forget about the holographic principle, and – imagine that we pixelate the whole Cosmos and there's a finite number of variables that you can specify with each pixelation. Now imagine that we take the digits of π in binary and we label each of those pixels by the numbers in the pixel so it's like a gigantic CA, we label each of them by the digits of π, so then you have a mess but then you go on to the next level of digits of π and it's still a mess, but when you're running out of digits – you will get this Universe, that's for sure: it's there, this Universe... Or any other Universe!

Stephen Wolfram: Well, but that's muddling some things.

Paul Davies: OK, what am I muddling? (Laughs).

Stephen Wolfram: Well, OK, so what you're muddling is this. So one thing you can say is that while you're looking for a fundamental theory of physics, once you have any universal computer if you believe that the universe is computable, then you've already got the answer.

Paul Davies: That's right. It's there, just don't know where it is.

Stephen Wolfram: So that's the muddle. What I am claiming is something much more extreme, I am saying that if we succeed in finding a fundamental theory of physics, I know this is going to give us some law of evolution, so I'm going to say: this initial condition, this law of evolution you will get precisely every detail of the Universe. So that means that as we work out the digits of π the first digit is what happens in the first 10^{-200} seconds, the next digit is what happens next ...

John Casti: Stephen, are you suggesting that the Universe is computable?

Stephen Wolfram: Yes.

John Casti: But then, I have to ask, do you think it is computable within every possible model of computation, or do you just confine yourself to the Turing model of computation?

Stephen Wolfram: I'm saying that it is computable in the Turing sense (at the Turing level of computation).

Cris Calude: So then randomness is just an illusion.

Stephen Wolfram: That's what I am saying – for our Universe. Randomness is a fascinating thing to study, no doubt, but we just don't have it ... we can't produce it. It just doesn't happen to be produceable in our Universe.

Greg Chaitin: But we have pseudo-randomness.

Stephen Wolfram: Yes, we have very good pseudo-randomness ...

Greg Chaitin: Like the digits of π. You look at them – it looks random.

Stephen Wolfram: The question is with your effort to compute Ω: is it actually possible to do that in our Universe? If you had more tools, any tools you can build in this Universe: is Ω actually there (for us to find it, and look at it) in our Universe? And I think, but I don't know, because we won't know until we have a fundamental theory of physics, yet my claim – my working hypothesis, if you will – is that there is no way you can extract Ω from our Universe, and that true randomness in Greg's sense simply does not exist in our Universe!

Cris Calude: How do you reconcile under this hypothesis the fact that in this Universe we cannot extract this kind of information with the fact that in this Universe we can talk about things that are so opaque?

Stephen Wolfram: I think it's a fascinating fact that we can talk about infinity even though we can't count to infinity and we can have, even in *Mathematica*, the symbol for infinity and it has various properties and we can reason about infinity. So I think it's a very interesting subject ...

Greg Chaitin: We can talk about Unicorns too! (everybody laughs.)

Stephen Wolfram: What I mean is that there is a practical matter to say our Universe it's somewhere in π, and to say where exactly it is located – for algorithmic information reason/purposes – the saying *where* is going to contain as much information as actually giving the whole Universe as a program ... What I am saying is that the whole Universe you could just hold it in your hand, as a little piece of *Mathematica* code and then

332 C. S. Calude et al.

you just run it and you make the whole Universe. And, it's a non-trivial question, whether that is conceivable or whether we need a higher level of computation beyond the Turing model of computation or something else ...

Greg Chaitin: It seems to me that your point of view is that Turing computation is *a priori* – that this notion of computation is given before we start discussing the physical Universe. David Deutsch has the position that what we consider to be a universal Turing machine or a possible computation depends utterly on the physics of our Universe and could be totally different from the Turing model.

Cris Calude: Then you have a circularity!

Greg Chaitin: Yes.

Stephen Wolfram: It is one of the remarkable – I mean, the fact that we in the last century happened to have discovered *the* invariant idea of what computation is ... well, if we're right about that idea, it's just a nice historical accident. I mean it could be the case that – and that's supported by what we seem to be learning in Physics which is that every sixty years somebody says that everything you knew in Physics so far is wrong and there is another level of stuff going on in Physics and from that intuition you might say: "Similarly, in sixty years, everything you know about computation is wrong and another level of computation will be possible" You might say: "Everything you knew about mathematics will have to change as well and in sixty years another kind of mathematics is going to be possible!" But we don't think that's going to happen, do we?

Paul Davies: Well, *wrong* will still be wrong, no? I hope... (laughter)

IS THERE A FUNDAMENTAL THEORY OF PHYSICS?

John Casti: It would be more accurate to say that everything you do is only an approximation to the real thing and the next level has all sorts of different properties and so on. But I think that that argument that you just made, if it's true that it will continue to have an approximation and then a better approximation, and then a better approximation, and so on – and you will never get to the end – then that's a pretty strong argument against the ultimate theory of physics, right?

Stephen Wolfram: No, that's what I am actually saying: the empirical observations of the last few centuries have been such that it seems to be a bottomless pit, that we don't seem to be able to ever hope that we will be done. I don't think that's correct. I don't think that that's the correct intuition. I think the reason for which people have tended to have that inaccurate intuition is that people have not seen examples where little simple things can build into things as rich as we might imagine the Universe to be.

Look, it's worth seeing the same thing in mathematics. And if we look at mathematics we could ask: is mathematics also the same kind of bottomless pit? That is, is it the case that the mathematics of today will necessarily be superseded by the mathematics of tomorrow and is it the case that the unsolved problems of today will necessarily be solved tomorrow - in the mathematics of tomorrow? And that seems to be the analogous issue and my claim would be (and I think Greg used to agree with this, at least, but I am not sure about the others) so, if you look at mathematics, and one my favorite example is the diophantine equations, equations of the integers: in antiquity they learned how to do linear diophantine equations. Later in 1800s they learned how to do quadratic diophantine equations. Later in the early part of 1900s they learned to do elliptic diophantine equations, which were leading towards cubic diophantine equations, and so on. But there are still plenty of diophantine equations which we don't know how to solve, and the question is: if we just project the future of mathematics, will we, just like (apparently) in the future of physics, will we, every fifty years, be able to get further into the world of diophantine equations? So my guess is: no, my guess is that we're going to run into a wall, that we're basically at the edge of where we can go and everything beyond this edge is likely to be undecidable. And that will be the case where the ultimate level of knowledge becomes something of a logical necessity, having to do with the ultimate limits of knowledge ... rather than ... this is a place where we really reach the edge ... and ... maybe one doesn't believe that ... so perhaps this would be a good time to see what others think.

Greg Chaitin: You've been using the word "muddle-headed." That's a very good word, because if you eliminate the muddle-headed-ness you force NKS!

Stephen Wolfram: That's an interesting claim.

Greg Chaitin: Let me say why. Because let's talk about the notion of real randomness in the physical universe. That means that there are things that you can't understand, things that are irrational. Now my understanding of you and the principle of sufficient reason of Leibniz (which Gödel believed in also) is this: if you believe that everything is rational and that everything in the Universe one should be able to understand using reason, then I think you're necessarily driven to look for a discrete Universe along the lines of ... a highly deterministic ...

Stephen Wolfram: Yes, it might be deterministic ... yes, but it might also be that there might be another level of reasoning. It may be that Turing level of reasoning wasn't the right level of reason for our Universe. Leibniz didn't know about the Turing level of reason, he just had an intuitive level of reason, we think we have a precise sense of reason which is the one that ... right now it seems to be all we need to describe what we do in computation. I think this point though that you can describe things that you can't actually make is an interesting point, which is kind of the key to the success of symbolic mathematics. But there's one question that is far from clear which is how far can you go in describing things that you cannot make? That is, is it the case – I know that in set theory people are always concerned whether as you start naming more and more and more infinite kinds of things, does it all eventually fall apart? Can you go on to even describing things as you get to more level of infinities so to speak?

Greg Chaitin: Maybe there won't be enough infinities to go around ... ?!

Paul Davies: What about hyper-computation, which is something we can imagine but can't ever do?

Stephen Wolfram: I think it's a question of physics, whether hyper-computation is something that we can really imagine but can't ever do, or whether the Universe does actually allow hyper-computation ...

Paul Davies: Oh, so we're back to this business of what is in principle doable but may in practice be not achievable because it's outside of the capacity of the Universe to provide us with whatever it is we need from the technology in order to ever be able to do it.

Stephen Wolfram: Right.

Paul Davies: So there are things which are logically possible, but it's not achievable because they're beyond the capacity of physics.

Stephen Wolfram: Right. So your question whether there are things that we can't achieve in our Universe – that would be an answer the question of what is achievable. But I think Karl (Svozil) has prepared some notes and wants to say something.

Karl Svozil: I would like first to comment on the issues that have been mentioned here before: finite versus infinite, continuum versus discrete, randomness versus deterministic. And in doing that I would like to use a concept (or an approach) that we can derive from the Freudian psychology and that they call 'evenly-suspended attentiveness' but before I even start doing I want to say that I am more on the side of Stephen's ideas, that is I too think so that the Universe is discrete because this idea is more iconoclastic. On the other hand I am also very interested in the technology of the continuum because there should be some sort of correspondence with the powers of the infinite which would then be in line with the continuum. Let's say for instance my perfect example from a long time ago would be the Banach-Tarski paradox, which allows the creation and morphing of objects by rearranging a finite number of parts through distance-preserving maps (essentially rotations and translations). This could be an example of how technologies might be some day. So, all in all, I think we should discuss this in threefold: scientific perspective, metaphysical perspective and the historical perspective. We should cover all three of them, but let me start by addressing one thing that is what interests me at the moment. I think that in order to go beyond what we are discussing at the moment the observer needs to be integrated into the system. I am so excited! I'd love to live for a period of much longer than I can and will, because I would like to see what happens – and I am looking forward to the new ideas and theories to what comes after that.

Stephen Wolfram: Can I try to ask a question that I am curious about? OK, so let's say that one day we have the fundamental laws of the Universe. We know them, we have them, we can write them out as code, in *Mathematica* or whatever. Then – what kind of thing can we conclude from that? That is, once we know that our Universe is six lines of *Mathematica* code and here they are – what would we then do with it? What will we then conclude from that?

Paul Davies: May I? I think that the first question to ask is clearly to be is that the only one, or is that just one of maybe an infinite number of different Universes that you can have in other words is there any other way or this is the only way? and I suspect that the answer has to be "No." So then the question is: is this the *simplest* way of doing it consistent with the existence of observers such as ourselves?

Stephen Wolfram: What does 'simplest' mean? Let me just say that we would feel very special if we were the simplest Universe that could be made.

Paul Davies: Right, it would be somewhat arbitrary ...

THE HUNT FOR OUR UNIVERSE

Stephen Wolfram: My guess is that out there in the universe of possible Universes there are zillions that are rich enough to support something kind of intelligent like us, and whether we write it in *Mathematica* or write it in *Java* or something else – which would end up with a different ordering, and there would be a different ordering as to which one is the simplest Universe, and whether there is any invariance, of course, remains to be seen. So one possibility would be that it will be a simple looking Universe because we live in that Universe and the primitives we put into our computer languages are such that our Universe will end up simple.

Paul Davies: That we couldn't discover our Universe if it's more complicated. In other words we will never complete the program you're talking about unless it is simple enough, of course, for us. So by definition either it's a mystery or it's simple.

John Casti: So the other real worlds are too complex?

Paul Davies: Right.

Stephen Wolfram: So that, OK – so one possibility would be that there is no other way ... no, but I don't think that's right, I think that we need to unpack a little more. I was hoping that you guys were going to try to help more ...

Cris Calude: Assuming, assuming by absurdity that this Universe is 60 lines of *Mathematica*, how would you realize that this is true?

Stephen Wolfram: In the same way that you would do it for 6 lines. You run it ...

Cris Calude: Oh, oh, OK ... then how would you do it for six lines?

Stephen Wolfram: ... (pause) ... so, that's a good question. But the fact is that it's hard! I'm in this business, I actually hunt for Universes, OK? I can tell you that being on the front line is quite a difficult business. You make something that is candidate Universe you can often tell that is hopeless, quite quickly. Sometimes there's a candidate, this thing we caught in the net, that is not obviously hopeless so, when that happens you start them off, you let them run, they run out of memory, etc. Now one's hope is that before they do that they appear to exhibit have some features that we can recognize some feature of our real Universe, like a finite dimensional space. As I keep on saying, my goal is to recapitulate the history of physics in a millisecond by being able to take the actual natural thing that is this Universe and be able to deduce from it the effective physical laws and then compare them with the other's ... That's the only way, but you're right, we will never know whether this Universe that we have is exactly our Universe. It could be that it would get, you know, we check off on the feature list we check off 20 features, and they all work perfectly. But unfortunately, it is exactly our Universe except the poor lepton is missing ... My own guess is that if the Universe is simple then then chances are we won't miss it by just a single bit. That it's very unlikely that among the simple Universes there's one that is exactly ours but for a small bug.

Karl Svozil: Stephen ... I think that you might have left out the possibility that there might be just one Universe that is consistent. And this might be our Universe, and this might be the only consistent Universe; a Universe created by universal computation; and maybe some interface to the beyond.

Stephen Wolfram: What does consistency mean?

Karl Svozil: Consistence in the mathematical logic sense.

Stephen Wolfram: I don't think that makes any sense.

Greg Chaitin: Not for programs.

Paul Davies: Not so for theoretical physics either, I think.

Stephen Wolfram: I don't think consistency has anything to do with programs. I mean: you take the thing and you run it, it may do something that you don't like, but there's no such notion of axiomatic consistency. I'd like to discuss this point a little bit further ...

Greg Chaitin: Well, one possibility: we have entire genomes now. We have the complete genome of hundreds of different creatures and people are comparing them, so one possible future is that there are lots of candidates that you can come up with, and it gets very very hard to see if they are right or wrong (because it's at the 10^{-200} scale or something) so you might end up having a future subject which is "comparative universality" – we have all these laws for the universe that cannot be eliminated as ours and you study them, you talk about them, you compare them, this could be a future subject. Students would be required to pass exams on their ten possible favorite universes ...

Stephen Wolfram: Right now we don't know the fundamental laws of physics. What you're saying is that in the future we may have this whole collection of possible fundamental or candidate Universes ...

Paul Davies: ... fairly good universes. And it's easy to imagine that you have two which are practically indistinguishable ...

Stephen Wolfram: But I don't think that it is that easy to imagine! Because if the laws are simple enough the distance between each candidate universe is small ...

Paul Davies: That's true... But you have to make the assumption of immutable laws – because you can always imagine tweaking things ...

Stephen Wolfram: I don't think that makes any sense.

Paul Davies: Well, then, I think it's horrible.

Stephen Wolfram: No, but it doesn't make any sense. Here's the thing with something like with cellular automata, people are always saying "Couldn't you get richer behavior if you let the rules change as evolve the CA (or as the CA is running)? That's a muddle. And the reason that's a muddle is because you could never get richer behavior by changing the underlying hardware because you can always emulate any other form of hardware within the hardware that you already have if your system is universal.

Paul Davies: But do you know about ... the Parrondo[j] games ...?

Stephen Wolfram: No.

Paul Davies: Well, you have a game of chance which has an expectation of loss. And you have another game of chance which also has an expectation of loss. You can play them in combination with an expectation of gain even if that combination is random. So it shows that randomly changing the rules as we go along actually has a creative aspect. So this is a bit like having your automata changing their rules, even randomly, may give you more, may give you ...

---

[j]http://en.wikipedia.org/wiki/J._M._R._Parrondo

Stephen Wolfram: Not so sure about the case where they change randomly! Because as soon as you talk about randomness you're talking about injecting some information from outside the system ...

Paul Davies: Well, you don't have to do it that way, you can actually do it systematically ...

Stephen Wolfram: But as soon as you do it systematically why isn't it just an extra bit in the rules?

Greg Chaitin: Can we go back to your original question? So let's say you have the laws of the Universe in front of you. Then what happens? One possible scenario is the following: let's say this law is very simple; then you might think we are the only possible Universe, that there is some kind of logical necessity that it should only be like this, etc. But let's say you find the laws of the Universe and they're actually rather complicated. Not too complicated that we never find them but that they turn out to be really very complicated. At that point you might suspect all possible universes might exist, otherwise why should we be in this complicated Universe, rather than in a simpler one that was a possibility but doesn't give us our Universe, it gives a different kind of Universe, but, you know, in some ways is a good Universe too – even if it's not ours. So, is that a conceivable ...

Stephen Wolfram: Yes, I think it is the case that if it seems like in our enumeration of Universes this would be the very first one that occurred it would indeed seem – even though it's kind of an illusion, because it depends on the kind of primitives we choose ...

Greg Chaitin: ... on the ordering of complexity, yes ...

Stephen Wolfram: But, in that case it would indeed seem like there's something logical and inevitable to that ...

Greg Chaitin: And maybe we're unique then, maybe this is the only Universe.

Stephen Wolfram: But if we are just random planet no. 3 on a star at the edge of a random galaxy, it would then seem that there should be a lot more like us, no? I don't know, I am just curious – because I just haven't figured this out yet, I haven't unpacked this problem yet. So let's say if the Universe is reasonably simple in our representation and suppose there are ten candidate universes sort of in the range of where one would think that our Universe would be, but they have eighteen and a half dimensions of space and suppose that they have various problems that they are not really quite right ... so what we would conclude, what would we say, would we then ...

Greg Chaitin: ... well we might believe that it's arbitrary and then the idea of an ensemble or landscape of the alternate universes acquires some persuasion, perhaps starts having some reality, starts to seem more persuasive. Whereas if we have a very simple universe and there are no other fairly good candidates nearby then we start to think that we are unique, that we're the only Universe, because this was the only possibility – in other words that there's a logical necessity ...

Stephen Wolfram: Can I try one more question – which I think might engage everybody here – which is about mathematics? I have a definite position on this question, of how inevitable the mathematics that we have is. I guess everybody here knows that I have a very definite position on this, which is that it is not inevitable at all, and it's just a historical accident basically. But I am just curious whether other people are sharing that point of view – or not.

John Casti: What do you mean by being inevitable – the inevitability of mathematics – in the axiomatic framework, or ... ?

The Future of Meta-Mathematics

Stephen Wolfram: So what I say is that the axioms of mathematics that have given us the more than $3 \cdot 10^6$ papers that have been written about mathematics, these are just a few axioms systems that are just one possible system of axioms that exist in a much bigger space of axiom systems and that there is nothing particularly special about them. I've done this thing with boolean algebra where I tried to determine which serial number of axiom system boolean algebra is and the answer was about the 50,000th axiom system. And so the question is that people nevertheless say: "God created the integers" for example – that's something somehow that seems more inevitable. But my point is ...

Cris Calude: But do you disagree with this?

Stephen Wolfram: Yes. I disagree with it.

Paul Davies: I think I disagree with this too. Because you can imagine a different kind of Universe where the natural numbers may not be natural at all.

Stephen Wolfram: Yes, that's what I think. So, Cris doesn't ...

Cris Calude: Well ... I don't know ... it's very unnatural for me to think that a natural number will come out in an unnatural way. I can fully agree with the fact that the various axiomatical systems we're using are entirely accidental and probably they will disappear very quickly. But there are some building blocks of mathematics which seem to have many more reasons to remain than the others. And the natural numbers are in that category.

Paul Davies: But that's only because of the physical structure of our Universe. I mean David Deutsch pointed this out a long time ago, that physics underlies everything. I mean a Turing machine is a physical machine. So if we imagine a Universe with a different set of goals and entirely different stuff in it – what is computable in that universe is not computable in this one!

Cris Calude: OK, so in this quest to find new Universes do you have an example of toy Universe strictly strictly simple where you know the mathematics is radically different?

Paul Davies: Topos theory?...

Cris Calude: ... you can't talk about topos theory without natural numbers.

Paul Davies: So you need natural numbers!

Cris Calude: Yes, you need natural numbers ... Actually I think this belief could be much more substantiated if you could construct a very simple Universe in which the natural mathematics would be unrecognizable from this Universe (or, from the mathematics in this Universe.)

Stephen Wolfram: So one question is: does our Universe give us our mathematics? I don't think that it does.

Paul Davies: Well, I think that it probably does. We live in an Universe that has a certain type of natural mathematics and in which there are certain types of processes that we call computable. It could be a completely different world ...

Stephen Wolfram: Yes, but the natural numbers remain invariant for all different mathematical objects that you have enumerated.

Paul Davies: All I'm saying is that as far as I understand it looks like you need natural numbers in all of these.

Stephen Wolfram: So imagine a world in which, the weather has a mind of its own so to speak, and we exist as a gaseous intelligence. Where there are things that are going on that but there's nothing really to count – you know, imagine something like that ...

Greg Chaitin: No fingers ... that's it! No fingers – no natural numbers! ... (everybody laughs)

Stephen Wolfram: But will it explain – so one possible claim would be that in any axiom system that is rich enough to describe something interesting there exists a copy, something analogous to natural numbers ...

Greg Chaitin: Euclidean geometry doesn't have it. And it's also complete.

Stephen Wolfram: But Euclidean geometry is decidable.

Greg Chaitin: Yes, it's complete. It's decidable. No natural numbers.

Stephen Wolfram: OK, so this will be interesting. There's a precise claim that can be made which is that (ponders, for a moment) : let's explain the technicalities because this is kind of interesting. In theories of mathematics you can have those that are closed and decidable, where truth statements could be established in finite ways and so on. So boolean algebra is an example of one of those, Euclidean geometry is another example of one of those, real algebra is another example of one of those, integer or Peano arithmetic integer equation (and so on) is not an example – it's an open thing that is undecidable and so is set theory and so are other axiom systems of mathematics. So the question might be (let's see, maybe this is obviously true) the claim might be that within any undecidable theory in mathematics you can create objects that are similar to integers (or which have the properties of integers). I think that is obviously true – because I think it's universality that supports this argument, I mean Peano axioms for arithmetic define a certain axiom system that allows you to prove things about the integers, but I think that as soon as you have an axiom system (actually it's probably not totally obvious) so if you have an axiom systems that is universal, in which you can essentially support Turing computation – then I claim that it follows that you can construct, you can emulate within that axiom system Peano arithmetic ...

Cris Calude: You can emulate natural numbers in various ways, so the difference between your Universe and what Greg said about geometry is that you can emulate natural numbers in both universes but with different properties. So the properties in geometry are much weaker than the ones in Peano arithmetic – so this is why you have this degree of decidable and undecidable.

Stephen Wolfram: But I think that the claim would be that as soon as the system is "universal" you will be able to emulate it or construct it with the Peano axioms. Now there is one thing which maybe is getting in technicalities of computability theory but in the case of systems that can be universal computers there is this possibility that there are things for which there are undecidable propositions. It is claimed – I don't think this is really right but – it is often claimed that there could be things where you have undecidability but not universality. So that's the Friedberg-Muchnik theorem[k] which is a rather abstruse piece of computability theory. So the question would be: "Could there be axiomatic systems where there is undecidability but where you cannot embed ... [doesn't finish because Chaitin starts talking with the answer]?"

Greg Chaitin: So I suspect that the answer is: "Yes" but that it requires, or relies on some other construction, like the Friedberg-Muchnik theorem. In fact doesn't that give it to you immediately?

Stephen Wolfram: Yes, it does.

Greg Chaitin: The Friedberg-Muchnik immediately gives us as a corollary that you have an axiom system that is undecidable but which is not equivalent to ...

Stephen Wolfram: I don't think that's right, because I think if you take apart – so my claim about the Friedberg-Muchnik theorem is that ... this is getting deeply technical! ...

Greg Chaitin: ... for most of the audience, yes...

Stephen Wolfram: ... so my claim is that if you look inside the construction there's a universal computer running around ...

Greg Chaitin: Well, yes, yes – but you can't get to it by that universal computer. That's the whole point. It is there but you can't get to it.

Audience Member: No, the Friedberg-Muchnik theorem says that [...] if you take two sets which are both intermediate – but if you're taking the two sets together they're complete ...

Stephen Wolfram: This is getting far too technical ...

---

[k]See page 1130 in the NKS book.

C. S. Calude et al.

Karl Svozil: It is worth mentioning that there is a sort of "computer-generated" undecidability related to Heisenberg complementarity. And it refers to a situation when you are inside a system and you decide you want to measure A then you can't measure B or, vice versa, you want to measure B you can't measure A ... this is something that is entirely impossible "from inside" ...

Stephen Wolfram: Is that really the same phenomenon at all?

Karl Svozil: Yes, it led to a model to formalize complementarity, and in doing so Edward Moore created finite automata theory, an entire field of computer science, in the 1950s ...

Greg Chaitin: 1956.

Stephen Wolfram: But automata theory went off in a totally different direction.

Karl Svozil: Yes, but this thing is robust and reminds me of Greg's remark regarding comparative universes: this should be a new research topic.

Greg Chaitin: Yes, this is E. F. Moore paper "Gedanken Experiments on Sequential Machines" published in 1956. Wonderful volume called "Automata Studies".

Cris Calude: Shannon was an editor.

Greg Chaitin: Yes, Shannon and McCarthy were editors, von Neumann had a contribution – this book is marvelous, it's the beginning of theoretic universality, it's the first book of its kind. I saw it reviewed in Scientific American – a very perceptive review – and I immediately ran to the bookstore, I had saved my lunch money and bought a copy and then I started doing computer experiments with Moore's stuff. Moore's book is epistemological, automata theory went off unfortunately in algebraical directions, you'll see from this that I don't like algebra but the original paper dealt with complementarity and models of induction [...] epistemology. It's a wonderful paper, and I met him once, he had another paper – I guess he's probably dead by now, but I remember him well, he was red headed, (to himself: I guess that's useful information (people laugh)) – and he had another paper on self-reproducibility of CAs ...

Stephen Wolfram: Ah, so he is the guy that had ...

Greg Chaitin: ... the Garden of Eden

Stephen Wolfram: Yes, but he also had the self-reproducing fishes in the ocean ...

Greg Chaitin: I think so.

Stephen Wolfram: He had this paper in Scientific American in which he talked how sometime in the future there will be self-reproducing ...

Greg Chaitin: ... artificial self-reproducible plants, yes! That then you harvest and they go all over the sea ...

Karl Svozil: For example I would like to run an experiment based on these computational complementarity issues. ... My suspicion, or my claim is that in a computer generated universe you would experience complementarity very much as you would experience the complementary of quantum mechanics.

Greg Chaitin: Very technical work.

Stephen Wolfram: Could degenerate quickly.

Greg Chaitin: But this discussion is still about the future of metamathematics?

Stephen Wolfram: Yes, this is related to this question whether our mathematics – that is, if integers are inevitable and what aspects of our mathematics are inevitable, and what aspects are just a feature of the Universe we happen to be in? What is the relationship between (a) the Universe we're in, (b) the mathematics that we make, (c) what we're like and (d) the space of possible mathematics – that's what this is all about.

Greg Chaitin: It's a complicated question ... Well, when you think that the mathematics is just axiomatic systems then the whole thing is very arbitrary, because it's very artificial, but if you believe in mathematics as a kind of description of a platonic world of ideas which has an independent existence, then mathematics tries to describe it and there's some necessity to it.

Stephen Wolfram: So your question is: is mathematics about something?

Greg Chaitin: Is it about something – exactly! Or is it just symbol manipulation? So I believe that there is a two level structure, and the way you are looking at it – it is a one level structure ...

Stephen Wolfram: So if it's about something maybe it's about our Universe.

John Casti: But if you look at Gödel's theorem, it says that number theory is about numbers – which are an irreducible semantic component ... it's not just symbolic.

Stephen Wolfram: But wait a minute ... isn't Gödel theorem ... doesn't it say that number theory as practiced in the Peano axioms ... it is effectively saying that it isn't necessarily about numbers ... because the whole point is that Gödel's theorem shows ...

John Casti: ... that numbers are just not part of a symbolic structure, that numbers have indeed a semantic content. Number theory *is* in fact about something. It's not ...

Greg Chaitin: ... otherwise.

Stephen Wolfram: But wait a minute! One of the results of Gödel's theorem is that – so Gödel theorem's says that given the Peano axioms (the axiomatic system that's supposed to describe numbers,) that in a sense any axiomatic system says that there are things out there, somewhere, in the platonic world (or wherever else), that have the following properties [...]. And then the purpose of an axiom system is to sculpt those properties so well that the only thing that the axiom system can't be describing ...

Greg Chaitin: But that's impossible.

Stephen Wolfram: Well, that's the whole point ... for logic it is possible.

Cris Calude: But axiomatic systems are themselves an accident of history. You know, this is what we have done in the last one hundred years and we are trying to turn it into a religion, but there is nothing intrinsic that should indicate that we must get stuck on these systems.

Greg Chaitin: Well, because they are self-destructive. I mean what Gödel did, in a way, formal axiomatic systems – he showed that they're useless, and that's Gödel's result. He is working from within Hilbert's program, and he explodes Hilbert's program. So where are we after that, and what are we left with ...

Cris Calude: Well, randomness – it's all random!

Greg Chaitin: ... and Gödel says "intuition".

Cris Calude: Randomness. It's one possibility. Again by all means not a unique possibility but there it's an alternative – a credible alternative ...

Stephen Wolfram: We should say for people that don't know, that this is a relatively interesting fact actually: what Gödel theorem actually establishes is that there exists – in addition to the ordinary integers that satisfy the axioms of Peano arithmetic – an infinite hierarchy, of bizarre kinds of non-standard arithmetics ... where even the operation of addition is not computable ...

Greg Chaitin: ... which is not ours but that satisfy the same axioms ...

Stephen Wolfram: ... so the axiom system is not an adequate way to say what you're talking about, so to speak. But your statement about axiomatic systems – I tend to agree with your statement that axiomatic theory systems are an unfortunate accident ...

Greg Chaitin: ... and one from which we will hopefully recover ...

Stephen Wolfram: ... and what I certainly think, and perhaps this is a very biased group to judge my statement – is that the programs and things that you can specify by computing rules are the things that we should look at, rather than these axiom systems used as ways to describe things ...

Paul Davies: In which case this raises an issue that has interested me for a long while, which is that if we assume that the laws of the Universe are computable then we can imagine different sorts of Universes, in which different types of mathematical functions would be computable functions but they didn't obey the rules of classical Turing machines. And then we'd have a different definition of computability. So then there is a self-consistent loop: if our laws are indeed computable, given a Universe which has the right type of physical structure to give the right type of computability to make the laws which describe it as computable (that's what I call a nice consistent loop) – is that loop unique? Or are there any other number of self consistent loops, where the physics and the stuff and laws and computability would be nice enough to exhibit something of the same kind?

Greg Chaitin: Well the answer is: probably. There probably are, but Stephen and I believe that there probably aren't. David Deutsch probably believes there are but Stephen and I seem to think that the current notion of Turing computability is a priori and is [indistinguishable] from the physical universe, while David Deutsch says : "No!" ...

Stephen Wolfram: No – the current notion of Turning computability is enough for our Universe. But your question is a good question and a precise question. Is there a way ... I mean – what seems to be the case is that Turing computability is a nice robust notion. All these different things that we come up with: potential models of computation, all seem to be equivalent. What has happened in the world of hyper-computation, or super-computation is that there are a variety of names and there are lots of various definitions and unlike what happened historically with Turing computability, which is that all definitions seem to converge in hyper-computation that is not what seems to be happening, I mean everybody will claim that their definition is the right definition but it seems to me a perfectly valid question ...

Greg Chaitin: ... you are offering them a number of alternatives. Add to a Turing machine an oracle for the halting problem. That gives you a new kind of Turing machine. It, in turn, has its own halting problem. If you add an oracle for that to a Turing machine you get a still higher Turing machine that can compute even more stuff and you have an infinite hierarchy – and in fact it's much more than ... just this complicated structure.

Stephen Wolfram: So: can you make a consistent model of the Universe in which you have a Turing machine plus oracle as the fundamental stuff running in the system?

Paul Davies: Remember it's machine ... a real machine with real laws ... made of real stuff ...

Greg Chaitin: Well, you can certainly have a consistent notion of computation, an infinite number of possibilities where you can compute things that are not computable in our world and it just keeps going on like that. So I suspect there are perfectly consistent physical universes in which each of these levels of uncomputability is the level of computability that corresponds to that next physical Universe. So we would take as the most natural thing being able to compute the halting probability for *our* machine but not for their own Turing machines.

Cris Calude: In this context an interesting question is how much knowledge do you get in these higher and higher models of computation. I can tell you that for instance the algorithmic randomness that we discussed here is level zero, and Ω numbers are in a sense the weakest random numbers that are random but they have the maximal compressed knowledge in them. You can construct higher and higher more random numbers, none of them having the knowledge compressed as much as the Ω number. So the question is not only how much you can compute but with what result. What knowledge can you deduce from more powerful computation?

Greg Chaitin: ... at our level? And the answer is: not much.

Cris Calude: ... as a level of randomness ...

Stephen Wolfram: No, but one issue is: you're asking what kind of computations can you do in a particular type of system or what kind of computations are easy to do with a particular kind of system. There's a question of whether in that kind of system you can do a lot of computations that you care about, or whether there are computations that you care about that you can't do. So when you say "not much knowledge is obtained" ... you've only got the Ω number...

Greg Chaitin: Well, what we can perceive at our level, you know ... if you saw an Ω at a higher level, Ω at our level is in principle giving you an immense amount of information. But given an Ω at a higher level (where there's an oracle for a halting problem for every Turing machine) that Ω (unlike our Ω down here, which gives a lot of information to us down here) that higher order Ω gives absolutely no information.

Stephen Wolfram: So you're saying that from our level we can't make much use of an Ω.

Greg Chaitin: ... a higher order Ω ...

Cris Calude: Let me give you some examples. With Ω for instance you need about three thousand eight hundred bits to solve the Goldbach conjecture. With about 8000 bits you can solve the Riemann hypothesis ...

Paul Davies: Where did you come up with ... these numbers?

Cris Calude: I computed them.

Paul Davies: But the Riemann hypothesis is not solved.

Stephen Wolfram: He's saying "to solve it – if you had the Ω number".

Cris Calude: And *if* you had the Ω number.

Paul Davies: Oh – if you had the Ω number!

Greg Chaitin: Cris has written a large program which holds if and if the Riemann hypothesis holds. So that's how ...

Cris Calude: But if you go to any higher parts of Ω you won't find any finite number of bits that would give you the solution for the Riemann hypothesis, for example – so, in spite of the fact that you get incredible more power, this power doesn't help you to solve anything ...

Stephen Wolfram: But it would help you if there were problems formulated at the next level up – then it would help you. But the point is – so this relates to what we consider to be interesting [and] at this point we don't have the possibility to even imagine [what is interesting to ask as a question] ...

Cris Calude: We don't know.

John Casti: So we come back to the problem of an observer and the complexity that is in fact relative to the observer.

Stephen Wolfram: That's true.

Greg Chaitin (to Wolfram): My bottom line is that I definitely agree with your statement that the notion of computation is more basic than logics and axiom systems and it's a better way to think about things. I may have formulated that in a different flavor than yours ... but I think we agree on that.

John Casti: So if everybody is in agreement we can bring this panel discussion to a close. Thank you all.

End of transcript.

Chapter 26

What is Computation? (How) Does Nature Compute?[a]

Cristian S. Calude, Gregory J. Chaitin, Edward Fredkin, Anthony J.
Leggett, Rob de Ruyter, Tommaso Toffoli and Stephen Wolfram

PANEL DISCUSSION, NKS MIDWEST CONFERENCE 2008. UNIVERSITY OF
INDIANA BLOOMINGTON, USA. NOVEMBER 2, 2008.[b]

QUANTUM VS. CLASSICAL COMPUTATION

Adrian German: About five years ago at MIT Stephen Wolfram was asked
a question at the end of his presentation:

Question (on tape): "How does your Principle of Computational Equivalence (PCE) explain the separation in complexity between a classical
cellular automaton (CA) and a quantum CA?"

Stephen Wolfram (answers the question on tape): "If you take the current formalism of quantum mechanics and you say: let's use it, let's just
assume that we can make measurements infinitely quickly and that the
standardized formalism of quantum mechanics is an exact description of
the actual world, not some kind of an idealization, as we know that it is
– because we know that in fact when we make measurements, we have
to take some small quantum degree of freedom and amplify it to almost
an infinite number of degrees of freedom, and things like that – but let's
assume we don't worry about that, let's take the idealization that, as
the formalism suggests, we just do a measurement and it happens. So
then the question is: how does what I am talking about relate to the
computation that you can get done in a quantum case vs. the classical

---

[a]Prof. Anthony Leggett has requested the editor to mention that he had no time to
proofread in full detail the transcription of his interventions.
[b]Transcript and footnote comments by Adrian German (Indiana University Bloomington) with some assistance from Anthony Michael Martin (UC Berkeley). Organized by Adrian German, Gerardo Ortiz and Hector Zenil. Moderated by
George Johnson assisted by Gerardo Ortiz and Hector Zenil. Plays recording from
http://mitworld.mit.edu/video/149/ in which the selected paragraph appears at the
1:28:28 mark. Tape starts.

case. And the first thing I should say is that, ultimately, the claim of my PCE is that in our universe it is in fact not possible to do things that are more sophisticated than what a classical CA, for example, can do. So, if it turned out that quantum CAs (or quantum computers of some kind) can do more sophisticated things than any of these classical CAs then this PCE of mine would just be wrong."

Question (on tape): "By more sophisticated do you mean with a better complexity then, or do you mean in a universal sense?"

Wolfram (on tape): "So – to be more specific, one question is: can it do more than a universal Turing machine, can it break Church's thesis? Another one is: can it do an NP-complete computation in polynomial time? And on the second issue I am not so sure. But on the first issue I certainly will claim very strongly that in our actual universe one can't do computations that are more sophisticated than one can get done by a standard classical Turing machine. Now, having said that there's already then a technical question which is: if you allow standard idealizations of quantum mechanics, with complex amplitudes that have an arbitrary number of bits in them, and things like that, even within the formalism can you do computations that are more sophisticated than you can do in a standard classical universal Turing machine? I think that people generally feel that you probably can't. But it would be interesting to be able to show that. For example, the following is true: if you say that you do computations just with polynomials then there is a result from the 1930s that says that every question you ask is decidable. This was Tarski's result. So, if the kinds of primitives that you have in your theory would be only polynomial primitives then it is the case that even using these arbitrary precision amplitudes in quantum mechanics you couldn't compute something more than what we can classically compute. But as soon as you allow transcendental functions that is no longer the case – because for example even with trigonometric functions you can easily encode an arbitrary diophantine equation in your continuous system. So if you allow such functions then you can immediately do computations that cannot be done by a classical Turing machine, and if there was a way to get quantum mechanics to make use of those transcendental functions – which certainly doesn't seem impossible – then it would imply that quantum mechanics in its usual idealization would be capable of doing computations beyond the Church's thesis limit. Then you can ask questions about the NP completeness level of things and, as I said, I'm less certain on that issue. My own guess, speculation if you want, is that if you try and eventually unravel the idealizations that are made in quantum mechanics you will find that in fact you don't succeed in doing things that are more sophisticated than you could do with a classical Turing machine type system – but I don't know that for sure!"

Adrian German: Where this answer ends, our conference starts. As quantum computation continues to generate significant interest while the interest in NKS grows steadily we decided to ask ourselves: Is there really a tension between the two? If there is – how is that going to inform us? What if the conflict is only superficial? It was then that we remembered a quote from Richard Feynman from the 1964 Messenger Lectures at Cornell, later published as the book "The Character of Physical Law":

> "It always bothers me that, according to the laws as we understand them today, it takes a computing machine an infinite number of logical operations to figure out what goes on in no matter how tiny a region of space, and no matter how tiny a region of time. So I have often made the hypothesis that ultimately physics will not require a mathematical statement, that in the end the machinery will be revealed, and the laws will turn out to be simple, like the chequerboard with all its apparent complexities[c]."

The seed of that hypothesis may have been planted by one of our guests today: Ed Fredkin[d] who says that in spite of his extensive collaboration with Feynman was never sure that his theories on digital physics had actually made an impact with the legendary physicist – until he heard it in the lectures as stated above.

THE IDEA OF COMPUTATION

George Johnson: Well, good morning, and I want to thank you all for coming out and also for inviting me to what's turned out to be a really, really fascinating conference. Friday night just after I got here I had dinner with my old friend Douglas Hofstadter who of course lives in Bloomington and I was thus reminded why (and how) I later became interested in computation in the abstract. I was already very interested in computers when Doug's

---

[c]In 1981, Feynman gave a talk at the *First Conference on the Physics of Computation*, held at MIT, where he made the observation that it appeared to be impossible in general to simulate an evolution of a quantum system on a classical computer in an efficient way. He proposed a basic model for a quantum computer that would be capable of such simulations.

[d]Richard Feynman's interaction with Ed Fredkin started in 1962 when Fredkin and Minsky were in Pasadena and one evening not knowing what to do with their time "sort of invited [them]selves to Feynman's house," as documented by Tony Hey. Twelve years later, in 1974, Fredkin visited Caltech again, this time as a Fairchild scholar and spent one year with Feynman discussing quantum mechanics and computer science. "They had a wonderful year of creative arguments," writes Hey, "and Fredkin invented Conservative Logic and the Fredkin Gate, which led to Fredkin's billiard ball computer model." (See http://www.cs.indiana.edu/~dgerman/hey.pdf)

book "Gödel, Escher, Bach" came out – and I remember in graduate school picking up a copy in a bookstore on Wisconsin Ave., in Washington, DC, and just being absolutely sucked into a vortex – and then eventually buying the book and reading it. And as I was reading it I thought "Well, this is great, because I am a science journalist," and at the time I was working for a daily newspaper in Minneapolis, "I think we really need to do a profile of this Hofstadter person." It was a great way to get introduced to this topic and I had my first of many other trips to Bloomington to meet Doug and over the years we got to know each other pretty well – in fact he copy-edited my very first book "The Machinery of the Mind."

And it was while writing that book that I came across a quote from one of tonight's guests Tommaso Toffoli that I just remembered while I was sitting here on these sessions and I looked it up the other night and I just wanted to use it to get us started. This was an article in 1984 in Scientific American by Brian Hayes[e] that mentions Stephen Wolfram's early work on CAs and mentions Norman Packard who was at the Institute for Advanced Studies (IAS) at the time (he later went on to co-found a prediction company in my hometown of Santa Fe and is now researching Artificial Life) and quotes[f] Dr. Toffoli as saying:

> "... in a sense Nature has been continually computing the 'next state' of the universe for billions of years; all we have to do – and actually, all we can do – is 'hitch a ride' on this huge ongoing computation."

It was reading that and things like that really got me hooked and excited about this idea of computation as a possible explanation for the laws of physics. We were talking about a way to get started here to have you briefly introduce yourselves, although most of us all know who you are and why you're important, but just to talk in that context of how you first got hooked to this idea and the excitement of computation in the abstract.

Greg Chaitin: Being a kid what I was looking for was the most exciting new idea that I could find. And there were things like Quantum Mechanics and General Relativity that I looked at along the way and Gödel's incompleteness theorem but at some basic level what was clearly the big revolution was the idea of computation, embodied in computer technology. Computer technology was exciting but I was even more interested in the idea of computation as a deep mathematical and philosophical idea. To me it was

[e]http://bit-player.org/bph-publications/AmSci-2006-03-Hayes-reverse.pdf
[f]http://www.americanscientist.org/issues/id.3479,y.0,no.,content.true,page.1,css.print/
issue.aspx

already clear then that this was a really major new mathematical idea – the notion of computation is like a magic wand that transforms everything you touch it with, and gives you a different way of thinking about everything. It's a major paradigm shift at a technological level, at the level of applied mathematics, pure mathematics, as well as the level of fundamental philosophy, really fundamental questions in philosophy. And the part about the physical universe – that part was not obvious to me at all. But if you want to discover the world by pure thought, who cares how this world is actually built? So I said: let us design a world with pure thought that is computational, with computation as its foundational building block. It's like playing God, but it would be a world that we can understand, no? If we invent it we can understand it – whereas if we try to figure out how this world works it turns into metaphysics again[g]. So, obviously, I've hitched a ride on the most exciting wave, the biggest wave I could see coming!

Ed(ward) Fredkin: When I was in the Air Force and was stationed at Lincoln Labs I had the good fortune to meet Marvin Minsky almost right away and not long afterwards I met John McCarthy and I'd get lots of advice from them. I don't know exactly when I first thought of the idea of the Universe being a simulation on a computer but it was at about that time. And I remember when I told John McCarthy this idea (that had to be around 1960) he said to me something that in our long time of talking to each other has probably said to me maybe a hundred times: "Yes, I've had that same idea," about it being a computer. And I said "Well, what do you think of it?" and he said "Yeah, well we can look for roundoff or truncation errors in physics..." and when he said it I immediately thought "Oh! He thinks I'm thinking of an IBM 709 or 7090 (I guess wasn't out yet) in the sky..." and I was thinking of some kind of computational process... I wasn't thinking of roundoff error... or truncation error. When I told this idea to Marvin he suggested that I look at cellular automata and I hadn't heard of cellular automata at that point so what I had to do was to find a paper by... or, find out what I could, I remember I couldn't find the paper

---

[g]Beginning in the late 1960s, Chaitin made contributions to algorithmic information theory and metamathematics, in particular a new incompleteness theorem. He attended the Bronx High School of Science and City College of New York, where he (still in his teens) developed the theories that led to his independent discovery of algorithmic complexity. Chaitin has defined Chaitin's constant Ω a real number whose digits are equidistributed and which is sometimes informally described as an expression of the probability that a random program will halt. Ω has the mathematical property that it is definable but not computable. Chaitin's early work on algorithmic information theory paralleled the earlier work of Kolmogorov and other pioneers.

that von Neumann had done and from that point on – which was around 1960 (say, 1959 or 1960) – I remained interested in it[h]. And on one of my first experiments on the computer I decided to find the simplest rule that is symmetrical in every possible way, so I came up with the von Neumann neighborhood and binary cellular automata and I thought "what function that's symmetrical is possible?" And, as it turned out it's XOR. And so I programmed that one up and I was kind of amazed by the fact that that simple rule was at least a little bit interesting.

Rob de Ruyter[i]: I also have to go back to my youth, or maybe high-school, when (in the time that I was in high-school, at least) there was a lot of mistery still around about biology and there were a number of wonderful discoveries made and are still being made about how life on the cellular scale works. So that was a kind of an incredibly fascinating world for me that opened as I got interested in biology and learned ever more intricate mechanisms. At the same time of course we learned physics – actually in Holland in high-school you learn them both at the same time, as opposed to in this country. And in physics you get all these beautiful laws, mathematical descriptions of nature and they give you a real sense that you can capture nature in pure thought. Now it was obvious then already that going from that physical picture where you can describe everything very precisely to the workings of a biological cell that there's an enormous range of things that you have to cover in order to understand how this biological cell works and we still don't know that in terms of underlying physical principles. At the same time at that age when you're in high school you also go through all kind of hormonal, etc., transformations and one of the things you start to realize is that (a) as you look at the world you take information in from the world outside but (b) you also have an own mind with which you can also think and do introspection. And you know that this introspection, your own thoughts, somehow have to reside, they have to be built out of matter – that probably, out there somewhere, somehow, that has to happen. Then that's a question that has always fascinated me tremendously and still really fascinates me which is (if you want to put it

---

[h]http://www.digitalphilosophy.org/
[i]Rob de Ruyter van Stevenick is a Professor at the Department of Physics and Program in Neural Science Indiana University Bloomington. He is a Fellow of the American Physical Society. He has a Ph.D. (1986) from the University of Groningen, the Netherlands, and has had postdoctoral positions at Cambridge in the UK and the Univ. of Groningen. His main interest is in understanding basic principles underlying coding, computation and inference in the sensory nervous system.

in shorthand) the question of how matter leads to mind, the mind-matter question and everything that's related to that.

I was lucky enough to find a place in the physics department where there they had a biophysics group and I could study questions of mind – probably a simple mind, the mind of an insect, or at least little bits of the mind of an insect – and try to study them in a very, highly quantitative way that harkens back to the nice, beautiful mathematical description that you get in physics. So I think this tension between matter and how you describe it in mathematical formalisms and thought – how you introspectively know what thought is, and presumably to some extent all animals have thoughts – and thought to some extent takes the form of computation, of course, is the kind of thing that drives me in my work[j]. And as an experimentalist I find it very pleasing that you can do experiments where you can at least take little little slivers off these questions and perhaps make things a little bit more clear about the mind.

(An)T(h)ony Leggett: I guess I am going to be "skunk" of the group, because I actually am not convinced that a computational approach is going to solve at least those fundamental problems of physics which I find most interesting[k]. I certainly think that the sort of ideas, like some of those that we've discussed at this conference, are extremely intriguing and they may well be right – I don't know. And one area in which I think that one could quite clearly demonstrate that this confluence of computational science and physics has been fruitful is of course the area of quantum information and in particular quantum computing. Certainly what's happened in that area is that, as I said earlier, an approach coming from computer science gives you a completely new way of looking at the problems, posing them as questions that you would not have thought about otherwise. But quite interesting actually that at least in the decade from 1995 to 2005, when quantum information clearly gained way, a lot of the papers which appeared in Physical Review Letters at that time in some sense could have easily appeared in 1964, but they hadn't. Why not? Because people then didn't

---

[j]Rob de Ruyter van Steveninck has co-authored *Spikes: Exploring the Neural Code* (MIT Press, 1999) with Prof. William Bialek of Princeton University and others

[k]Sir Anthony James Leggett, KBE, FRS (born 26 March 1938, Camberwell, London, UK), is the John D. and Catherine T. MacArthur Chair and Center for Advanced Study Professor of Physics at the University of Illinois at Urbana-Champaign since 1983. Dr. Leggett is widely recognized as a world leader in the theory of low-temperature physics, and his pioneering work on superfluidity was recognized by the 2003 Nobel Prize in Physics. He set directions for research in the quantum physics of macroscopic dissipative systems and use of condensed systems to test the foundations of quantum mechanics.

have the knowledge (or intuition) to ask such particular type(s) of questions. So certainly I think that was a very very useful and fruitful interaction.

But I think that if someone told me, for example, that such and such a problem which can be easily posed in physics cannot be answered by a computer with a number of bits which is greater or equal to the total number of particles in the universe: I don't think I would be too impressed. And my reaction would probably be: "All right, so what?" I don't think that's useful, I don't think that nature goes around computing what it has to do – I think it just doesn't! But I think perhaps a little more fundamentally, one reason I'm a little skeptical about the enterprise is that I actually think that the fundamental questions in physics[l] have much more to do with the interface between the description of a physical world and our own consciousness[m]. One of the most obvious cases is the infamous quantum measurement problem[n], which I certainly do regard as a very

---

[l]In 1987 Oxford University Press published "The Problems of Physics" by Anthony J. Leggett in which he wrote: "Although [the so-called anthropic principle, the 'arrow of time', and the quantum measurement paradox] are associated historically with quite different areas of physics — cosmology, statistical mechanics, and quantum mechanics, respectively — they have much in common. In each case it is probably fair to say that the majority of physicists feel that there is simply no problem to be discussed, whereas a minority insist not only on the problem's existence, but also on its urgency. In no case can the issue be settled by experiment, at least within the currently reigning conceptual framework -a feature which leads many to dismiss it as 'merely philosophical'. And in each case, as one probes more deeply, one eventually runs up against a more general question, namely: In the last analysis, can a satisfactory description of the physical world fail to take explicit account of the fact that it is itself formulated by and for human beings?"

[m] "The anthropic principle in cosmology has a long and venerable history — in fact it goes back effectively to long before the birth of physics as we know it. Most modern versions start from two general observations. The first is that, in the current formalism of particle physics and cosmology, there are a large number of constants which are not determined by the theory itself, but have to be put in 'by hand'. The second observation is that the physical conditions necessary for the occurrence of life — still more for the development of life to human stage — are extremely stringent. It is easy to come to the conclusion that for any kind of conscious beings to exist at all, the basic constants of nature have to be exactly what they are, or at least extremely close to it. The anthropic principle then turns this statement around and says, in effect, that the reason the fundamental constants have the values they do is because otherwise we wouldn't be here to wonder about them." (The Problems of Physics, by Anthony J. Leggett, 1987 Oxford University Press, pp. 145-146)

[n] "Within the formalism of quantum mechanics in its conventional interpretation, it looks as if in some sense a system does not possess definite properties until we, as it were, force it to declare them by carrying out an appropriate measurement. But is this the only possible interpretation of the experimental data? We can summarize the situation succinctly by saying that in the quantum formalism things do not 'happen', while in everyday experience they do. This is the quantum measurement paradox. As indicated

serious problem; and secondly the question of the arrow of time°: that we can remember the past, that the past influences the future and vice-versa. It's something that underlines some of the most fundamental aspects of our existence as human beings and I don't think we understand them at all, at least I don't think we understand them in physical terms, and I personally find it very difficult to imagine how a computational approach could enhance our approach to any of these problems.

Now one of the reasons I come to conferences like this is that I hope that I may in the future somehow be convinced. So we'll have to wait and see.

---

above, it is regarded by some physicists as a non-problem and by others as undermining the whole conceptual basis of the subject. Various resolutions of the paradox have been proposed. There is one so to say exotic solution that needs to be mentioned namely the interpretation variously known as the Everett-Wheeler, relative state, or many-worlds interpretation. The basis for this alleged solution is a series of formal theorems in quantum measurement theory which guarantee that the probability of two different measurements of the same property yielding different results (of course, under appropriately specified conditions) is zero The many worlds interpretation, at least as presented by its more enthusiastic advocates, then claims, again crudely speaking that our impression that we get a particular result in each experiment is an illusion; that the alternate possible outcomes continue to exist as 'parallel worlds', but that we are guaranteed, through the above formal theorems, never to be conscious of more than one world, which is moreover guaranteed to be the same for all observers. Let me allow myself at this point the luxury of expressing a strong personal opinion . It seems to me that the many-worlds interpretation is nothing more than a verbal placebo, which gives the superficial impression of solving the problem at the cost of totally devaluing the concepts central to it, in particular the concept of 'reality'. I believe that our descendants two hundred years from now will have difficulty understanding how a distinguished group of scientists of the late twentieth century, albeit still a minority, could ever for a moment have embraced a solution which is such manifest philosophical nonsense."

° "It is almost too obvious to be worth stating that the world around us seems to exhibit a marked asymmetry with respect to the 'direction' of time; very many sequences events can occur in one order, but not in reverse order. Why is this observation problematic? As a matter of fact, there is a great deal more to the problem than the above discussion would indicate. Indeed, in general discussions the subject it is conventional to distinguish (at least five different arrows of time. One is the 'thermodynamic' arrow, a second is the 'human' or psychological arrow of time, determined by the fact that we remember the past but not the future. A third arrow is the 'cosmological' one determined by the fact that the Universe is currently expanding rather than contracting. A fourth arrow is the so-called electromagnetic one, and this requires a little explanation. Finally, for completeness I should mention the fifth arrow, which is associated with a class of rare events in particle physics — the so-called CP-violating decays of certain mesons — and probably with other very small effects at the particle level. This arrow is rather different from the others in that the relevant equations are themselves explicitly asymmetric with respect to the sense time, rather than asymmetric being imposed by our choice of solution. At the moment the origins of this asymmetry are very poorly understood, and

Cris(tian) Calude: I guess I am probably the most conventional and ordinary person[P] here. My interests are in computability, complexity and
randomness, and I try to understand why we can do mathematics and
what makes us capable of understanding mathematical ideas. I am also a
little bit skeptical regarding the power of quantum computing. I've been
involved in the last few years in a small project in dequantizing various
quantum algorithms, i.e., constructing classical versions of quantum algorithms which are as quick as their quantum counterparts. And I have come
to believe that quantum computing and quantum information say much
more about physics than they could become useful tools of computation.
Finally, I am very interested to understand the nature and the power of
quantum randomness, whether this type of hybrid computation where you
have your favorite PC and the source of quantum randomness can get you
more. Can we surpass the Turing barrier, at least in principle, with this
kind of hybrid computation or not?

---

the general consensus is that it probably does not have anything much to do with the
other arrows; but it is too early to be completely sure of this. Regarding the first four
arrows a point of view which seems free from obvious internal inconsistencies at least is
that one can take as fundamental the cosmological arrow; that the electromagnetic arrow
is then determined by it, so that it is not an accident that, for example, the stars radiate
light energy to infinity rather than sucking it in; that, because radiation is essential to life,
this then uniquely determines the direction of biological differentiation in time, and hence
our psychological sense of past and future; and finally, that the thermodynamic arrow is
connected in inanimate nature with the electromagnetic one, and in the laboratory with
the psychological one in the manner described above. However, while it is easy to think,
at each step, of reasons why the required connection *might* hold, it is probably fair to say
that in no case has a connection been established with anything approaching rigor, and
it would be a brave physicist who would stake his life on the assertion, for example, that
conscious life must be impossible in a contracting universe. This fascinating complex of
problems will probably continue to engage physicists and philosophers (and biologists
and psychologists) for many years to come." The Problems of Physics, 1987 Oxford
University Press, pp. 148-157
[P]Cristian S. Calude is Chair Professor at the University of Auckland, New Zealand.
Founding director of the Centre for Discrete Mathematics and Theoretical Computer
Science. Member of the Academia Europaea. Research in algorithmic information theory
and quantum computing.

Tom(maso) Toffoli[q]: Well, I was born by a historical accident in a post office and that may have something to do with all of this (laughter) because it was exactly organized like Ethernet: press a key and you bring a whole lot of lines (40 miles) from a high value to ground, so you have to listen to what you are typing and if what you get is different from what you are typing, or what you're expecting to hear, then it means that someone else is on the line too. So you both stop. And then both of you start again, at random – exactly in the spirit of Ethernet. But anyway, when I was five, it was right after the end of the war, my mother already had four kids and she was very busy trying to find things for us to eat from the black market and such – so I was left off for the most part of the day in a place on the fifth or sixth floor. Where we lived we had long balconies with very long rails, for laundry etc., and the rail was not raised in the middle, it was just hooked, stuck in the wall at the endpoints on the two sides. So I was like a monkey in a cage in that balcony and I had to think and move around and find a way to spend time in some way. So one of the things I noticed was that as I was pulling and shaking this thing, the rope, somehow I discovered that by going to more or less one third of its length and shaking it in a certain way, eventually the thing would start oscillating more and more and more. And I discovered what you could call resonance, phase locking and other things like that – and I felt really proud about this thing, about this power unleashed into your [tiny] hands. But I also got very scared because – as a result of my early scientific activity – the plaster had come out at the place where the rail was hooked in the wall and there was even plaster outside on the sidewalk if you looked down from the sixth floor balcony where I was, so I was sure I was going to get into trouble when my mother would come home later that day.

---

[q]Prof. Tommaso Toffoli's interests and expertise include among others: Information Mechanics, Foundations and physical aspects of computing, Theory of cellular automata, Interconnection complexity and synchronization, Formal models of computation consistent with microscopical physics (uniformity, locality, reversibility, inertia and other conservation principles, variational, relativistic, and quantum aspects of computation). He proved the computation-universality of invertible cellular automata (1977); formulated the conjecture (later proved by Kari) that all invertible cellular automata are structurally invertible (1990). Introduced the Toffoli gate (1981), which was later adopted by Feynman and others as the fundamental logic primitive of quantum computation. Proposed, with Fredkin, the first concrete charge conserving scheme for computation (1980) – an idea that has been taken up by the low-power industry in recent years. Proved that dissipative cellular automata algorithms can be replaced by non dissipative lattice gas algorithms (2006).

In any event, the point that I'd like to make is that "I cannot understand why people cannot understand" (as Darwin wrote to a friend) "that there's no such thing as simply experiments – there are only experiments 'for' or 'against' a subject." In other words you have to ask a question first, otherwise the experiment is not a very useful thing. Now you will say: "Fine, but how are you going to answer the question that you're making the experiment for?" And what we do with computation, cellular automata and so on, is that we just make our own universe. But the one that we make, we know it, we make the rules so the answers that we get there are precise answers. We know exactly whether we can make life within this CA (like von Neumann) and we know whether we can compute certain things, so it's a world that we make so that we have precise answers – for our limited, invented world – and that hopefully can give us some light on answers for the questions in the real world. And so we have to go continually back and forth between the worlds that we know because we have complete control [over them] and we can really answer questions and the other worlds about which we would really like to answer questions. And my feeling is that computation, CA and all these things that Stephen has called our attention to are just attempts to get the best of the two worlds, essentially. The only world we can answer questions for is the one we make, and yet we want to answer the questions about the world in which we are – which has all of its imperfections, and where resonance does not give you an infinite peak but still gives you crumbled plaster on the floor near the wall and so on.

But I think that we should be able to live with one foot in one world and the other in the other world.

Stephen Wolfram: Let's see, the question I believe was how did one get involved with computation, and how did one get interested in computation and then interested in these kind of things. Well, the story for me is a fairly long one by now: when I was a kid I was interested in physics. I think I got interested in physics mainly because I wanted to understand how stuff works and at the time, it was in the 1960s, roughly, physics was the most convincing kind of place to look for an understanding of how stuff works. By the time I was 10-11 I started reading college physics books and so on and I remember a particular moment, when I was 12, a particular physics book that had a big influence on me: it was a book about statistical mechanics which had on the cover this kind of a simulated movie strip that purported to show how one would go from a gas of hard spheres or something like that, that was very well organized, to a gas of hard spheres that looked completely random. This was an attempt to illustrate the second law of thermodynamics. And I thought it was a really interesting picture, so I read the whole mathematical explanation of the book as to why it was this way.

And the mathematical explanation to me was pretty unconvincing, not at all in the least because this was one of those explanations which – it's about actually the analog of time basically – ended with a statement that said "well, the whole explanation could be just as well run in reverse, but somehow, that isn't how it works." (General laughter) Not very convincing, so I decided: all right, I will figure out on my own what was really going on there, I would make my own version of this movie strip. At the time I had access to a simple computer, it was an English computer: Eliott 903, computers we used mostly to operate tanks and such. It had eight kilowords of eighteen bit memory. It was otherwise a fine computer and I started trying to program this thing to work out the second law of thermodynamics. In some respects the computer was fairly primitive especially when doing floating point operations so I tried to simplify the model that I had for these hard sphere gas bouncing around. What I realized many years later was that the model that I had actually come up with was a two dimensional cellular automata system. The thing that went wrong was: (a) I simulated this 2D CA system and (b) I found absolutely nothing interesting and (c) I could not reproduce the second law of thermodynamics. So that was when I was about 12 years old. And at that point I said: maybe there was something else going on, maybe I don't understand this so I'm going to work in an area where I could learn about stuff, where I think I can understand better what's going on. And the area that I got involved with was particle physics. And so I got to know all sorts of stuff about particle physics and quantum field theory and figured out lots of things that made me kind of a respectable operative in the particle physics business at the time.

And doing particle physics something that happened was that we had to do these complicated algebraic calculations, something that I was never particularly good at. I also have the point of view that one should always try to figure out the best possible tools that one can build to help one out. And even though I wasn't very good myself at doing things like algebra I learned to build computer tools for doing algebra. And eventually I decided that the best way to do the kinds of computations I needed to do was to build my own tool for doing those computations. And so I built a system called SMP which was the precursor of Mathematica – this was around

1981. In order to build this computer system I had to understand at a very fundamental level (a) how to set up a wide range of computations and (b) how to build up a language that should describe a wide range of computations. And that required inventing these primitives that could be used to do all sorts of practical computations. The whole enterprise worked out rather well, and one of my conclusions from that was that I was able to work out the primitives for setting things up into a system like SMP. And then for a variety of reasons when I got to started thinking about basic science again my question was: "Couldn't I take these phenomena that one sees in nature and kind of invent primitives that would describe how they work – in the same kind of way that I succeeded in inventing primitives for this computer language that I built?" I actually thought that this would be a way of getting to a certain point in making fairly traditional physics models of things, and it was only somewhat coincidentally that later I actually did the natural science to explore abstractly what the computer programs that I was setting up did. And the result of that was that I found all sorts of interesting phenomena in CAs and so on.

That got me launched on taking more seriously the idea of just using computation, computer programs, as a way to model lots of things. I think that in terms of the conviction for example that the universe can be represented by computation I certainly haven't proved that yet. I've only become increasingly more convinced that it's plausible – because I did use to say just like everybody else: "Well, all these simple programs and things: they can do this amount of stuff but there'll be this other thing (or things) that they can't do." So, for instance, say, before I worked on the NKS book one of the things that I believed was that while I could make simple programs representing the standard phenomena in physics, that standard phenomena – not at the level of fundamental physics, but at the level of things like how snowflakes are formed, and so on – I somehow believed that, for example, biological systems with adaptation and natural selection would eventually lead to a higher level of complexity that couldn't be captured by these very simple programs that I was looking at. And I kind of realized as a result of working on the NKS book and so on that that's just not true. That's a whole separate discussion, but I kind of believed that that there would be different levels of things that could and couldn't be achieved by simple programs. But as I worked on more and more areas, I got more and more convinced that the richness of what's available easily in the computational universe is sufficiently great that it's much less crazy to think that our actual universe might be made in that kind of way. But we still don't know if that's right until we actually have the final theory so to speak.

IS THE UNIVERSE A COMPUTER?

George Johnson: Very good. And now we have a fairly formal linear exercise for you. Let me preface it by saying that early in the conference there was, I thought, this very interesting exchange between Tom Toffoli and Seth Lloyd, who's another old Santa Fe friend. And I thought it was funny that he said he had to leave to help his kids carve jack-o-lanterns, because shortly after I met Seth he invited me to his house in Santa Fe that he was renting and he was carving jack-o-lanterns then as well – this was before he was married and had children. Our other jack-o-lantern carver then was Murray Gell-Mann. I was writing a biography of Murray at the time and he had not agreed to cooperate, in fact he was being somehow obstructive, but somehow sitting there with a Nobel prize winner carving jack-o-lanterns helped break the ice.

Seth of course has described his grand vision of the universe as a quantum computer – the universe *is* a quantum computer, he said, and then Dr. Toffoli had an interesting subtle objection and basically I think you said that it was this idea of the computational relation between the observer and the system, that somebody has to be looking at it. And later, actually – to skip out to another context during the same session – when Seth made some joke about the election and you said, quoted Stalin saying that it doesn't matter who votes, it's who counts the votes, that kind of seemed to connect back to it and got me thinking about this whole idea of endo-physics of looking at the universe from within or from without. So what I am wondering is this: does Seth's view of looking at the universe not *as* a computer but saying that it *is* a computer, implies a God's eye view of somebody who's watching the computation? What is the universe computing, if it is computing? Is the universe *a* computer, or is the universe *like* a computer– or do the two state the exact same thing?

Tom Toffoli: Are you asking anyone in particular?

George Johnson: Anyone. Please jump in.

Stephen Wolfram: I can think of a few fairly obvious things to say. You know, there's the question of: "What are models?" In other words, if you say, in the traditional physics view of things: "Is the motion of the Earth a differential equation?" No, it's not – it's *described* by a differential equation, it isn't a differential equation. And I don't think, maybe other people here think differently about it, but my view of computation as the underlying thing in physics is that computation is a *description* of what happens in physics. There's no particular sense in which it is useful to say the universe is a computer. It's merely something that can be described as a computation[al process] or something that operates according to the rules of some program.

Cris Calude: I agree and I would add one more idea. When you have a real phenomenon, and a model – a model typically is a simplified version of the reality. In order to judge whether this model is useful or not you have to get some results (using the model). So I would judge the merit of this idea about the universe being modeled faithfully by a gigantic computer primarily by saying: "Please tell me three important facts that this model reveals about the universe that the classical models can't."

Ed Fredkin: I will tell you one in a minute. The thing about a computational process is that we normally think of it as a bunch of bits that are evolving over time plus an engine – the computer. The interesting thing about informational processes (digital ones) is that they're independent of the engine in the sense of what the process is: any engine that is universal (and it's hard to make one that isn't) can – as long as it has enough memory – exactly produce the same informational process as any other. So, if you think that physics is an informational process then you don't have to worry about the design of the engine – because the engine isn't here. In other words, if the universe is an informational process then the engine, if there is one, is somewhere else.

Greg Chaitin: George you're asking a question which is a basic philosophical question. It's epistemology versus ontology. In other words when you say the Universe looks like a computer, that this is a model that is helpful – that's an epistemological point of view, it helps us to understand, it gives us some knowledge. But a deeper question is what is the universe really. Not just we have a little model, that sort of helps us to understand it, to know things. So, that is a more ambitious question! And the ancient Greeks, the pre-Socratics had wonderful ontological ideas: the world is number, the world is this, the world is that. And fundamental physics also wants to answer ontological questions: "What is the world really built of at the fundamental level?" So, it's true, we very often modestly work with models, but when you start looking at fundamental physics and you make models for that and if a model is very successful – you start to think [that] the model *is* the reality. That this is really an ontological step forward. And I think modern philosophy doesn't believe in metaphysics and it certainly doesn't believe in ontology. It's become unfashionable. They just look at epistemological questions or language, but mathematicians and physicists we still care about the hard ontological question: If you're doing fundamental physics you *are* looking for the ultimate reality, you *are* working on ontology. And a lot of the work of lots of physicists nowadays really resembles metaphysics– when you start talking about all possible worlds, like Max Tegmark does or many other people do, or David Deutsch for that matter. So philosophers have become very timid, but some of us here, we are continuing in the tradition of the pre-Socratics.

Ed Fredkin: The problem – one problem – we have is the cosmogony problem which is the origin of the universe. We have these two sort of contradictory facts: one is that we have a lot of conservation laws, in particular mass energy is conserved and we have the observation that the universe began, it seems, with a big bang not so long ago, you know, just thirteen or fourteen billion years ago. Well, basically there's a contradiction in those two statements: because if something began and you have a conservation law where did everything come from, or how did it come about? And also the idea that it began at some time is problematic – the laws of physics can't help us right there because, if you think a lot about of those details it's not so much why matter suddenly appeared but why is there physics and why this physics and stuff like that.

And there is a way to wave your hands, and come up with a kind of answer that isn't very satisfying. Which is that you have to imagine that there is some other place – I just call it 'other' – and in this other place for whatever reason there is an engine and that engine runs an informational process. And one can actually come to some kind of feeble conclusions about this other place, because there are some numbers that we can state about how big that computer must be. In other words if we ask the question what would it take to run a computer that emulated our universe, well – we can guess some quantitative numbers if we can figure out a few things, and so you can make a few statements [educated guesses] about that other kind of place. The point about 'other' is that it is a place that doesn't need to have conservation laws, it is a place that doesn't need to have concepts such as 'beginnings' and 'ends' – so there aren't that many constraints. And one of the wonderful things about computation is that it is one of the least demanding concepts, if you say: well, what do you need in order to have a computational engine? Well, you need a space of some kind. What kind of space? How many dimensions does it have to have. Well, it could have three, it could have two, it could have one, it could have seven – it doesn't matter. They can all do the same computations. Of course, if you have a one-dimensional space you can spend a lot of time overcoming that handicap. But does it need the laws of physics as we know them? No, you don't need to have the laws of physics as we know them. In fact, the requirements are so minimal for having a computation compared to the wonderful rich physics we have that it's very, very simple. I am, in fact, reminded of a science-fiction story, by a Polish author where there's a robot that could make everything that started with the letter 'n' and in Polish, like in Russian, or in English, the word 'nothing' starts with an 'n' so someone bored said: "OK, make nothing." And the robot started working and where the sky had previously been white with so many stars and galaxies just minutes earlier, it slowly started to fade away, little by little, galaxy by galaxy. Admittedly, it's just a science fiction story, but the point is that one could even inquire as to what would be the motivations to create an emulation like this? You can imagine that there is some question, and [then] one needs to think about: what could the question be?

We can speculate about that. But the point is that this is an explanation that says: well there's this thing called 'other' that we don't know anything about – as opposed to all other explanations that imply that some kind of magic happened. Well: I don't like magic, myself.

Rob de Ruyter: A couple of sentences from a biological perspective: Let's take a naive standpoint that there's a world out there and that there's a brain and this brain needs to understand what's happening in the world, what's going around and unfortunately, maybe – or fortunately for us – this brain is an engine that is really well adapted to information processing in the savannah, or in the trees, or wherever. I don't think that the brain itself is a universal computational engine – at least I don't see it that way – but it's a device that is extremely well adapted to processing information that comes to our sensory organs, in from the world that we happen to inhabit. So if we want to start thinking about more complex things or deeper things we need to develop tools that allow us translate our thoughts about the phenomena that we observe in the world. Or the other way around: just like we developed hammers and pliers, in dealing with phenomena in the world we need to develop tools to think [about them]. I have no idea of what the limitations that the structure of our brain are, and that impose – I mean, there must be limitations (in the way we think) that impose structure on those tools that we develop to help think about things. But computation, I think, in a sense, is one of the tools [that] we tried to develop in dealing with the world around us. And what computation allows us to do, like mathematics, is to be able to develop long chains of reasoning that we normally don't use, but that we can extend to reason about very long series, sequences of complicated observations about phenomena in the world around us. So what interests me is this relationship between the way we think – and the way we have evolved to think about the world around us – and the things that we think about now in terms of scientific observations and the origin of the universe. To what extent does the hardware that we carry around inform and determine the kinds of tools and computations and the strategies that we're using?

Tom Toffoli: I would like to give some examples to illustrate why the question of whether the universe is a computer, is a really hard question. In some sense it resembles the question of what is life. Let's take for example the concept of randomness: say you buy a random number generator and you start producing numbers with it. First number you get out is 13, then you get 10, 17, and so on – and then you start asking yourself about the numbers that you obtained: how random are they? What is random? Is 10 a random number? Is 13 a random number? How about 199999 – is it a random number? Then you realize that, of course, randomness is not a property of the number, it is a property of the process. You pay for a random number generator because you want to be surprised. You want not to know what will come out. If you knew it – it would not be random to you because it would be perfectly predictable. So we use that term 'a random number' as an abbreviation for whatever [sequence] is produced by a random number generator. I'll give you another example: somebody tried to trademark icons. When they were invented icons were small: 16 by 16 pixels black and white, and you can draw, you know, some simple things with those 16 by 16 bits. So you may want to trademark them. And some greedy businessman said: look I will go to the judge and I will try to trademark *the entire set* of 16 by 16 pixel icons, all of them, and everybody has to pay me. So now if you are the judge and you have to decide whether you can or cannot allow to someone the right to trademark not just one icon, but all of the icons that can be made that way. And if you are the judge you really have two ways: you say (a) either you have to give me a reason why this icon is really something interesting or (b) you pay 5 cents for each one of the icons that you register and being that there are 22^{56} items you know, you don't have to exercise any judgment, it would just turn into a big contribution to the community.

In other words, the question is: what makes an icon what it is? Is it the fact that it is 16 by 16 bits or that you have reason to believe that there is something useful in it? Brian Hayes, whom you mentioned a moment ago, once said: "What surprises me is that most people don't use the computer for what makes it unique and powerful – which is that it is a programmable machine." My partial definition of a computer is: something that can compute (evaluate) a lot of different functions. If it can evaluate just one function, then I wouldn't call it a computer I would call it a special purpose machine or whatever it is. So we may get the surprise that if we find the formula that gives us the universe as a computer, then at that very point the universe itself becomes a special purpose machine. I mean, we know the formula, we know the machine, we know the initial conditions, and we just go: tick, tick, tick, tick. And it's the largest computer but nobody can program it – if this is the universe, we cannot program it because we are inside of it.

So, the definition of a computer is a bit like the definition of life and the definition of evolution or being adaptive: if there isn't a component of adaptiveness, I wouldn't call the thing a computer. Now the thing can be formalized better, I just said it in an intuitive way, but I'm asking some of these questions to try to clarify a bit what exactly it was that we wanted.

Greg Chaitin: Look, I'd like to be aggressive about this – the best way for me to think about something is to make claims that are much too strong (at least it brings out the idea). So the universe *has to* be a computer, as Stephen said, because the only way to understand something is to program it. I myself use the same paradigm. Every time I try to understand something the way I do it, is: I write a computer program. So the only possible working model of the universe has to be a computer – a computational model. I say a working model because that's the only way we can understand something: by writing a program, and getting it to work and debugging it. And then trying to run it on examples and such. So you say that you understand something only if you can program it. Now what if the universe decides however that it's not a – that you *can't* do a computational model about it. Well, then: no problem. It just means we used the wrong computers. You know, if this universe is more powerful than a com-

puter model of it can be, that means that our notion of what a computer
is is wrong and we just need a notion of computer that is more powerful,
and then things are in sync. And by the way, there is a way to define the
randomness of individual numbers based on a computer [we hear: infinite
ones, from Toffoli] well, anyway, but that's another issue.

Stephen Wolfram: One point to make, in relation to using computation as
a model of a universe: we're used to a particular thing happening when we
do modeling, we're used to models being idealizations of things. We say
we're going to have a model of a snowflake or a brain or something like
that, we don't imagine that we're going to make a *perfect* model! It's a
very unusual case that we're dealing with in modeling fundamental physics
(perhaps modeling isn't the right term, because what we imagine is that
we're going to actually have a precise model that reproduces our actual
universe in every detail.) It's not the same kind of thing as has been the
tradition of modeling in natural science, it's much more. So when you
say, when you talk about what runs it and so on, it's much more like
talking about mathematics: you wouldn't ask when you think about a
mathematical result, [if you] work out some results in number theory, for
example, one wouldn't be asking all the time "Well, what's *running* all
these numbers?" It's just not a question that comes up when one is dealing
with something where what you have is a precise model of things.

One other point to make regarding the question of to what extent our
efforts of modeling relate to what our brains are good at doing and so on.
One of the things I am curious about is: if it turns out to be the case that
we can find a precise representation, a new representation (better word
than model) for our universe in terms of a simple program and we find
that it's, you know, program number such and such – what do we conclude
in that moment? It's a funny scientific situation, it kinds of reminds one
of a couple of previous scientific situations, like for instance Newton was
talking about working out the orbits of planets and so on and made the
statement that once the planets are put in their orbits then we can use the
laws of gravity and so on to work out what would happen – but he couldn't
imagine what would have set the planets originally in motion in their orbit
in the first place. So he said, "Well, the Hand of God must have originally
put the planets in motion. And we can only with our science figure out
what happens after that." And it's the same with Darwin's theories: once
we have life happening then natural selection will lead us inexorably to all
the things that we see in biology. But how to cause life in the first place

– he couldn't imagine. So some of the things I'd be curious about would be: if in fact we do come up with a precise representation of the universe as a simple program – what do we do then and can we imagine what kind of a conclusion we can come to, about why this program and not another program and so on? So one of the possibilities would be that we find out that it's, you know, program number 1074 or whatever it is. The fact that it is such a small number might be a consequence of the fact that our brains are set up because they are made from this universe in such a way that it is inevitable, and in [all] the enumerations that we [might] use our universe will turn out to be a small number universe. I don't think that's the case – but that's one of those self fulfilling prophecies: because we exist in our universe our universe will have to have laws that will somehow seem simple and intuitive to us. I think it's more clear than that, but that's one of the potential resolutions of this question: so now we have our representation of the universe, what do we conclude metaphysically from the fact that it is this particular universe representation and not the other one.

Tony Leggett: I think with regard to the strong and forceful statement that could be related to Seth Lloyd's argument namely that the universe is a computer, I just have a little very naive and simple question: it seems to me that if a statement is called for then the converse is not called for. So my question to Seth Lloyd is: "What would it be like for the universe *not* to be a computer?" And so far I fear I don't find that particular statement truly helpful. I do find quite helpful the thesis that it may be useful to look at the universe in the particular framework of computational science and to ask different questions about it, although I have to say that I'm not yet convinced that looking at it in this way is going to help us to answer some very obvious questions, some of which are usually met with a certain amount of friction, namely is the anthropic principle physically meaningful? That is, why do the constants of nature as we know them have the particular values that they have? Is it perhaps for some arbitrary reason, or maybe for some other deeper reason. Now, of course, there have been plenty of arguments and speculations here about all sorts of things but as far as I can see they don't seem to give any direct relationship (of the universe with a specific computational model, or the universe as a computer.) But I would really like to hear a plausible argument as to why this point of view takes us further on these types of questions that I just mentioned.

Ed Fredkin: I want to react a little bit to what Stephen was saying. There exist areas where we use computers to write programs that are *exact* models, exact and perfect in every possible way – and that is when you design a program to emulate another computer. This is done all the time both for writing a trace program map and for debugging, where you write an emulator for the computer that the software is running on. Or you want to run software that is for another computer like the Mac did when it switched CPUs from the 68000 to the PowerPC: they made an emulator, which is an exact implementor of the software on another computer.

This relates to something that I used to call 'The Tyranny of Universality.' Which is: "Gee, we can never understand the design of the computer that runs physics since any universal computer can do it." In other words if there's a digital computer running all physics of course then any computer can do it, but then, after convincing myself that that point of view made sense a long time later I came up with a different perspective: that if the process that runs physics is digital and it is some kind of CA there will exist a model that's one to one onto in terms of how it operates. And it would probably be possible to find in essence the simplest such model so that if some kind of experimental evidence showed us that physics is some kind of discrete, digital [physical] process like a CA I believe we should be able to find the exact process (or, you know, one of a small set of processes) that implement it exactly.

IS THE UNIVERSE DISCRETE OR CONTINUOUS?

George Johnson: Maybe a good way to get to Tony Leggett's question that he raised at the end is just to ask the same question that our host Adrian asked after we saw that brief film that he showed us first thing this morning which is: is there a fundamental difference between a computational physics, or a computational model, or emulation of the universe and quantum mechanics? Or is there a fundamental distinction between a discrete and a continuous physics? Does anyone have a reaction to that?

Stephen Wolfram: (speaking to George Johnson): So if you're asking is there some definitive test for whether the universe is somehow discrete or somehow fundamentally continuous...

George Johnson: Yeah – if there's a conflict that it could possibly be both at a deeper level.

Stephen Wolfram: If you're asking for that – for example in the kinds of models that I made some effort to study, there are so many different ways to formulate these models that this question "Is it discrete or is it continuous?"

becomes kind of bizarre. I mean, you could say: we'll represent it in some algebraic form in which it looks like it's talking about these very continuous objects – and what matters about it may yet turn out to be discrete, it may turn out to be a discrete representation (which is much easier to deal with). So I think that at the level of models that I consider plausible the distinction between continuity and discreteness is much less clear than we expect. I mean, if you ask this question I think you end up asking 'very non-physics questions' like, for example, how much information can in principle be in this volume of space. I'm not sure that without operationalizing that question that it's a terribly interesting or meaningful question.

Tom Toffoli: I would like to say something that will be very brief. Look at CAs, they seem to be a paradigm for discreteness. But as it turns out one of the characterizations of CAs is that they are a dynamical system [that perform certain kinds of translations] and they are continuous with respect to a certain topology which is identical to the Cantor set topology, continuous in exactly that very sense of that definition of continuity that is studied in freshman calculus. But the interesting thing is that this is the Cantor set topology, invented by Cantor (the one with the interval where you remove the middle third, etc.) And as it turns out, this topology for CAs is – not in the sense of geometrical topology, but in the sense of set topology of circuits with gates that have a finite number of inputs and a finite number of outcomes outputs – that is exactly the topology of the Cantor set, so it's sort of a universal topology for computation. And so we come full circle that (a) something that was not invented by Cantor to describe computers in fact represents the natural topology to describe discrete computers and (b) the moment you take on an in[de]finite lattice then you have continuity exactly the kind of continuity, continuity of state, of the dynamics that you get when you study continuous functions. So these are some of the surprises that one gets by working on things.

Stephen Wolfram: I just want to say with respect to the question of how do we tell, you know, the sort of thing that Tony is saying: "How do we tell that this is not all just complete nonsense?" Right?

Tony Leggett: Refutation of the negative statement.

Stephen Wolfram: Yes, right. We're really only going to know for sure if and when we finally get a theory, a representation that is the universe and that can be represented conveniently in computational form. Then people will say: "Great! This computational idea is right, it was obvious all along, everybody's thought about it for millions of years..."

Tom Toffoli: At that point they will probably say: "In fact it's trivial!"

Stephen Wolfram (agrees laughing, everybody laughs): And I think that until that time one could argue back and forth forever about what's more plausible than what and it's always going to be difficult to decide it based on just that. Yet these things tend to be decided in science in a surprisingly sociological way. For example the fact that people would seriously imagine that aspects of string theory should be taken seriously as ways to model the reality of a physical universe it's – it's interesting and it's great mathematics – but it's a sociological phenomenon that causes [or forces] that to be taken seriously at the expense of other kinds of approaches. And it's a matter of history that the approach we're using (computational ideas) isn't the dominant theme in thinking about physics right now. I think it's purely a matter of history. It could be that in place of string theory people could be studying all kinds of bizarre CAs or network systems or whatever else and weaving the same kind of elaborate mathematical type web that's been done in string theory and be as convinced as the string theorists are that they're on to the right thing. I think at this stage until one has the definitive answer, one simply doesn't know enough to be able to say anything with certainty and it's really a purely sociological thing whether we can say that this is the right direction or this isn't the right direction. It's very similar actually to the AI type thing, people will argue forever and ever about whether it's possible to have an AI and so on – and some of us are actually putting a lot of effort into trying to do practical things that might be identified as relevant to that. I think that actually the AI question is harder to decide than the physics question. Because in the physics case once we'll have it it's likely (it seems to me) that we'll be able to show that the representation is of the universe that is obviously the actual universe and the question will be closed. Whereas the question of AI will be harder to close.

Cris Calude: Apparently there is an antagonism between the discrete and continuous view. But if we look at mathematics there are mathematical universes in which discrete and continuous are co-existing. Of course, what Tom said, the Cantor space is a very interesting example, but it might be too simple for the problem that we are discussing. For instance, non-standard analysis is another universe where you find [this same phenomenon] and you have discreteness and you have continuity and maybe, to the extents that mathematics can say something about the physical universe, it could be just a blend of continuous and discreteness and some phenomena may be revealed through discreteness and some others will be revealed through continuity and continuous functions.

Greg Chaitin: Again I am going to exaggerate – on purpose. I think the question is like this: discreteness vs. continuity. And I'm going to say why I am on the side of discreteness.

The reason is this: I want the world to be comprehensible! Now there are various ways of saying this. One way would be: God would not create a world that we couldn't understand. Or everything happens for a reason (the principle of sufficient reason). And other ways. So I guess I qualify as a neo-Pythagorean because I think the world is more beautiful, if it is more comprehensible. We are thinkers, we are rationalists – we're not mystics. A mystic is a person that gets in a communion with an incomprehensible world and feels some kind of community and is able to relate to it. But we want to understand rationally so the best universe is one that can be completely understood and if the universe is discrete we can understand it – it seems to me. This is something that you said, at one point, Ed – it is absolutely totally understandable because you run the model and the model is exactly what is happening.

Now a universe which uses continuity is a universe where no equation is exact, right? Because we only have approximations up to a certain order. So I would also say: a universe would be more beautiful if it were discrete! And although we now end up in aesthetics, which is even more complicated, I would still say that a discrete universe is more beautiful, a greater work of art for God to create – and I'm not religious, by the way. But I think it's a very good metaphor to use – or maybe I am religious in some sense, who knows?

Another way to put it is let's say this universe does have continuity and messy infinite precision and everything – well, too bad for it. Why didn't God create as beautiful a universe as he could have?

Tom Toffoli: He should have asked you, Greg!

Greg Chaitin (laughs): What? ... No... No ... maybe at this point I think Stephen is the leading candidate for coming up with a ... [everybody is still laughing, including Chaitin who continues] ... so that would be more beautiful it seems to me. You see, it would be more understandable it would be more rational it would show the power of reason. Now maybe reason is a mistake as may be to postulate that the universe is comprehensible – either as a fundamental postulate or because you know God is perfect and good and would not create such a universe, if you want to take an ancient theological view. Maybe it's all a mistake, but this one of the reasons that I'm a neo-Pythagorean, because I think that would be a more beautiful, or comprehensible universe.

Stephen Wolfram: I have a more pragmatic point of view, which is that if the universe is something that can be represented by something like a simple discrete program, then it's realistic to believe that we can just find it by searching for it. And it would be embarrassing if the universe would indeed be out there in the first, you know, billion universes that we can find by enumeration and we never bothered to even look for it. [There's a very sustained reaction from the rest of the round table members, especially Greg Chaitin, whom we hear laughing.] It may turn out that, you know, the universe isn't findable that way – but we haven't excluded that yet! And that's the stage we're at, right now. Maybe in, you know, 10-20-50 years we will be able to say: yes we looked at all the first I don't know how many – it will be like looking for counterexamples of the Riemann hypothesis, or something like that – and we'll say that we've looked at the first quadrillion possible universes and none of them is our actual universe, so we're beginning to lose confidence that this approach is going to work. But right now we're not even at the basic stage of that yet.

Ed Fredkin: If this were a one dimensional universe, Steve (Wolfram), you would have found the rule by now, right? Because you've explored all of them...

Tony Leggett: Well, George I think, raised the question whether quantum mechanics has any relevance to this question, so let me just comment briefly on that. I think if one thinks about the general structure of quantum mechanics, and the ways in which we verify its predictions, you come to the conclusion that almost all the experiments (and I'll stick my neck out and say *all* the experiments) which have really shown us interesting things about quantum mechanics do measure discrete variables in fact. Experiments on the so-called macroscopic quantum coherence, experiments on Bell's theorem, and so forth – they all basically use discrete variables in practice. Now of course the formalism of quantum mechanics is a continuous formalism. You allow amplitudes to have arbitrary values, but you never really

measure those things. And I think that all one can say when one does
sometimes measure things like position and momentum which are [appar-
ently] continuous variables – if you look at it hard you'll see that the actual
operational setup is such that you are really measuring discrete things. So
measurements within the framework of quantum mechanics which claim to
be of continuous variables usually are of discrete variables. So I think from
the, as it were, the ontological point of view, one can say that quantum
mechanics does favor a discrete point of view.

IS THE UNIVERSE RANDOM?

George Johnson: When Hector[r], Gerardo[s] and I were talking about good
questions that would stimulate debate we thought that perhaps we should
ask something that would be really really basic about randomness – and
the great thing about being a journalist particularly a science journalist
is that you get this license of asking really really smart people questions
about things that have been puzzling you. And this is something that
has always kind of bugged me – the SETI (Search for ExtraTerrestrial
Intelligence) where we get these signals from space which are then analyzed
by these computers, both by super computers and by SETI at home, where
you donate some CPU time on your PC, computer cycles etc. They're
looking for some structure in what appears to be random noise. And I was
wondering we're getting a signal that seems to be pure noise but to some
extent – as I think Tomasso Toffoli has suggested – perhaps the randomness
is only in the eye of the beholder.

If, for example we're getting this noisy signal that just seems to be static
– how do we know we're not getting the ten billionth and fifty seventh digit
of the expansion of π forward? How do we know that we're not getting
line ten trillion quadrillion stage forward of the computation of the rule 30
automata? So I am wondering if you can help me with that.

Tom Toffoli: This is not an answer. It's just something to capture the
imagination. Suppose that people are serious about computing and they
say: "Look: you're not using your energy efficiently because you're letting
some energy – that has not completely degraded – out." So they start to
make better re-circulators, filters and so on and now whatever thermal energy comes out is as thermalized as possible. Because if it's not, they would
have committed a thermodynamical sin. But this is exactly what happens

---

[r]Hector Zenil
[s]Gerardo Ortiz

when you look at the stars. They are just sending close to thermodynamical equilibrium a certain temperature – so you can say well probably then this is prima facie evidence that that there are people there computing and they are computing so efficiently that they are just throwing away garbage, they're not throwing away something that is still recyclable! And this could be an explanation as to why we see all these stars with all these temperatures.

Stephen Wolfram: You know I think the question about SETI and how it relates to the type of things we're talking about – I think it gets us into lots of interesting things. I used to be a big SETI enthusiast and because I'm a practical guy I was thinking years ago about how you could make use of unused communication satellites and use them to actually detect signals and so on. And now I have worked on the NKS book for a long time, and thought about the PCE and I have became a deep SETI non-enthusiast. Because what I realized is that it goes along with statements like "the weather has a mind of its own". There's this question of what would constitute – you know, when we say that we're looking for extra terrestrial intelligence – what actually is the abstract version of intelligence? It's similar to the old question about what is life, and can we have an abstract definition of life that's divorced from our particular experience with life on the Earth. I mean, on the Earth it is pretty easy to tell whether something – reasonably easy to tell whether something – is alive or not. Because if it's alive it probably has RNA it has some membranes it has all kinds of historical detail that connects it to all the other life that we know about. But if you say, abstractly: what is life? It is not clear what the answer is. At times, in antiquity it was that things that can move themselves are alive. Later on it was that things that can do thermodynamics in a different way than other things do thermodynamics are alive. But we still – we don't have – it's not clear what the abstract definition of life is divorced from the particular history. I think the same is true with intelligence. The one thing that most people would (I think) agree with – is that to be intelligent you must do some computation. And with this principle of computational equivalence idea what one is saying is that there are lots of things out there that are equivalent in the kind of computation that they can do ...

Tom Toffoli: But you can also do computation without being intelligent!

Stephen Wolfram (replying to Toffoli): That's precisely the question: can you – what is the difference, what is the distinctive feature of intelligence?

If we look at history, it's a very confusing picture: a famous example that
I like is when Marconi had developed radio and (he had a yacht that he
used to ply the Atlantic with, and) at one point he was in the middle of the
Atlantic and he had this radio mast – because that was the business that
he was in – and he could hear these funny sounds, you know: ... wooo ...
ooooeeo ... eeooo ... woooo ... this kind of sounds out in the middle of the
Atlantic. So what do you think he concluded that these sounds were? He
concluded that they must be radio signals from the martians! Tesla, was in
fact more convinced that they were radio signals from the martians. But,
what were they in fact? They were in fact some modes of the ionosphere on
the Earth, they were physical processes that – you know, something that
happens in the plasma. So the question was, how do you distinguish the
genuinely intelligent thing, if there is some notion of that, from the thing
that is the ... the computational thing that is.

The same thing happened with pulsars, when the first pulsars were
discovered. In the first days of discovery it seemed like this periodic millisecond thing must be some extraterrestrial beacon. And then it seemed
like it was too simple. We now think it's too simple to be of intelligent
origin. It also relates to this question about the anthropic principle and
the question of whether our universe is somehow uniquely set up to be capable of supporting intelligence like us. When we realize that there isn't
an abstract definition of actual intelligence, it is (as I think) just a matter
of doing computation. Then the space of possible universes that support
something like intelligence becomes vastly broader and we kind of realize
that this notion of an anthropic principle with all these detailed constraints
– just doesn't make much sense.

There's so much more we can say about this, and I'll let others do so.

George Johnson: Randomness: is it in the eye of the beholder?

Greg Chaitin: George, I suppose it would be cowardly of me not to defend
the definition of randomness that I have worked on all my life, but I think
it is more fun to say (I was defending rationalism, you know) that a world
is more understandable because it is discrete, and for that reason it is more
beautiful. But in fact I've spent my life, my professional life, working on
a definition of randomness and trying to find, and I think I have found,
randomness in pure mathematics which is a funny place to find something
that is random. When you say that something is random you're saying
that you can't understand it, right? So defining randomness is the rational
mind trying to find its own limits, because to give a rational definition to

randomness is odd there's something paradoxical in being able to know *that*, you know, being able to define randomness, or being able to know that something is random – because something is random when it escapes ... I'm not formulating this well, actually improvising it, but there are some paradoxes involved in that. The way it works out in these paradoxes is that you can define randomness but you can't know that something is random, because if you could know that something is random then it wouldn't be random. Randomness would just be a property like any others, and it could be used would enable you to classify things. But I do think that there is a definition of randomness for individual numbers and you don't take into account the process by which the numbers are coming to you: you can look at individual strings of bits – base 10 number – and you can at least mathematically say what it means for this to be random. Now, although most numbers or most sequences of bits are random according to this definition, the paradoxical thing about it is that you can never be sure that one individual number is random – so I think it is possible to define a notion of randomness which is intrinsic and structural and doesn't depend on the process from which something comes but there is a big problem with this definition which is: it's useless. Except to create a paradox or except to show limits to knowledge, or limits to mathematical reason. But I think that's fun, so that's what I've been doing my whole life.

So I don't know if this is relevant to SETI? I guess it is, because if something looks random it then follows that it probably doesn't come from an intelligent source. But what if these superior beings remove redundancy from their messages? They just run it through a compression algorithm because they are sending us enormous messages, they're sending us all their knowledge and wisdom and philosophy everything they know in philosophy, because their star is about to go nova, so this is an enormous text encompassing all their accomplishments of their thinking and civilization – so obviously they think any intelligent mind would take this information and compress it, right? And the problem is, we're getting this LZ compressed message and we think that it's random noise and in fact it's this wonderfully compact message encapsulating the wisdom and the legacy of this great civilization?

George Johnson: Oh, but do they include the compression algorithm?

Greg Chaitin: Well, they might think that a priori this is the only conceivable compression algorithm, that it is so simple that any intelligent being would use this compression algorithm – I don't know ...

Stephen Wolfram: I think it's an interesting question – about SETI. For example, if you imagine that there was a sufficiently advanced civilization that it could move stars around, there's an interesting kind of question: in what configuration would the stars be moved around and how would you know that there is evidence of intelligence moving the stars around? And there's a nice philosophical quote from Kant who said "if you see a nice hexagon drawn in the sand you know that it must come from some sort of intelligent entity [that has created it]" And I think that it's particularly charming that now in the last few years it's become clear that there are these places in the world where there are hexagonal arrangements of stones that have formed and it is now known that there is a physical process that has causes a hexagonal arrangement of stones to be formed. That's sort of a charming version of this ...

Ed Fredkin: One of the poles of Saturn has this beautiful hexagon – at the pole and we have pictures of them.

Stephen Wolfram (continues): ... right, so the question is what do you have to see to believe that we have evidence that there was an intention, that there was a purpose. It's just like Gauss for example, he had the scheme of carving out in the Syberian forest the picture of the Pythagorean theorem, because that would be the thing that would reveal the intelligence. And if you look at the Earth now a good question to ask an astronaut is: "What do you see on the Earth that makes you know that there is some sort of a civilization?" And I know the answer: the thing that is most obvious to the astronauts is – two things, OK? One is: in the great salt lake in Utah [there is] a causeway that divides a region which has one kind of algae which tend to be of orangeish color, from a region that has another kind of algae that tend to be bluish, and there's a straight line that goes between these two colored bodies of water. It's perfectly straight and that is thing number one. Thing number two is in New Zealand. There's a perfect circle that is visible in New Zealand from the space. I was working on the NKS book and I was going to write a note about this particularly thing. We contacted them, this was before the web was as developed as it is today, so we contacted the New Zealand Geological Survey to get some information about this perfect circle and they said: "If you are writing a geology book (the circle is around a volcano,) please *do not* write that this volcano produces this perfect circle, because it isn't true." What's actually true is that there is a national park that was circumscribed around the volcano and it happens to be perfectly circular and there are sheep that have grazed inside the national park but not outside, so it's a human produced circle. But it's interesting to see what is there on the Earth that sort of reveals the intelligence of its source.

And actually, just to make one further comment about randomness, and integers – just to address this whole idea of whether there are random integers or not, and does it matter, and how can you tell, and so on – we have a little project called 'integer base' which basically is a directory of integers. And the question is to find is the simplest program that makes each of these integers. And it's interesting, it's a very pragmatical project actually trying to fill in actual programs that make integers. We have to have some kind of metric as to what counts as simple. When you use different kinds of mathematical functions, you use different kinds of programming constructs, you actually have to concretely decide a measure to quantify simplicity. And there are lots of ways, for example: how many times does this function appear is referenced on the web, that could be a criterion as to how much weight should be given; or how long is this function's name in ; or other kinds of criteria like that. So it's kind of a very concrete version of this question about random integers.

Tom Toffoli: I'd like to say something that throws out another corollary. There's this self-appointed guru of electronics, Don Lancaster, he's very well known in circles and he said something that is very true. He said: the worst thing that could happen to humanity is to find an energy source that is inexhaustible and free. You know, we are all hoping that we will find something like that, but if we found it it would be a disaster, because then the Earth would be turned into a cinder in no time.

If you don't have any of the current constraints it can turn very dangerous. For example, you have a house there, and you have a mountain, and in the afternoon you would like to have sunshine. And in the morning you would like to have shading from the cold or whatever, so if energy is free you just take the mountain from where it is in the morning you take it away and you plant it back in the evening. And this is what we're doing in essence when we're commuting from the suburbs to the center of Boston. You see this river of cars that is rushing in every day, and rushing out every day, with energy that is costing quite a bit. Imagine if it was completely free. So, again, let's try to think up what the answer to our question would be if we really got it and then see the consequences of that first.

And, again: this comment is in relation to what I said earlier about the randomness of the stars if it's an indication of a super-intelligence, or super-stupidity. Who knows, it could be that they're one and the same thing?

Rob de Ruyter: Just to take your question completely literally: there's a lot of randomness in our eyes, as we look around the world. And especially outside in moonlight conditions there are photons flying around, but they are not that many and we are very aware of the fact that information that we're getting into our visual system, information that we have to process in order to navigate successfully, is of low quality and the interesting thing is that we as organisms are used to walking around in a world that is random and we're very conscious of it. Yesterday I spoke about how flies cope with this – we cope with this too and we cope with it at all levels, from adaptation in photoreceptors in our eyes to the adaptation in the computational algorithms that our brain is using to where you are in an environment where you are subject to large levels of noise, because there are not that many photons around. In that case you tend to move very cautiously – you don't start running, unless maybe the tiger is just following you, but that is a very rare situation – so, I think, in a lot of senses we're used to the measurements that our sensors make being more or less random depending on how the situation is at the moment. And so as computational engines we are very well aware of that.

Cris Calude: I am interested in the quality of quantum randomness. We were able to prove that quantum randomness, under some mild assumptions on the quantum model of physics we agree on, is not computable. So this means no Turing machine can reproduce the outcome of a quantum generated sequence of bits (finitely many) and this gives you a weak form of relation between Greg's theory, Greg's definition of algorithmic randomness and what one would consider to be the best possible source of randomness in this this universe, i.e., quantum randomness. And one of the things that is delicate as we are thinking and experimenting is a way to distinguish quantum randomness from generated randomness. Is it possible, by using finitely many well-chosen tests, to find a mark of this distinction you know between something that is computer computably generated from something that is not generated in that way?

So whereas here we have some information about the source, you know, like Tom said – we know very well that there is an asymptotic definition of the way that these bits can be generated – it is still very difficult to account in a finite amount of tests for that difference.

Ed Fredkin: Just a funny story about random numbers: in the early days of computers people wanted to have random numbers for Monte Carlo simulations and stuff like that and so a great big wonderful computer was being designed at MIT's Lincoln laboratory. It was the largest fastest computer in the world called TX2 and was to have every bell and whistle possible: a display screen that was very fancy and stuff like that. And they decided they were going to solve the random number problem, so they included a register that always yielded a random number; this was really done carefully with radioactive material and Geiger counters, and so on. And so whenever you read this register you got a truly random number, and they thought: "This is a great advance in random numbers for computers!" But the experience was contrary to their expectations! Which was that it turned into a great disaster and everyone ended up hating it: no one writing a program could debug it, because it never ran the same way twice, so ... This was a bit of an exaggeration, but as a result everybody decided that the random number generators of the traditional kind, i.e., shift register sequence generated type and so on, were much better. So that idea got abandoned, and I don't think it has ever reappeared.

Stephen Wolfram: Actually it has reappeared, in the current generation of Pentium chips there's a hardware random generator that's based on double Johnson noise in the resistor. But in those days programs could be run on their own. In these days there are problems in that programs can no longer be run on their own: they are accessing the web, they're doing all sorts of things, essentially producing random noise from the outside, not from quantum mechanics but they're producing random noise from the outside world. So the same problem has come up again.

George Johnson: This reminds me that in the dark ages before the published this huge tome – I found a reprint of it called "One hundred thousand random numbers" (Cris Calude corrects out loud: "One *million* random numbers") in case you needed some random numbers – and Murray Gell-Mann used to tell a story that he was working at RAND and at one time, I think when they were working on the book, they had to print an errata sheet! (There is laughter)

Stephen Wolfram: Well, the story behind the erratum sheet I think is interesting because those numbers were generated from (I think a triode, or something) some vacuum tube device and the problem was that when they first generated the numbers, they tried to do it too quickly, and basically didn't wait for the junk in the triode to clear out between one bit and the next. This is exactly the same cause of difficulty in randomness that you get in trying to get perfect randomness from radioactive decay! I think the null experiment for quantum computing, one that's perhaps interesting to talk about here, is this question of how do you get – I mean, can you get – a perfect sequence of random bits from a quantum device? What's involved in doing that? And, you know, my suspicion is the following: my suspicion would be that every time you get a bit out you have to go from the quantum level up to the measured classical level, you have to kind of spread the information about this bit out in this bowl of thermodynamic soup of stuff so that you get a definite measurement; and the contention, or my guess, would be that there is a rate at which that spreading can happen and that in the end you won't get out bits that are any more random than the randomness that you could have got out just through the spreading process alone without kind of the little quantum seed. So that's an extreme point of view, that the extra little piece (bit) of *quantumness* doesn't really add anything to your ability to get out random bits. I don't know if that is correct but you know I, at least a long time ago, I did try looking at some experiments and the typical thing that's found is that you try to get random bits out of a quantum system quickly and you discover that you have $\frac{1}{f}$ noise fluctuations because of correlations in the detector and so on. So I think that the minimal question from quantum mechanics is: can you genuinely get sort of random bits and what's involved in doing that? What actually happens in the devices to make that happen? I'd be curious to know the answer to this [question].

Tom Toffoli: I know the answer, Intel already says: yes, you can get good quantum numbers if you have a quantum generator of random numbers. Just generate one random number, then throw away your generator and buy a new one, because the one that you have used is already entangled with the one it generated. So you buy a new one and you solved the problem. [Wolfram says: this is also a good commercial strategy ... people laugh]

Ed Fredkin: There's a great story in the history of – back in the '50s people doing various electronic things needed noise. They wanted noise, random noise so they thought: what would be a good source of it. And so it was discovered that a particular model of photomultiplier, if you covered it up and let no light into it gave beautiful random noise. And as a result various people conducted experiments, they characterized this tube and it was essentially like a perfect source of random noise, and the volume of sales started to pick up. Various people started building these circuits and using them all over. Meanwhile, back at the tube factory which was RCA, someone noticed: "Hey, that old noisy photomultiplier that we had trouble selling lately, sales are picking up, we better fix that design so it isn't so noisy!" So they fixed it and that was the end of that source of random noise.

INFORMATION VS. MATTER

George Johnson: I think I'll ask another question about one other thing that has been bothering me before we start letting the audience jump in. I first ran across this when I was writing a book called "Fire of the mind" and the subtitle was 'Science, faith and the search for order.' And I was re-reading a book that I read in college that really impressed me at the time, and seeing that it still stood up, which was Robert Pirsig's book "Zen and the art of motorcycle maintenance" – and it did stand up in my opinion.

There's a scene early on in the book the protagonist who called himself Phædrus after Plato's dialogue is taking a motorcycle trip around the country with his son Chris – who in real life later was tragically murdered in San Francisco where he was attending a Zen monastery, which is neither here or there – but, in the book this person's running around with Chris and they're sitting around the campfire at night and drinking whisky and talking and telling ghost stories and at one point Chris asks his father: "Do you believe in ghosts?" And he says "Well, no, of course I don't believe in ghosts because ghosts contain no matter and no energy and so according to the laws of physics they cannot exist." And then he thinks for a moment and says: "Well, of course the laws of physics are also not made of matter or energy and therefore they can't exist either." And this really seemed like an interesting idea to me, and when I was learning about computational physics, this made me wonder: *where* are the laws of physics? Do you have to be a Platonist and think that the laws of physics are written in some theory realm? Or does this computational physics gives us a way to think of them as being embedded within the very systems that they explain? [waits a little, sees Chaitin wanting to answer, says:] Greg!

Greg Chaitin: George, we *have* ghosts: information! Information is non-material.

George Johnson: Information is physical, right? Well, wouldn't Landauer say that?

Greg Chaitin: Well, maybe Rolf would say that, but ontologically we've come up with this new concept of information and those of us that do digital physics somehow take information as more primary than matter. And this is a very old philosophical debate: is the world built of spirit or mind or is it built of matter? Which is primary which is secondary? And the traditional view of what we see as the reality, is that everything is made of matter. But another point of view is that the universe is an idea and therefore (and information is much closer to that) made of spirit, and matter is a secondary phenomenon. So, as a matter of fact, perhaps everything is ghost. If you believe in a computational model informational model of the universe, then there is no matter! It's just information – patterns of information from which matter is built.

George Johnson: Does that sound right to you, Tony?. Does it sound right to you that information is more fundamental than matter or energy? I think most of us – you know, the average person in the street – asked about information would think about information as a human construct that is imposed on matter or energy that we make. But I really like the idea that Gregory has suggested – and I actually wrote about it quite a bit in this book – that information is actually in the basement there, and that matter and energy are somehow built out of information.

Tom Toffoli [starts answering the question]: Well, ideally, I mean... You can ask the same thing about correlation rather than information because it conveys the same meaning. Furthermore one can bring up and discuss the notion of entanglement, and in the same spirit: it's not here nor there. Where is it? Very related issues. That's the point I wanted to make.

Tony Leggett: Well I think I would take the slightly short-sighted point of view that information seems to me to be meaningless – unless it is information *about something*. Then one has to ask the question: "what is the something?" I would like to think that *that something* has something to do with the matter of interest, and the matter and energy that's involved.

Stephen Wolfram: This question about whether abstract formal systems are *about something* or not is a question that obviously has come up from mathematics. And my guess about the answer to this question: is information the primary thing or is matter the primary thing? I think that the answer to that question would probably end up being that they are really the same kind of thing. That there's no difference between them. That matter is merely our way of representing to ourselves things that are in fact some pattern of information, but we can also say that matter is the primary thing and information is just our representation of that. It makes little difference, I don't think there's a big distinction – if one's right that there's an ultimate model for the representation of universe in terms of computation.

But I think that one can ask this question about whether formal systems are about something – this comes up in mathematics a lot, we can invent some axioms system and then we can say: is this axiom system describing something really, or is it merely an axiom system that allows us to make various deductions but it's not really about anything. And, for example, one of the important consequences of Gödel's theorem is that you might have thought that the Peano axioms are really just about integers and about arithmetic but what Gödel's theorem shows is that these axioms also admit different various non-standard arithmetics, which are things that are not really like the ordinary integers, but still consistent with its axioms. I think it's actually a confusion of the way in which mathematics has been built in its axiomatic form that there is this issue about 'aboutness' so to speak – and maybe this is getting kind of abstract. But when we think about computations we set things up, we have particular rules, and we just say "OK, we run the rules, you know, and what happens – happens." Mathematics doesn't think about these things in those terms, typically. Instead, it says: let's come up with axiom systems which constrain how things could possibly work. That's a different thing from saying let's just throw down some rules and then the rules run and then things just happen. In mathematics we say: let's come up with axioms which sort of describe how things have to work, same thing was done in physics with equations – the idea of, you know, let's make up an equation that describe what's possible to happen in the world – and not (as we do in computation) let's do something where we set up a rule and then the rule just runs. So, for example, that's why in gravitation theory there's a whole discussion about "What do the solutions to the Einstein's equations look like?" and "What is the set of possible solutions?" and not (instead) "How will the thing run?" but the question traditionally asked there is: "What are the possible things consistent with these constraints?"

So in mathematics we end up with these axiom systems, and they are trying to sculpt things, to ensure that they're really talking about the thing that you originally imagine[d you were] talking about, like integers. And what we know from Gödel's theorem and so on, is that that kind of sculpting can never really work. We can never really use this constraint-based model of how to understand things to actually make our understanding be about a certain thing. So, I think that's kind of the ultimate problem with this idea of whether the laws that one's using to describe things and the things themselves are one and the same or they are different, and if so what the distinction really is.

Greg Chaitin: There's this old idea that maybe the world is made of mathematics and that the ultimate reality is mathematical. And for example people have thought so about continuous mathematics, differential equations and partial differential equations, and that view was monumentally successful, so that already is a non-materialistic view of the world. Also let me say that quantum mechanics is not a materialistic theory of the world; whatever the Schrödinger wave equation is it's not matter, so materialism is definitely dead as far as I am concerned. The way Bertrand Russell put it is: if you take the view that reality is just what the normal day appearances are, and modern science shows that everyday reality is not the real reality, therefore – I don't know how he called this ... 'naive realism' I think – and if naive realism is true then it's false, therefore it's false. That's another way to put it. So the only thing we're changing in this view that the actual structure of the world is mathematical, the only thing new that we're adding to this now is we're saying mathematics is ultimately computational, or ultimately it is about information, and zeroes and ones. And that's a slight refinement on a view that is quite classical.

So I am saying in a way we're not as revolutionary as it might seem, this is just a natural evolution in an idea. In other words, this question of idealism vs. materialism, or "Is the world built of ideas or is the world built of matter?" it might sound crazy, but it's the question of "Is the ultimate structure of the world mathematical?" versus matter. That sounds less theological and more down to Earth. And we have a new version of this, as ideas keep being updated, the current version of this idea is just: "Is it matter or is it information?" So these are old ideas that morph with time, you know, they evolve, they recycle, but they're still distinguishably not that far from their origin.

Cris Calude: Information vs. matter, discrete vs. continuous: interesting philosophical contrasts and I found Stephen's description to be extremely interesting, because I too think that these views should in fact coexist in a productive duality. And it depends on your own abilities, it depends on your own problems if one of them would be more visible or useful. And at the end of the day what really counts is – if you have a view that in a specific problem information prevails – what can you get from that? Can you prove a theorem, can you get some result, can you build a model which you know answers an important question or not? In some cases one view may be the right one, in other cases the other one is, so from my point of view, which is, I would guess, more pragmatic, I would say: look, you choose whatever view you wish in order to get a result and if you get the result that means that for this specific problem that choice was the correct one.

Ed Fredkin: But in fact the world either is continuous or discrete, and we can call it anything we want, to get results and so on, to add convenience – but there really is an answer to it and one answer is right and the other is wrong. I go with Kronecker who said "God invented the integers and all else is the work of man." So you can do anything with discrete models and/or continuous models but that doesn't mean that the world is both, or can be, or could be either. No, the world is either one – or the other.

Stephen Wolfram: This whole question about mechanism is kind of amusing. For example, with CAs models, people in traditional physics (not so in other areas but people in traditional physics) have often viewed this kinds of models with a great deal of skepticism. And it's kind of an amusing turn of historical fate, because, in the pre-Newtonian period people always had mechanistic models for things – whether there were angels pushing the Earth around the orbit, or other kinds of mechanistic type of things. And then along came this kind of purely abstract mathematical description of law of gravity and so on and everybody said – but only after a while! – everybody said: "Well, it's all just mathematics and there isn't a material reality, there isn't a mechanism behind these things!" And so, when one comes with these computational models which seem to have much more of a tangible mechanism, that is viewed as suspicious and kind of non-scientific by people who spend their lives working in the mathematical paradigm, that it can't be simple enough if there's an understandable mechanism to behind things. So it's kind of an interesting turning around of the historical process. As you know I work on various different kinds of things and, as I said, I've not been working much on finding a fundamental theory of physics lately. I actually find this discussion as an uptick in my motivation and enthusiasm to actually go and find a fundamental theory of physics.

Because I think, in a sense, what with all of these metaphysical kinds of questions about what might be there, what might not be there and so on: damn it, we can actually answer these things!

Tom Toffoli: I know a way out of this! It is similar to the one that Ed proposed a long time ago. He said (about computability, referring to exponential, polynomial problems) he said that one can turn all problems into linear problems. You know, *all* the exponential problems! And people of course said: "Ed you are an undisciplined amateur you say these things without knowing what you're talking about." But he said: "Look, we have Moore's law! And with Moore's law everything doubles its speed every so many years. So we just wait long enough and we get the solution – as long as it takes."

So the key then, is to wait, as long as we need to. With this, I am now giving a simpler solution to this problem, of finding a fundamental theory of physics, starting from the observation that domesticated animals become less intelligent than wild animals. This has been proven recently with research on wolves. And maybe domesticated animals are somewhat less intelligent in certain ways but they can see what humans want and they obey. But then some more experiments were run and the findings were that wild wolves once they put them in an environment with humans they learn humans faster than domesticated animals to anticipate the wills of their trainers. It's not that they follow their will, but they anticipate it faster.

Now we are doing a big experiment on ourselves, on humanity – humanity is domesticating itself. And there was a time when people said: we discovered differential equations and very few people can understand them so we have the monopoly, we are the scientists. And eventually somebody said: "Well, wait a second, but why can't we find a model like, you know, Ed or Stephen – that is, sort of computational, discrete so that everyone can own it and possess it and so on?" But [one forgets that] we are domesticating ourselves! To the point that the computer that – according to Brian Hayes the first thing about a computer is that you can program it, make it do whatever we want – now most people don't even know that that is possible, they just think of the computer as an appliance. And soon even the computer will be a mystery to most people! The programmable digital computer [like] the differential equations were a generation ago – so we solved the problem, in that sense, just wait long enough and nobody will even be able to care about these things, it will just be a mystery for us.

Cris Calude: Well, I would like to just add a small remark, suggested by your idea about physical computation. So, this is essentially a personal remark about the P vs. NP problem: I believe this is a very challenging and deep and interesting mathematical question, but I think one that has no computer science meaning whatsoever. For the simple fact that P is not an adequate model of feasible computation, and there are lots of results – both theoretical and experimental – which point out that P does not model properly what we understand as feasible computation. Probably the simplest example is to think about the simplex algorithm which is exponentially difficult, but works much better in practice than all the known polynomial solutions.

UNMODERATED AUDIENCE QUESTIONS

George Johnson: This is a good time to move on to questions from the audience.

Jason Cawley: So I have a question for Greg Chaitin but for everyone else as well. You said that the world would be more intelligible and prettier if it were discrete, which was very attractive to me. But how intelligible would it be, really, even if I grant you finiteness and discreteness, even if we find the rule? Won't it have all these pockets of complexity in it, won't it be huge compared to us, finite, wouldn't it last much longer than us – and then we still have all kinds of ways that would be mysterious to us in all the little detail?

Greg Chaitin: Well, I didn't catch all of that... (collects his thoughts then proceeds to summarize the question as best as he heard it because Jason had no microphone) ... Oh, I guess the remark is disagreeing with what I said that a discrete universe would be more beautiful ... no, no, more comprehensible ... right? ... and you gave a lot of reasons why you think it would be ugly, incomprehensible, disgusting (audience laugher) – and I can't argue, if you feel that way! But in that case I don't think that's a question, I view that as a comment that doesn't need to be answered ...

Jason Cawley: Sorry. The question is "How intelligible is a perfectly discrete universe?" The reason I am concerned about this is: I happen to like rationalism too, but I don't want people concluding, when they see non intelligibility in the universe, that it is evidence against the rational.

Greg Chaitin refines his answer: Oh, Okay, great! Well, in that case, as Stephen has pointed out in his book, it could be that all the randomness in the world is just pseudo randomness, you know, and things only *look* unintelligible, but they are actually rational. The other thing is he's also pointed out – and this is sort of his version of Gödel's incompleteness theorem – is that something can be simple and discrete and yet we would not be able to prove things about it. And Stephen's version of this (which I think is very interesting) is that in the way that the universe is created, because you have to run a computation, in general you have to run a physical system to see what it will do, you can't have a shortcut to the answer[t]. So that can be viewed as bad, but it also means that you could have a simple theory of the world that wouldn't help us much to predict things. And you can also look at it as good, because it means that the time evolution of the universe is creative and surprising, it's actually doing something that we couldn't know in advance – by just sitting at our desks, and thinking – so I view this as fundamental and creative! And in regards to creativity Bergson was talking 100 years ago about "L'Evolution Créatrice[u]" – at this point, this would be a new version of that. But over aesthetics one can't ultimately argue too much. But still, I think it was a good question.

New question from the audience: First a question about SETI: I may be naive, but it seems strange to me why we would like to look at perfect circles and dividing lakes when one can simply look at Chicago and New York emanating light from Earth at night. If I were a Martian that's what I would do. But what interests me more and that's the question for Sir Leggett is this: suppose that there were phase transitions from quantum to classical. Would building a quantum computer – would the approach to build a quantum computer be different than trying to reduce decoherence as it's being done presently?

Tony Leggett: I'm not entirely clear what you mean by postulating that there was a phase transition from quantum to classical. Could you elaborate a bit?

Question: Ah, well, I am a bit ignorant about your theory but, if there is a theory that it's not just decoherence but in fact there are phase transitions from quantum to classical at some level – that's why we don't see

---

[t]Via the Principle of Computational Irreducibility (which is a corollary of the PCE).
[u]1907 book by French philosopher Henri Bergson. Its English translation appeared in 1911. The book provides an alternate explanation for Darwin's mechanism of evolution.

Schrödinger's cat after some. Would this imply perhaps a different approach of building a quantum computer?

Tony Leggett: If you mean – if you're referring to theories, for example, of the GRWP type which postulate that there are physical mechanisms systems which will meet the linear formalism of quantum mechanics which will have to be modified – and it will have to be modified more severely as it goes from the microscopic to the macroscopic – then I think the answer to your question is that to the extent that we want to use macroscopic or semi-macroscopic systems as qubits in our quantum computer, it wouldn't work. On the other hand I don't think that theories of the GRWP – scenarios of the GRWP type – necessarily against an attempt to build a quantum computer always keeping the individual bits at the microscopic level. You have to look at it in, of course, in detail in a specific context of a particular computer built around a particular algorithm, such as, say, Shor's algorithm. But I don't think it is a priori essential that a GRWP type scenario would destroy the possibility of it.

New question from the audience: Well, we still haven't answered the question of "How does nature compute?" We are just discussing differences between discrete and continuous, classical vs. quantum computation but if we see mathematics as a historical accident and we try to push it aside, shouldn't we try to look at nature, and try to understand how nature in fact computes? And not just try to translate it into a mathematical context. For example one can watch plants and see that there is some kind of parallel type of computation that is going on, that is, not based on our mathematics, but like *they* do it, the plants ... you know... Do you have you any thoughts on that?

Stephen Wolfram answers it: I think that one of the things that makes that somewhat concrete is the question of how we should build useful computational devices. Whether our useful computational devices should have ALUs (Arithmetic Logic Unit) inside them or not. The very fact that every CPU that's built has a piece that's called "the arithmetic logic unit," tells you that in our current conception of computation we have mathematics somewhere in the middle of it. So an interesting thing is: can we achieve useful computational tasks without ever having an ALU in the loop, so to speak. I think the answer is: definitely yes, but as the whole engineering development of computers has been so far we've just been optimizing this one particular model that's based on mathematics. And as we try to build computers that are more at the molecular scale, we could, actually, use the same model: we could take the design of the Pentium chip and we can shrink it down really really small and have it implemented in atoms. But an alternative would be that we can have atoms do things that they are more naturally good at doing. And I think the first place where this would come up are things like algorithmic drugs, where you want to have something that is in essence a molecule operating in some biological, biomedical context and it wants to actually do a computation as it figures out whether to bind to some site or not, as opposed to saying "I am the right shape so I'm going to bind there!" So that's a place where computation might be done. But it's not going to be computation that will be done through arithmetic but we'll be forced to think about computation at a molecular scale in its own terms, simply because that's the scale at which the thing has to operate. And I'm going to guess that there will be a whole series of devices and things, that – mostly driven by the molecular case – where we want to do computation but where the computation doesn't want to go through the intermediate layer of the arithmetic.

Cris Calude: Yes! There is a lot of research that we have started about ten years ago in Auckland, and a series of conferences called "Unconventional Computation." And this is one of the interesting questions. You know, there are basically two streams of thought: one is quantum computation, and the other one is molecular computing. And in quantum computing, you have this tendency of using the embedded mathematics inside. But in molecular computing, you know, you go completely wild because mathematics is not there so you use all sorts of specific biological operations for computation. And if you look at the results, some of them are quite spectacular.

Small refinement: Yeah, but they are still based on logic gates and ... you know ... molecular computing is still based on trying to build logic gates and ...

Cris Calude: No, no! There is no logic gate! That's the difference, because the philosophy in quantum computation is: you do these logical gates at the level of atoms or other particles – but in molecular computation there is no arithmetical instruction, there are no numbers, you know, everything is a string, and the way they are manipulated is based exactly on biological type of processing. No arithmetic.

Greg Chaitin: No boolean algebra, not even *and*'s and *or*'s?

Cris Calude: No, nothing! And then this is the beauty, and in a sense this was the question that I posed to Seth Lloyd in '98. I said you know, why don't you do in quantum computing something similar? Why don't you try to think of some kind of rules of processing – not imposed from the classical computation, from Turing machines, but rules which come naturally from the quantum processes – just like the typical approach in molecular computation.

Tom Toffoli: I would like to give a complementary answer to this. I've been teaching microprocessors and microcontrollers – you have them in your watches and cellphones. They're extremely complicated objects. And you would say, I mean given the ease with which we can fabricate these things, whenever we want to run a program or algorithm, we could make a very special purpose computer rather than using a microprocessor to program it. Apparently it turns out that it's much more convenient, if someone had designed a microprocessor with an ALU and a cache and the other things in between, to just take it as a given and then the process is complete. If we look at biology, biology has done the same thing. We have, at a certain moment, hijacked mitochondria that do the conversion of oxygen and sugar into recharging the ATP batteries. That was a great invention. And now, after probably three billion years, or something like that we still keep using that instead of inventing a method that is, maybe a little more efficient, but would have to be a very special choice for every different circumstance. Essentially we could optimize more, but we would do that at the cost of losing flexibility, modularity and so on but, apparently it's much more convenient. For three billion years we've kept using this kind of energy microprocessor, that worked fairly well... So there is, essentially, the flexibility or modular evolution that really suggests that choices like the ALU... is not an optimal choice, but – empirically – is a very good choice. This is my viewpoint.

Stephen Wolfram: Well – as is always the case in the history of technological evolution it is inconceivable to go back and sort of restart the whole design process. Because there is just far too much investment in that particular thing. The point that I want to make is that – what will happen is that there will be certain particular technological issues, which will drive different types of computing. My guess is that the first ones that will actually be important are these biomedical ones, because they have to operate on this molecular scale, because that's the scale on which biomedicine operates. And, you know, one can get away with having much bigger devices for other purposes – but biomedicine is a place where potentially a decision has to be made by a molecule. And whether – maybe there will be ways of hacking around that but in time, it won't be very long before the first applications of this kind will be finalized.

And what's interesting about [this] is that if you look at the time from Gödel and Turing to the time when computers became generic and everybody had one, and the time from Crick and Watson and DNA to the time when genomics becomes generic – it's about the same interval of time. It hasn't yet happened for genomics, it has more or less happened for computers. Computers were invented, the idea of computers is twenty-three years earlier or something like that, than the DNA idea. Anyway, it will soon happen for genomics as well, and in time we will routinely be able to sequence things, in real time, from ourselves, and we'll do all kinds of predictions that yes, you know we detect that today you have a higher population of antibodies that have a particular form so we'll be able to run some simulation that this means that you should go and by a supply of T-cells, that has this particular characteristic and so on. And there's a question whether the decisions about that will be made externally by ALU based computers, or whether they will be made internally by some kind of molecular device – more like the way biology actually does it. And if it ends up that they are made externally then there won't be any drive from the technology side to make a different substructure for computing.

New question from the audience: I study computer science. And one of the ideas I find truly fascinating and I think it really is – is this thing called the Curry-Howard isomorphism, which basically relates propositions to types and rules to terms and proof normalization to program evaluation. And since you're discussing models of computation, I was wondering if you have encountered something similar for cellular automata and such. I think that the classification, this particular relation is very explicit in the simply typed lambda calculus where program terms, which are lambda expressions, can be given types – and the types are pretty much propositional logic expressions. And if you can prove a certain proposition the structure of the proof in natural deduction style will actually look like a program type and if the proof is not a normal proof then the process of proof normalization is basically the process of the evaluation of the term into a normal form of the term, which basically means that if you have this computational model which is the lambda calculus, reductions of the calculus will correspond to normalizations of the proofs, and the types serve as a way of classifying

programs, types are a way of saying: these particular programs that behave
in such and such a way don't have such and such properties. And I feel that
this might be something that carries over to other notions of computations
as well, because there's nothing intrinsic about the lambda calculus that
makes this uniquely applicable to it.

Stephen Wolfram: Let me start by saying that I'm a very anti-type person.
And it turns out that, you know, in the history types were invented as
a hack by Russell basically to avoid certain paradoxes – and types then
became this kind of "great thing" that were used as an example and then
as a practical matter of engineering in the early computer languages there
was this notion of integer types versus real types and so on, and the very
idea of types became very inflated – at least that's how I see it.

So, for example in *Mathematica* there are no types. It's a symbolic system where there is only one type: a symbolic expression. And in practical
computing the most convincing use of types is the various kinds of checking
but in a sense when something is checkable using types that involves a certain kind of rigidity in programs that you can write, that kind of restricts
the expressivity of the language that you have. And what we found over
and over again in *Mathematica*, as we thought about putting things in that
are like types, that to do that would effectively remove the possibility of all
sorts of *between paradigm* kinds of programming, so to speak, that exist
when you don't really have types.

So, having said that, this question of the analogy between proof processes and computation processes is an interesting one. I've thought about
that a lot, and there's more than just a few things to say about it. But
one thing to think about is: "What is a proof?" and "What's the point
of doing a proof?" I mean, the real role of a proof is as a way to convince (humans, basically) that something is true. Because when we do a
computation, in the computation we just follow through certain steps of
the computation and assuming that our computer is working correctly the
result will come out according to the particular rules that were given for
the computation. The point of a proof is somehow to be able to say to a
human – look at this: you can see what all the steps were and you can verify
that it's correct. I think the role of proofs in modern times has become,
at best, a little bizarre. Because, for example, so here's a typical case of
this: when *Mathematica* first existed twenty years ago one would run into
mathematicians who would say "How can I possibly use this, I can't prove
that any of the results that are coming out are correct!" OK? That's what

they were concerned about. So, I would point out sometimes that actually when you think you have a proof, in some journal for example, it's been maybe checked by one person – maybe – if you're lucky. In *Mathematica* we can automate the checking of many things and we can do automatic quality assurance, and it's a general rule that – in terms of how much you should trust things – the more people use the thing that you're using the more likely it is that any bugs in it will have been found by the time you use it. So, you know, if you say: "Well, maybe it's a problem in the software that I myself am writing, maybe it's a problem in the system (like *Mathematica*) that I am using, maybe it's a problem in the underlying hardware of the computer" – it gets less and less plausible that there's a problem, the broader the use of the thing is.

So I think as a practical matter when people say: "I want a proof!" that the demand for proof, at least in the kind of things that *Mathematica* does, decayed dramatically in the first few years that *Mathematica* existed because it became clear that most likely point of failure is where you as a human were trying to explain to the computer what to do. Now, having said that it's interesting [that] in *Mathematica* we have more and more types of things that are essentially proof systems – various kinds of things, for example proof systems for real algebra, we just added in the great new *Mathematica* 7.0 all sorts of stuff of doing computation with hundreds of thousands of variables and so on. Those are effectively places where what we've done was to add a proof system for those kinds of things. We also added a general equational logic proof system, but again I think that this question whether people find a proof interesting or whether they just want the results – it seems that the demand for presentation of proofs is very low.

Tom Toffoli (wants to add something, Wolfram uses the break to drink some water): If you would give me the microphone for one second: coming back to Russell when he published Principia Mathematica – most of the theorems were right, but a good fraction of the proofs were found wrong. I mean this was Russell, OK? He was wrong, but there was no problem, because he was still convinced, by his own ways he was convinced of the theorems. So he put together some proofs (arguments) to try to convince the readers that the theorems were right, and he convinced them. But the proofs, as actual mechanical devices, were not working. So his proofs were coming out of just a heuristic device and he derived, and you can always derive the right theorem with the wrong proof. What are you going to do in that case?

Stephen Wofram: No, I think it's actually interesting this whole analogy between proof and computation and so on. One of the things that I have often noticed is that if you look at people's earlier attempts to formalize mathematics – the thing that they focused on formalizing was the process of doing proofs, and that was what Whitehead, Russell, Peano before him and so on worked on. It turned out that that direction of formalization was fairly arid. Not much came from it. The direction that turned out to be the most interesting direction of formalization was, in fact, the formalization of the process of computation. So, you know, in the construction of *Mathematica*, what we were trying to do was to formalize the process of computing things. Lots of people used that and did various interesting things with it. The ratio of people who do computation with formalized mathematics to the number who do proofs with formalized mathematics is a huge ratio. The proof side turned out not to be that interesting. A similar kind of thing, and an interesting question, is how one would go about formalizing every day discourse: one can take (the) everyday language and one can come up with a formalized version of it that expresses things in a sort of formal, symbolic way. But the thing that I've not figured out – actually I think that I've now figured it out, but I *hadn't* figured it out – was "So, what's the point of doing that?" In other words the Russell-Whitehead effort of formalizing proofs turned out not not lead to much. The right idea about formalizing mathematics was the one about formalizing the process of computation. Similarly formalizing everyday discourse as a way to make the semantic web or some such other thing, probably has the same kind of issue as the kind of formalization of proof as was done in mathematics and I think that maybe there is another path for what happens when you formalize everyday discourse, and it's an interesting analogy to what would happen in the mathematics case. You know: what's the point of formalization and what can you do with a formalized system like that.

Greg Chaitin: Let me restate this very deep remark that Stephen has just made about proofs versus computation: if you look at a[ny] formal system for a mathematical formal theory, Gödel shows in 1931 that it will always be incomplete. So any artificial language for doing mathematics will be incomplete, will never be universal, will never have every possible mathematical argument. There is no formal language for mathematics where every possible mathematical argument or proof can be written in.

Zermelo-Fraenkel set theory, you know, as a corollary of Gödel – may be wonderful for everything we have but it is incomplete. Now the exact opposite – the terminology is different – when you talk about a programming language you don't talk about completeness and incompleteness, you talk about universality. And the amazing thing is that the drive for formalism, and Russell-Whitehead is one data point on that, another data point is Hilbert's program, the quest for formalization started off in mathematics, and the idea was to formalize reasoning; and the amazing thing is that this failed. Gödel in 1931 and Turing in 1936 showed that there are fundamental obstacles– it can't work! But the amazing thing is that this is a wonderful failure! I mean what can be formalized beautifully, is not proof, or reasoning but: *computation*. And there almost any language you come up with is universal, which is to say: complete – because every algorithm can be expressed in it. So this is the way I put what Stephen was saying.

So Hilbert's dream, and Russell and Whitehead failed gloriously – is not good for reasoning but it's good as a technology, is another way to put it, if you're trying to shock people. The quest for a firm foundation for mathematics failed, but gave rise to a trillion dollar industry.

Tom Toffoli: Let me add something to this. You've heard of Parkinson's law, the one that says: "a system will use as many resources as are available." It was formulated because it was noticed that the whole British Empire was run by essentially a basement of a few dozen people for two hundred years. And then, the moment the British started losing their empire, they started de-colonizing and so on, then they had a ministry of the colonies and this ministry grew bigger and bigger and bigger as the colonies became fewer and fewer and fewer. And it was not working as well as before. And I think that formalization is often something like that. You can think about it, but you don't want to actually do it, even von Neumann, you know, the moment he decided that CAs were, sort of, plausible to give life – he didn't go through the process of developing the whole thing. The point was already made.

New question from the audience: If I may, can I add something about the Curry-Howard isomorphism before I ask my question? Yes? Maybe ... I think that the revolution that we are living in physics is only part of a wider revolution and there are many questions in physics to which the answer uses the notion of algorithm. For instance: what are the laws of nature? They are algorithms. So you may give this answer, only because you have the notion of algorithm. And for centuries we didn't have it. So, to these questions we had either no answer or ad-hoc answers. For instance: what are the laws of nature? Compositions. When we had compositions we didn't have algorithms. It was a good way to answer it. And there are many many areas in knowledge where there are many questions to which now we answer: it is an algorithm. And one of the very first questions on which we changed our mind, was the question: what is a proof? And: what is a proof? The original answer was a sequence of formulas verifying deduction rules and so on. And starting with Kolmogorov – because behind the Curry-Howard isomorphism there is the Kolmogorov interpretation – is this idea that proofs, like the laws of nature are in fact algorithms. So they are two facets, or two elements of a wider revolution, and I think that they are connected in this way.

Now consider this question: Does the Higgs boson exist? Today, I guess, there are people who believe that the Higgs boson exists, there are people who believe that it doesn't exist, but anyone – you can take the electron, if you want (instead of the Higgs boson) [which] most people I guess, believe it exists – but I guess that everyone agrees that we have to find some kind of procedural [means of verifying such a prediction]. It can be an experiment, it may be anything you want – in this case – that will allow eventually, if we are lucky enough, to solve this question. And I would be very uncomfortable if someone told me that the Higgs boson exists in the eye of the beholder, or that Higgs boson has to exist because then the theory would be more beautiful, or if the Higgs boson does not exist then it's actually a problem of the Universe and not ours and we can continue to postulate that it exists because it would be nicer. So I wonder if the two questions we have (not?) been discussing today are not of the same kind: we should try to look for a kind of procedure to answer the question at some point. One: is the universe computable? The other: is the universe discrete? In some sense are these questions only metaphysical questions or are they questions related to experiments that we could and should, in fact carry out?

George Johnson: I'm sorry we won't going to have time to get into that... I want to thank the speakers. Perhaps we could resume at the next conference and this would be another reason to look forward to it.

End of transcript.

Author Index

Abbott, A., 181
Agafonov, V.N., 184
Agrawal, 238
Albeth, 235
Aldous, 47
Allender, E., xiii, 232, 239, 267
Alton, A., 267
Aristotle, 109, 185
Arora, 237
Arruda, A., 188
Arslanov, A., 181
Avigad, J., 234

Büchi, J.R., 293
Babai, L., 237
Bach, E., 301
Baker, T., 267
Becher, V., 253
Bell, J., 188
Bennett, C., 105, 133, 247, 288,
   289, 296
Bergson, H., 394
Bernoulli, J., 24
Bialek, W., 357
Bienvenu, L., 227, 248
Blokh, 293
Blum, M., 268, 303
Boethius, vii
Boltzmann, L., 22, 27, 173
Born, M., 110
Bourbaki, 187
Brattka, 217, 235
Braverman, M., 235
Bridson, M., 234

de Brogli, L., 110
Buhrman, H., 269
Burks, A., 8
Busi, 79
Buzeţeanu, 179

Câmpeanu, C., 179
Cai, J-Y., 301
Calude, C.S., xiii, 75, 78, 106, 115,
   179, 207, 226, 294, 295, 309, 351
Candide, 30
Carnap, R., 150
Casti, J., 309
Catuscia, P., 73
Cawley, J., 393
Chaitin, G., xiii, 75, 78, 82, 93, 95,
   106, 121, 181, 197, 200, 207, 227,
   229, 230, 311
Chazelle, 285
Chiţescu, I., 179
Chistyakov, 44
Cholak, 215
Chomsky, N., 129, 150
Chuaqui, R., 188
Church, A., 121
Codd, E.F., 260
Coles, R., 181, 226
Colmerauer, A., 257
Conversi, M., 7
Conway, J., 224
Coron, 48
Cournot, 165
Crick, F., 80, 398

Subject Index